Aircraft Gas Turbine Engines

Operation, Components and Systems

Wexford
2008

Preface

This book provides the reader with a working understanding of modern aircraft gas turbine engines, with frequent mention made of military applications, particularly Army jets and helicopters.

The text concludes with several chapters on specific makes and models of turbines, some of which are used in military aircraft.

J. Vennard,

Editor

CONTENTS

	Paragraph	Page
INTRODUCTION		1
CHAPTER 1. INTRODUCTION TO GAS TURBINE ENGINES	1.1	3
Section I. Theory of Gas Turbine Engines......................	1.2	3
II. Principles of Operation.......	1.7	12
III. Major Engine Sections.........	1.15	21
CHAPTER 2. SYSTEMS AND ACCESSORIES	2.1	46
Section I. Fuel Systems and Components....................	2.2	46
II. Lubrication Systems...........	2.12	58
III. Ignition Systems and Engine Instrumentation........	2.21	62
CHAPTER 3. TESTING, INSPECTION, MAINTENANCE, AND STORAGE PROCEDURES	3.1	68
CHAPTER 4. LYCOMING T53	4.1	89
Section I. Operational Description of the T53 Gas Turbine Engine....	4.2	89
II. Major Engine Systems and Assemblies................	4.9	99
CHAPTER 5. LYCOMING T55	5.1	146
Section I. Operational Description of the T55 Gas Turbine Engine........	5.2	146

		Paragraph	Page
	Section II. Major Engine Sections and Systems..................	5.8	152
CHAPTER 6.	SOLAR T62 AUXILIARY POWER UNIT ...	6.1	174
CHAPTER 7.	ALLISON T63	7.1	188
CHAPTER 8.	PRATT AND WHITNEY T73	8.1	210
CHAPTER 9.	PRATT AND WHITNEY T74	9.1	224
	Section I. Operational Characteristics and Description..............	9.2	224
	II. Major Engine Systems..........	9.13	240
CHAPTER 10.	GENERAL ELECTRIC T700	10.1	256

INTRODUCTION

Aircraft designers have always been limited by the efficiency of available powerplants. Their constant pleas have been for:

- Higher power
- Less weight
- Lower frontal area
- Better cooling characteristics
- Lower fuel consumption

Designers of reciprocating engines were able to address these problems to some extent during the first half of the twentieth century. The design of the piston engine was eventually refined to the point that the only way to obtain a further increase in power was to add more cylinders. This is not possible without adding significant weight and additional other complex problems.

The aircraft designers' pleas were finally answered with the development of the gas turbine engine. Since the end of World War II, progress in the gas turbine field has been rapid. Development of improved materials, high temperature metals, and better fuels should expedite further progress in this held. The gas turbine engine's greatest contribution to aviation is that it has lifted all previous limits that were imposed by the reciprocating engine.

This text describes the operation, components, and systems of the gas turbine engine. The first chapter includes an introduction to gas turbine engines. Chapter 2 discusses the systems and accessories such as fuel, oil, and electrical. Chapter 3 covers testing, inspection, maintenance, and storage procedures. Chapters 4 through 10 describe in detail the gas turbine engines used in jet aircraft, including military aircraft.

Chapter 1

INTRODUCTION TO GAS TURBINE ENGINES

1.1. INTRODUCTION

This chapter introduces the theory and operating principles of gas turbine engines. Gas turbine engines can be classified according to the type of compressor used, the path the air takes through the engine, and how the power produced is extracted or used. The chapter is limited to the fundamental concepts of the three major classes of turbine engines, each having the same principles of operation.

Chapter 1 is divided into three sections; the first discusses the theory of turbine engines. The second section deals with principles of operation, and section III covers the major engine sections and their description.

Section I. Theory of Gas Turbine Engines

1.2. GENERAL

Section I covers the laws of physics and fundamentals pertaining to the theory of jet propulsion. The gas turbine engines used to power Army aircraft are turboshaft powerplants. The energy produced drives the power shaft. Energy is generated by burning the fuel-air mixture in the engine and accelerating the gas tremendously. These high-velocity gases are directed through turbine wheels which convert the axial movement of the gas to a rotary motion. This rotary power is used to drive a powershaft, which drives a propeller or a rotor transmission.

1.3. LAWS OF MOTION

The theory of gas turbine engines is based on the laws and principles of physics discussed in the subparagraphs that follow.

<u>Newton's First Law of Motion</u>. The first law states that a body in a state of rest remains at rest, and a body in motion tends to remain in motion at a constant speed and in a straight line, unless acted upon by some external force.

Newton's Second Law of Motion. The second law states that an imbalance of forces on a body produces or tends to produce an acceleration in the direction of the greater force, and the acceleration is directly proportional to the force and inversely proportional to the mass of the body.

Newton's Third Law of Motion. The third law states that for every action there is an equal and opposite reaction, and the two are directed along the same straight line.

Bernoulli's Principle. This principle states that if the velocity of a gas or liquid is increased its pressure will decrease. The opposite is also true. If the velocity of a gas or liquid is decreased its pressure will increase. This fact relates directly to the law of conservation of energy.

Einstein's Law of Conservation of Energy. This law states that the amount of energy in the universe remains constant. It is not possible to create or destroy energy; however, it may be transformed.

Boyle's Law. This law states that if the temperature of a confined gas is not changed, the pressure will increase in direct relationship to a decrease in volume. The opposite is also true -- the pressure will decrease as the volume is increased. A simple demonstration of how this works may be made with a toy balloon. If you squeeze the balloon, its volume is reduced, and the pressure of air inside the balloon is increased. If you squeeze hard enough, the pressure will burst the balloon.

Charles' Law. This law states that if a gas under constant pressure is so confined that it may expand, an increase in the temperature will cause an increase in volume. If you hold the inflated balloon over a stove, the increase in temperature will cause the air to expand and, if the heat is sufficiently great, the balloon will burst. Thus, the heat of combustion expands the air available within the combustion chamber of a gas turbine engine.

Pressure and Velocity. Air is normally thought of in relation to its temperature, pressure, and volume. Within a gas turbine engine the air is put into motion so now another factor must be considered, velocity. Consider a constant airflow through a duct. As long as the duct cross-sectional area remains unchanged, air will continue to flow at the same rate (disregard frictional loss). If the cross-sectional area of the duct should become smaller (convergent area), the airflow must increase velocity if it is to continue to flow the same number of pounds per second of airflow (Bernoulli's Principle). In order to obtain the necessary velocity energy to accomplish this, the air must give up some pressure and temperature energy (law of conservation of energy). The net result of flow through this restriction would be a decrease in pressure and temperature and an increase in velocity. The opposite would be true if air were to flow from a smaller into a larger duct (divergent area); velocity would then decrease, and pressure and temperature would increase. The throat of an automobile carburetor is a good example of the effect of airflow through a restriction (venturi); even on the hottest day the center portion of the carburetor feels cool. Convergent and divergent areas are used throughout a gas turbine engine to control pressure and velocity of the air-gas stream as it flows through the engine.

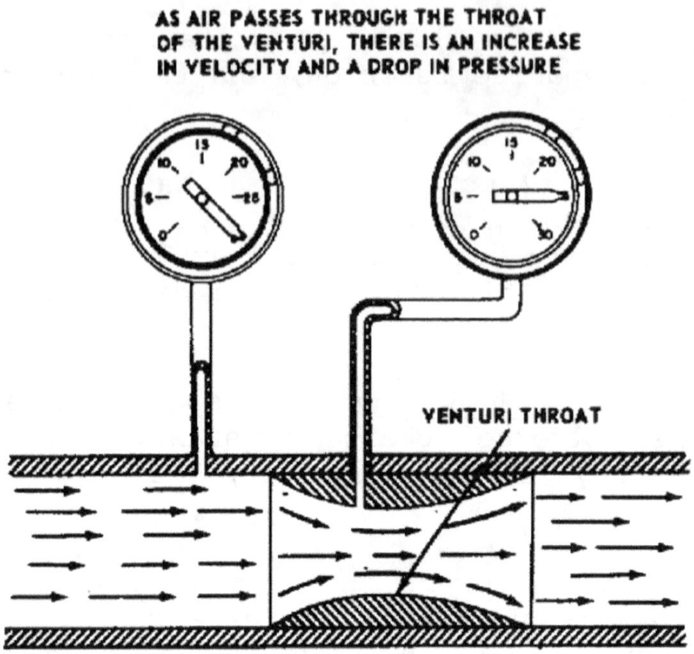

1.4. THEORY OF JET PROPULSION

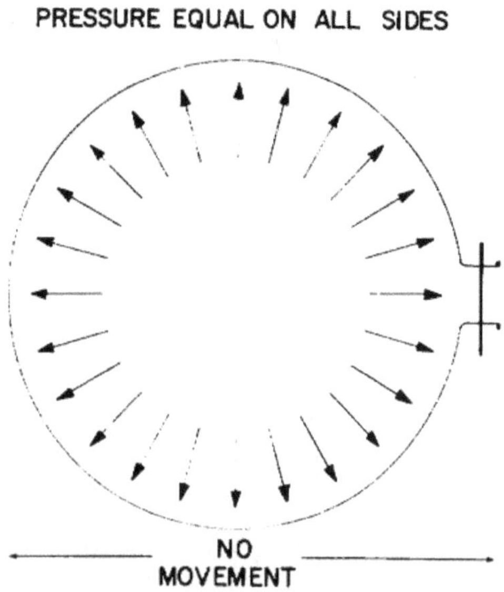

The principle of jet propulsion can be illustrated by a toy balloon. When inflated and the stem is sealed, the pressure is exerted equally on all internal surfaces. Since the force of this internal pressure is balanced there will be no tendency for the balloon to move.

If the stem is released the balloon will move in a direction away from the escaping jet of air. Although the flight of the balloon may appear erratic, it is at all

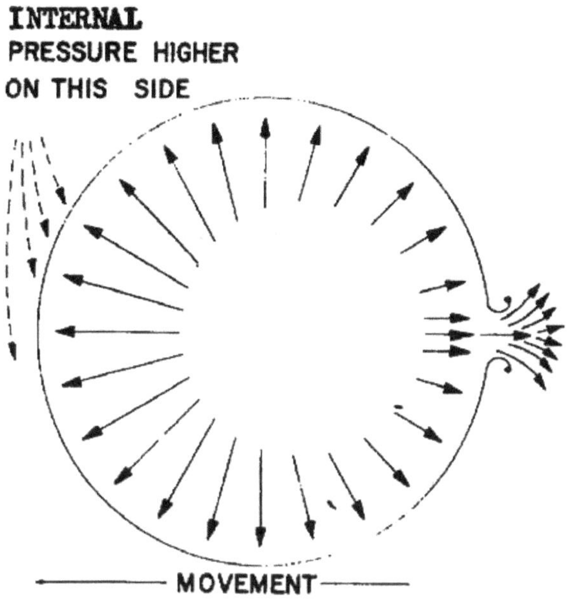

times moving in a direction away from the open stem.

The balloon moves because of an unbalanced condition existing within it. The jet of air does not have to push against the outside atmosphere; it would function better in a vacuum. When the stem area of the balloon is released, a convergent nozzle is created. As the air flows through this area, velocity is increased accompanied by a decrease in air pressure. In addition, an area of skin against which the internal forces had been pushing is removed. On the opposite internal surface of the balloon, an equal area of skin still remains. The higher internal pressure acting on this area moves the balloon in a direction away from the open stem. The flight of the balloon will be of short duration, though, because the air in the balloon is soon gone. If a source of pressurized air were provided, it would be possible to sustain flight of the balloon.

1.5. THEORY OF THE GAS TURBINE ENGINE

If the balloon were converted into a length of pipe, and at the forward end an air compressor designed with blades somewhat like a fan were installed, this could provide a means to replenish the air supply within the balloon.

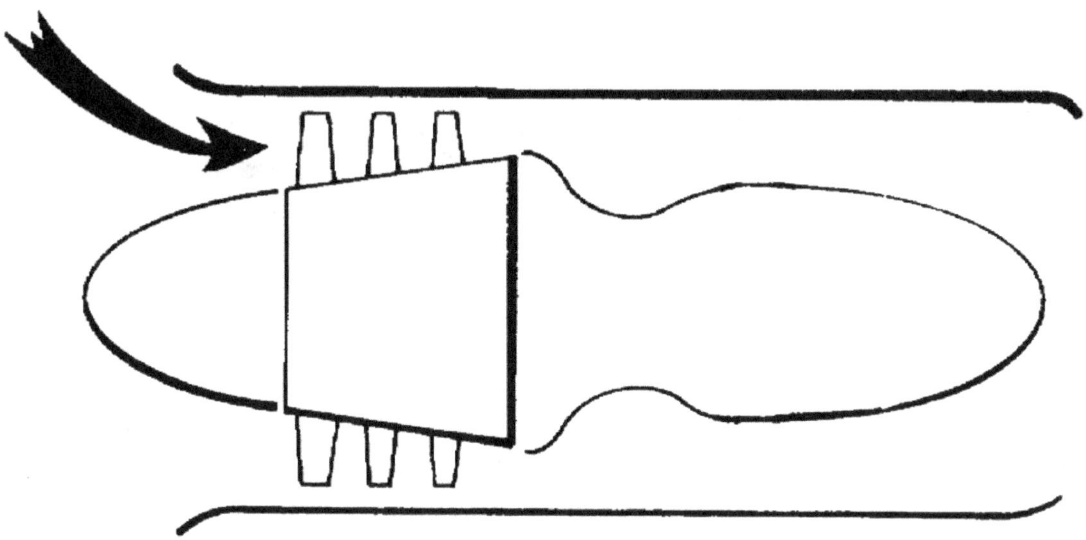

A source of power is now required to turn the compressor. To extend the volume of air, fuel and ignition are introduced and combustion takes place. This greatly expands the volume of gas available.

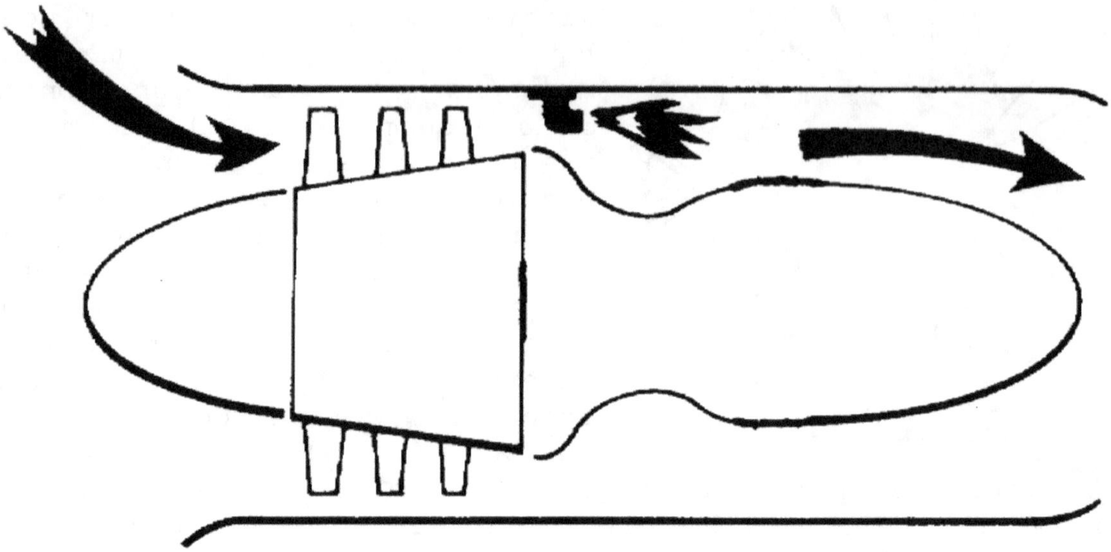

In the path of the now rapidly expanding gases, another fan or turbine can be placed. As the gases pass through the blades of the turbine, they cause it to rotate at high speed. By connecting the turbine to the compressor, we have a mechanical means to rotate the compressor to replenish the air supply. The gases still possessing energy are discharged to the atmosphere through a nozzle that accelerates the gas stream. The reaction is thrust or movement of the tube away from the escaping gas stream. We now have a simple turbojet engine.

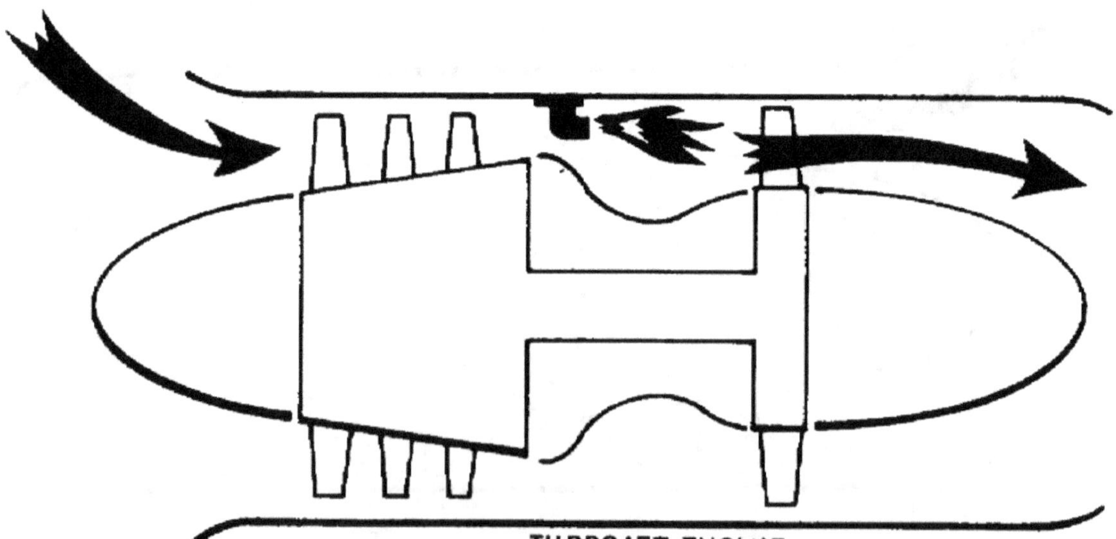

TURBOJET ENGINE

The turbojet engine is a high-speed, high-altitude powerplant. The Army, at present, has no requirement for this type of engine. Because it is simple and easy to operate and maintain, however, the Army does use the gas turbine engine. The simple turbojet engine has primarily one rotating unit, the compressor/turbine assembly. The turbine extracts from the gas stream the energy necessary to rotate the compressor. This furnishes the pressurized air to maintain the engine cycle. Burning the fuel-air mixture provides the stream of hot expanding gas from which approximately 60 percent of the energy is extracted to maintain the engine cycle. Of the total energy development, approximately 40 percent is available to develop useful thrust directly.

If we had ten automobile engines that would equal the total shaft horsepower of a turbine engine, it would take six of these engines to turn the compressor, and the other four would supply the power to propel the aircraft. The amount of energy required to rotate the compressor may at first seem too large; however, it should be remembered that the compressor is accelerating a heavy mass (weight) of air towards the rear of the engine. In order to produce the gas stream, it was necessary to deliver compressed air by a mechanical means to a burner zone. The compressor, being the first rotating unit, is referred to as the N_1 system.

With a requirement for an engine that delivers rotational shaft power, the next step is to harness the remaining gas stream energy with another turbine (free turbine). By connecting the turbine to a shaft, rotational power can be delivered to drive an aircraft propeller, a helicopter rotor system, a generator, a tank, an air cushion vehicle (ACV), or whatever is needed. The power shaft can extend from the front, back, or from an external gearbox. All of these locations are in use on various types of Army engines at present.

The following sketch shows a turboshaft engine with the power shaft extended out the front. The bottom sketch shows the same engine with the power shaft extending out the back.

The basic portion of the turbine engine, the gas producer, extracts approximately 60 percent of the gas stream energy (temperature/pressure) to sustain the engine cycle. To develop rotational shaft power, the remaining gas stream energy must drive another turbine. In Army engines today, a power turbine that is free and independent of the gas producer system accomplishes this task. The power turbine and shaft (N_2 system) are not mechanically connected to the gas producer (N_1 system). It is a free turbine. The gas stream passing across the turbines is the only link between these two systems. The free-turbine engine can operate over wide power ranges with a constant output-shaft speed.

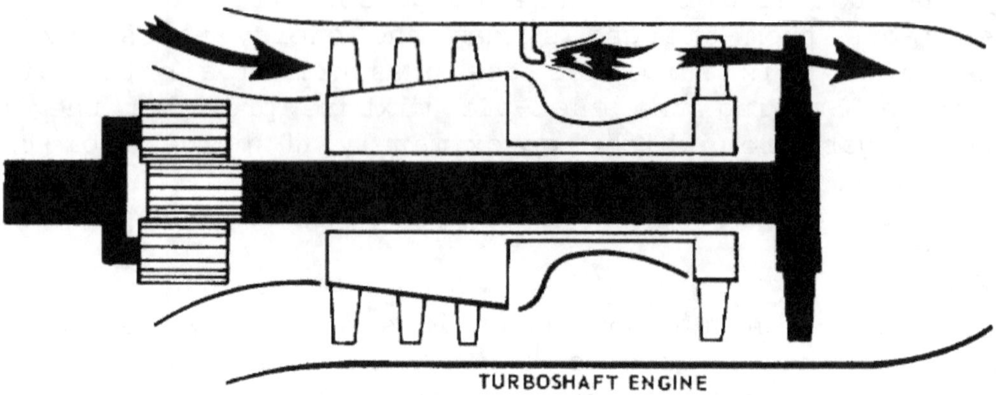

TURBOSHAFT ENGINE

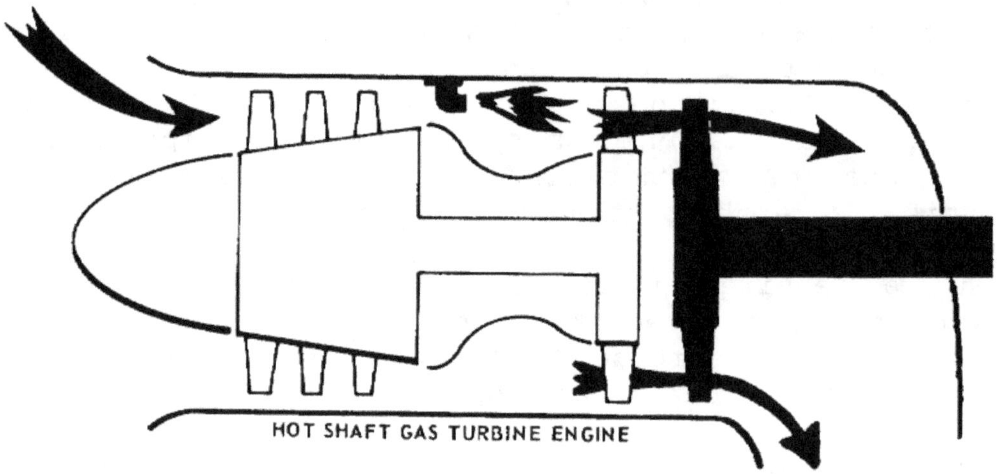

HOT SHAFT GAS TURBINE ENGINE

In operation, the gas producer (N_1) system automatically varies its speed, thereby controlling the intensity of the gas stream in relation to the load applied to the power (N_2) shaft. This is accomplished by a fuel metering system that senses engine requirements. The free turbine design has revolutionized the methods of application of shaft turbine engines. Why a shaft turbine? Why is a perfectly good jet engine used to drive a propeller? Because in the speed range that Army aircraft operate, the propeller or helicopter rotor is more efficient. With a turbojet engine, power (thrust) produced is roughly the difference between the velocity of the air entering the engine and the velocity of the air exiting from the engine. Efficiency of the engine (power producer versus fuel consumed) increases with speed until it is 100 percent efficient when the forward speed of the engine is equal to the rearward speed of the jet. It is this low efficiency at takeoff and at low cruising speed (i.e., 400 mph) that makes the turbojet engine unsuitable for use in Army aircraft. The propeller does not lack efficiency at low speed; the reverse is true, in that efficiency falls off at high speed. The result is to harness the jet engine's gas stream energy to drive a propeller or helicopter rotor system, thereby taking advantage of the best features of both.

Aircraft reciprocating engines operate on the four-stroke, five-event principle. Four strokes of the piston, two up and two down, are required to provide one power impulse to the crankshaft. Five events take place during these four strokes: the intake, compression, ignition, power, and exhaust events. These events must take place in the cylinder in the sequence given for the engine to operate.

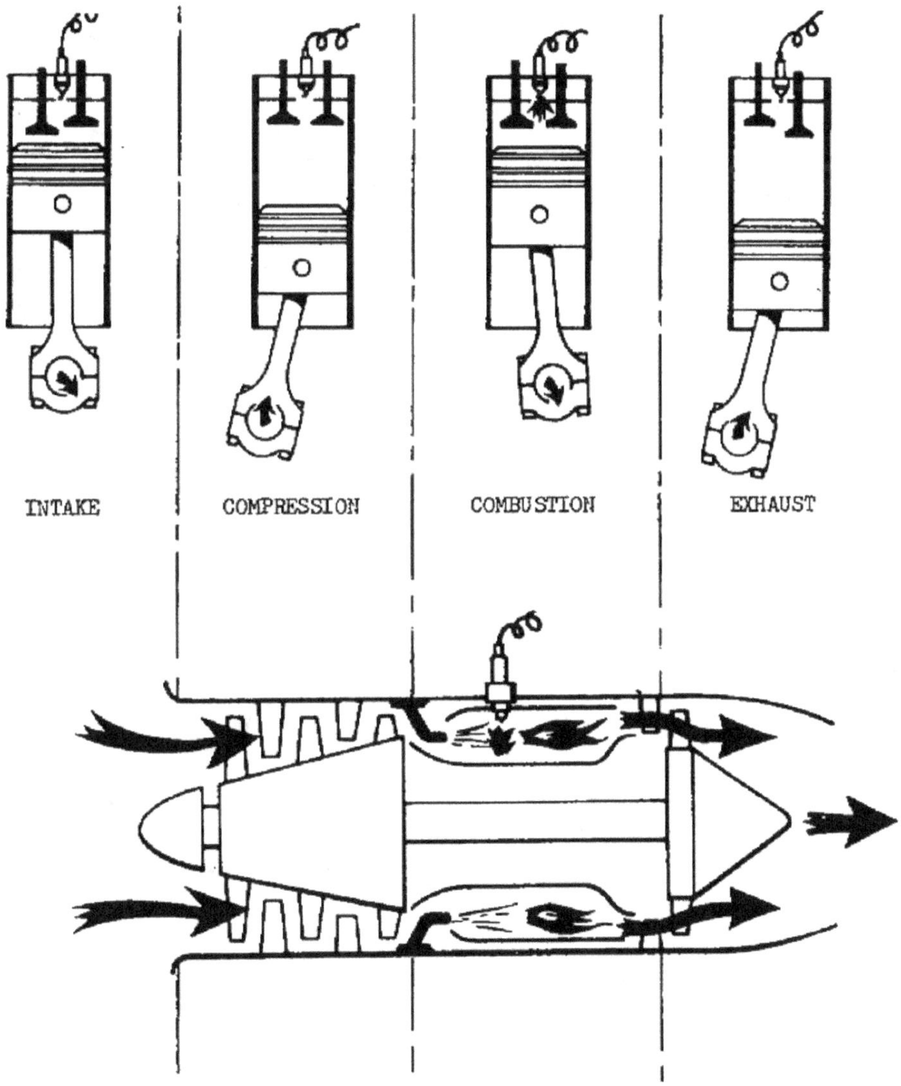

Although the gas turbine engine differs radically in construction from the conventional four-stroke, five-event cycle reciprocating engine, both involve the same basic principle of operation. In the piston (reciprocating) engine, the functions of intake, compression, ignition, combustion, and exhaust all take place in the same cylinder and, therefore, each must completely occupy the chamber during its respective part of the combustion cycle. In the gas turbine engine, a separate section is devoted to each function, and all functions are performed at the same time without interruption.

1.6. SUMMARY

The theory of gas turbine engine operation is based on the laws or principles of physics. The principle of jet propulsion can be illustrated by a toy balloon. When the balloon is inflated and the stem is unsealed the balloon will move in a direction away from the escaping jet of air. If the balloon is converted into a length of pipe, and at the forward end an air compressor is installed to supply air for combustion, and to expand the volume of air, fuel and ignition are introduced and combustion takes place. Then, in the path of the expanding gases a turbine rotor is installed. As the gases pass through the turbine blades, the turbine rotor is rotated at high speed. This turbine rotor is connected to the compressor shaft, and we now have a means to rotate the compressor to replenish the air supply. The remaining gases are discharged to the atmosphere. The reaction of these gases is thrust, or movement of the tube away from the escaping gases. This is a simple turbojet engine. At present the Army has no requirement for this high-speed, high-altitude powerplant. However, if we install another turbine rotor after the rotor that drives the compressor, we have a turboshaft engine that can be used to drive a transmission in a helicopter or a propeller on a fixed-wing aircraft.

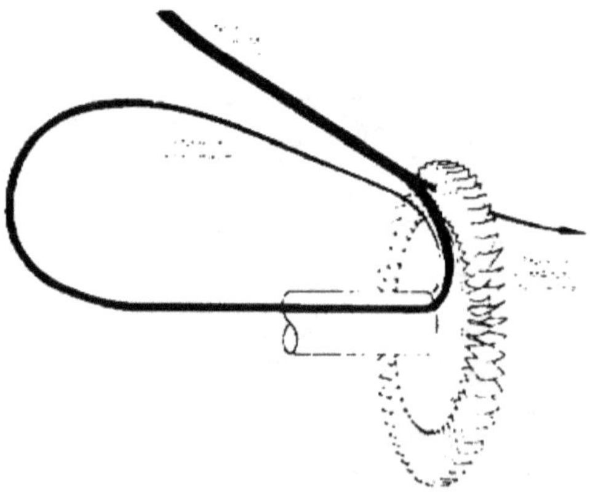

-Power transformed by turbine wheel is fed back to maintain engine cycle

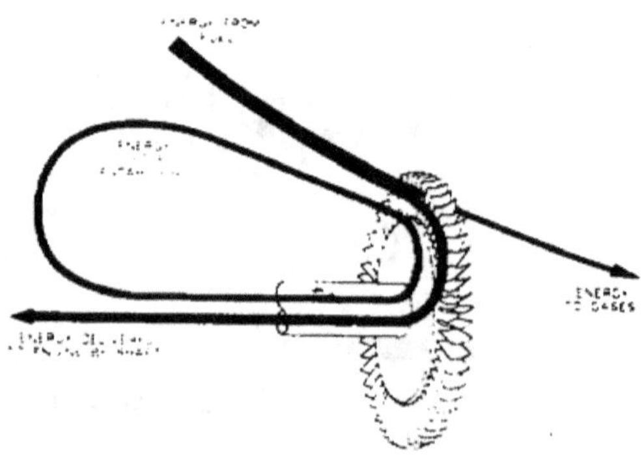

-Energy drawn off for purposes other than maintaining engine cycle

In the turbojet engine, approximately 60 percent of the energy is extracted to rotate the compressor, while the remaining 40 percent is used to develop thrust. In the turboshaft engine, the remaining energy is used to drive a turbine rotor attached to a transmission or propeller. On a free-turbine engine, the gas stream passing across the turbines is the only link between the two turbine rotors. One turbine drives the compressor and the other turbine propels the aircraft. The free-turbine engine is used in Army aircraft.

The gas turbine engine differs radically in construction from the reciprocating engine in that the turbine engine has a separate section for each function, while in the reciprocating engine all functions are performed in the same cylinder.

Section II. Principles of Operation

1.7. GENERAL

This section covers the principles of turbine engine operation. The three classifications of turbine engines are turbojet, turboshaft, and ramjet. The term "turbo" means "turbine." Therefore, a turboshaft engine is one which delivers power through a shaft.

1.8. OTTO AND BRAYTON CYCLES

There is an element of similarity to both the reciprocating and jet engines, but the thermodynamic cycle of each is different from the other. The reciprocating engine operates on the Otto cycle, a constant volume cycle, consisting of four distinct operations. These operations are performed intermittently by a piston reciprocating in an enclosed cylinder. It is important to remember that the piston in a reciprocating engine delivers power only during one of its four strokes.

The turbine engine operates on the Brayton cycle, a constant pressure cycle containing the same four basic operations as the Otto cycle, but accomplishing them simultaneously and continuously so that an uninterrupted flow of power from the engine results. Figure 1.1 shows a graph display of the Otto and Brayton cycles.

1.9. BRAYTON CYCLE OF OPERATION

Ambient air is drawn into the inlet section by the rotating compressor. The compressor forces this incoming air rearward and delivers it to the combustion chamber at a higher pressure than the air had at the inlet. The compressed air is then mixed with fuel that is sprayed into the combustion chamber by the fuel nozzles. The fuel and air mixture is then ignited by electrical igniter plugs similar to spark plugs. This ignition system is only in operation during the starting sequence, and once started, combustion is continuous and self-sustaining as long as the engine is supplied with the proper air-fuel ratio. Only about 25 percent of the air is used for combustion. The remaining air is used for internal cooling and pressurizing.

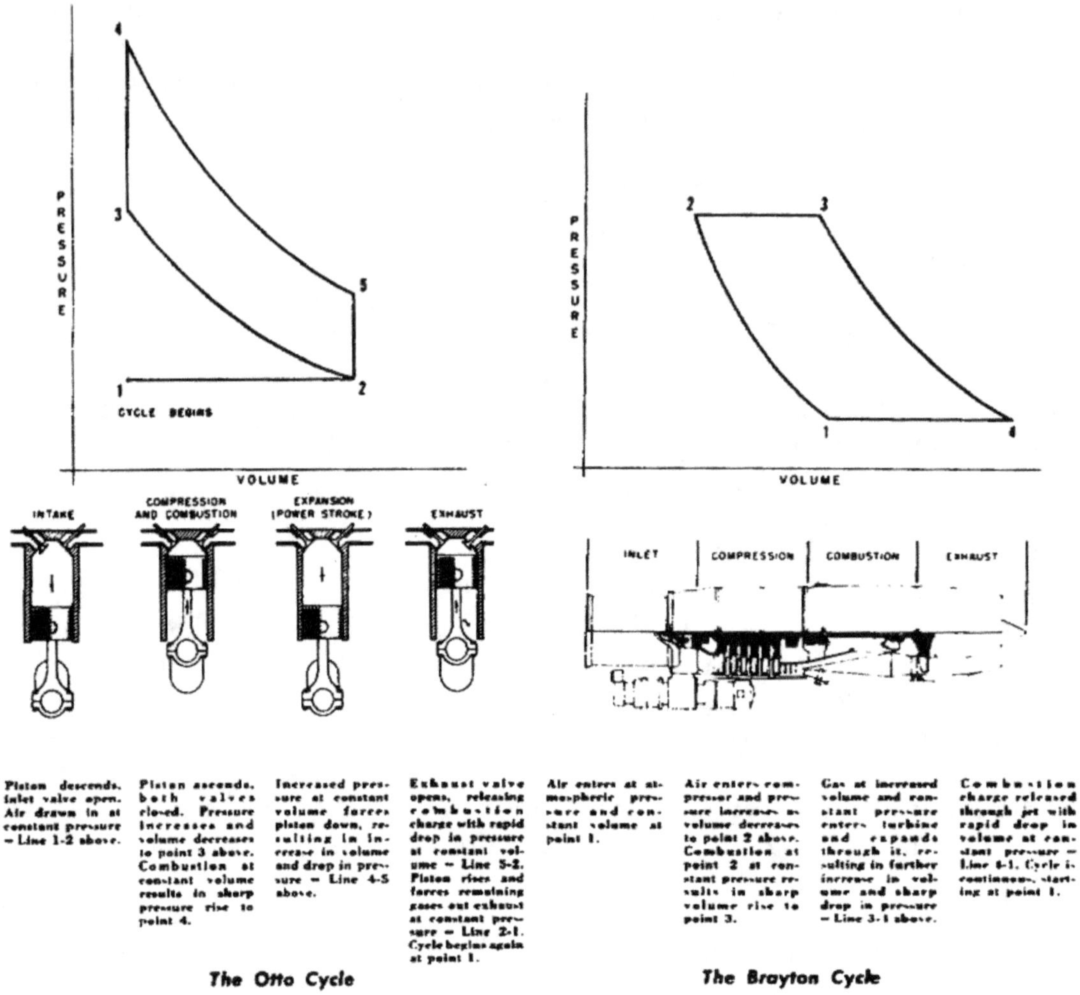

Figure 1.1. Otto and Brayton Cycles.

The turbine engines in the Army inventory are of the free-power turbine design, as shown in figure 1.2. In this engine, nearly two-thirds of the energy produced by combustion is extracted by the gas producer turbine to drive the compressor rotor. The power turbine extracts the remaining energy and converts it to shaft horsepower (shp), which is used to drive the output shaft of the engine. The gas then exits the engine through the exhaust section to the atmosphere. Army helicopters use a divergent duct to eliminate the remaining thrust. The various kinds of exhaust ducting are discussed in detail with the engine using that particular ducting.

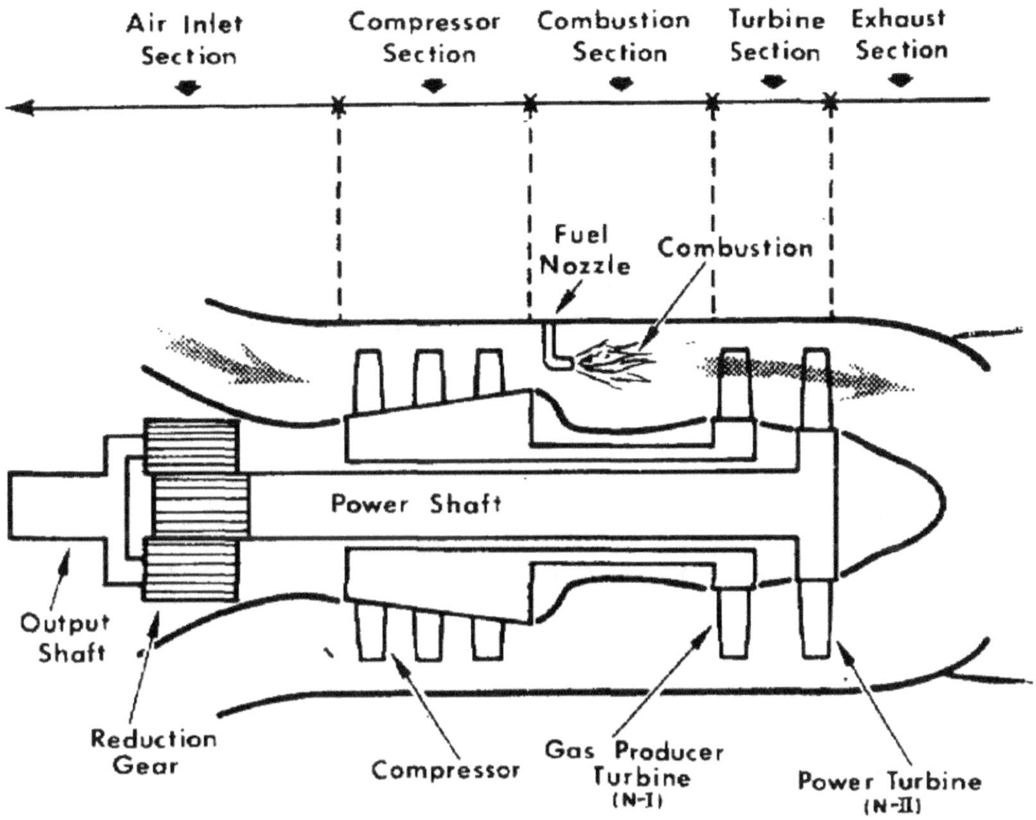

Figure 1.2. Typical Free-Power Turboshaft Engine.

1.10. TURBOJET

The turbojet is the engine in most common use today in high-speed, high-altitude aircraft, not in Army aircraft. With this engine, air is drawn in by a compressor which raises internal pressures many times over atmospheric pressure. The compressed air then passes into a combustion chamber where it is mixed with fuel to be ignited and burned. Burning the fuel-air mixture expands the gas, which is accelerated out the rear as a high-velocity jet-stream. In the turbine section of the engine, the hot expanded gas rotates a turbine wheel which furnishes power to keep the compressor going. The gas turbine engine operates on the principle of intake, compression, power, and exhaust, but unlike the reciprocating engine, these events are continuous. Approximately two-thirds of the total energy developed within the combustion chamber is absorbed by the turbine wheel to sustain operation of the compressor. The remaining energy is discharged from the rear of the engine as a high velocity jet, the reaction to which is thrust or forward movement of the engine. The turbojet is shown schematically in figure 1.3.

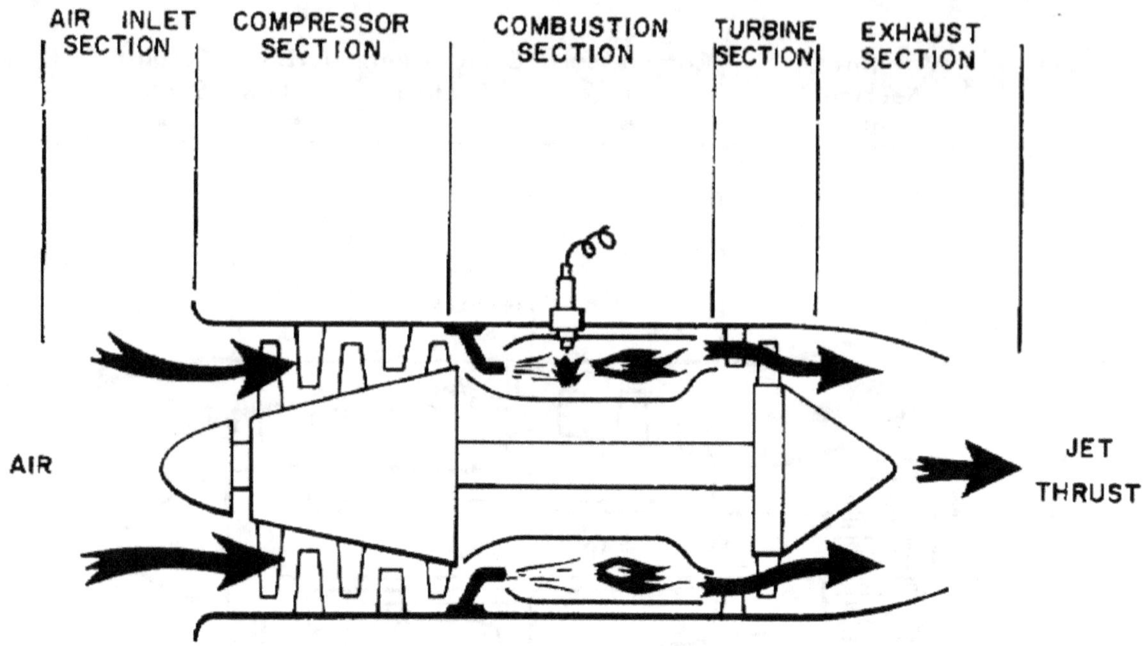

Figure 1.3. Axial-Flow Turbojet Engine.

1.11. TURBOPROP ENGINE AND TURBOSHAFT ENGINE

The turboprop engine and turboshaft engines, shown in figures 1.4 and 1.5, are of the same basic type as the turbojet. Instead of ejecting high-velocity exhaust gases to obtain thrust, as in the turbojet, a turbine rotor converts the energy of the expanding gases to rotational shaft power. A propeller or helicopter transmission can be connected to the engine through reduction gearing. This energy may be extracted by the same turbine rotor that drives the compressor, or it may be a free-power turbine which is independent of the compressor turbine and only linked to it by the expanding gases.

The free-power turbine is the type used in Army aircraft to harness the energy of the gases and convert this energy to rotational shaft power. This feature of having a free-power turbine enables the power output shaft to turn at a constant speed while the power producing capability of the engine can be varied to accommodate the increased loads applied to the power output shaft. Turbine engines may be further divided into three general groups, centrifugal-flow, axial-flow, and axial-centrifugal-flow, depending upon the type of compressor. Figure 1.4 shows an axial-flow turboprop engine, figure 1.5 shows a centrifugal-flow turbojet engine, and figure 1.5a shows an axial-centrifugal-flow compressor.

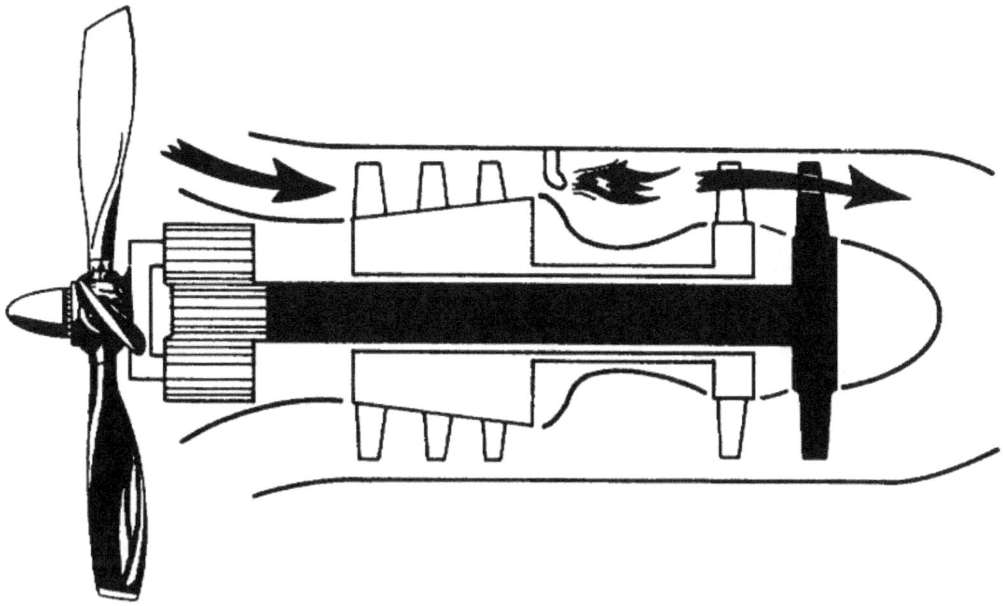

Figure 1.4. Axial-Flow Turboprop Engine.

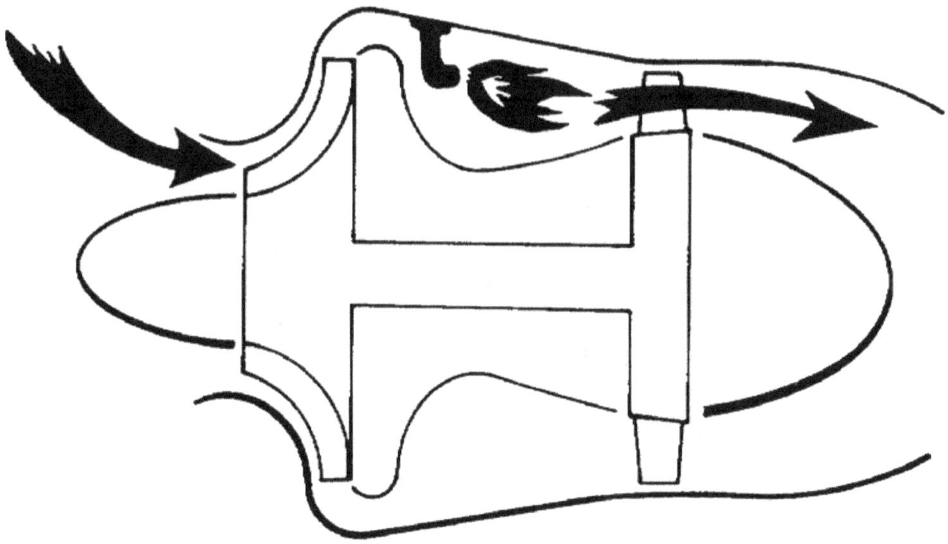

Figure 1.5. Centrifugal-Flow Turbojet Engine.

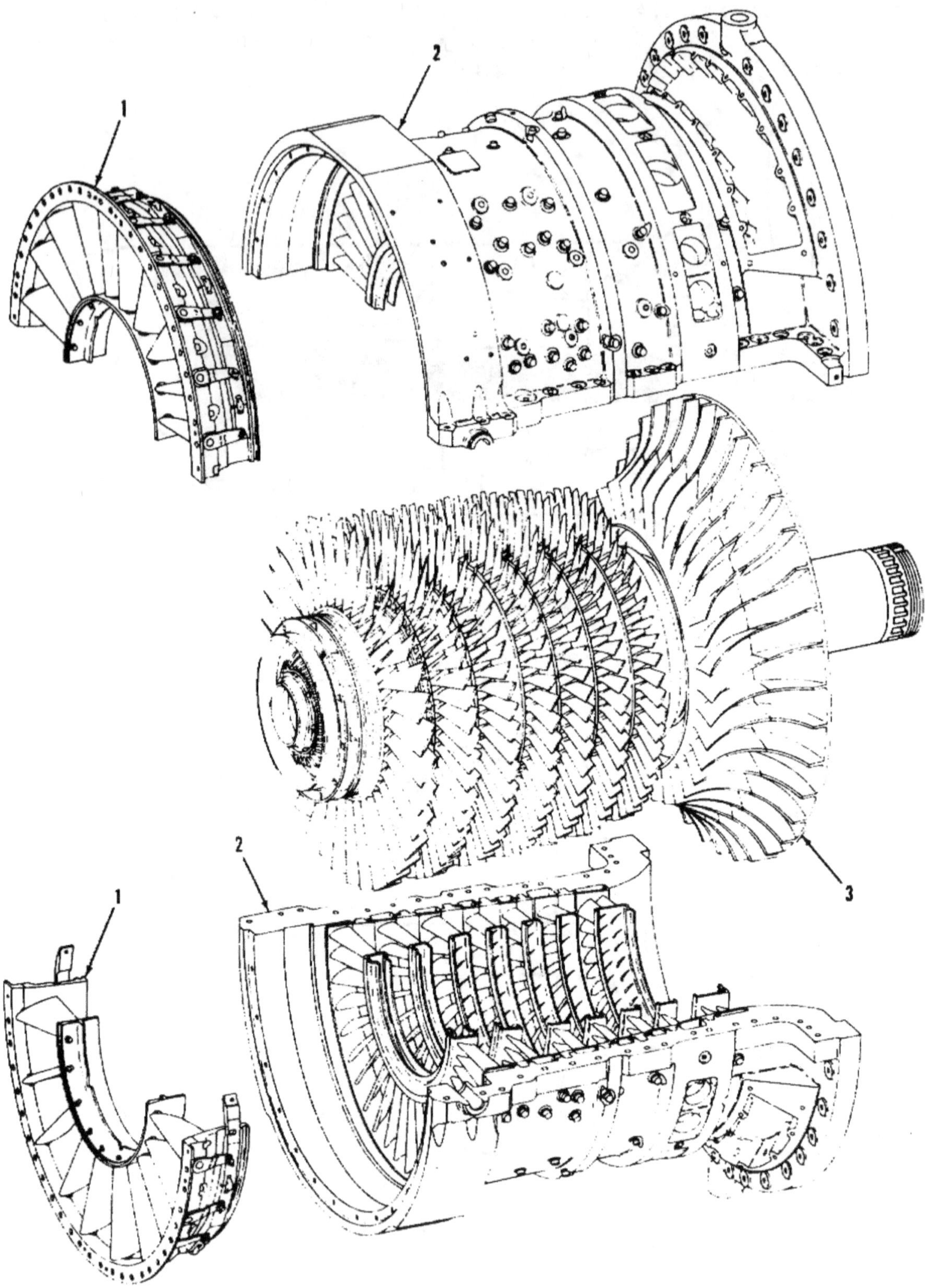

Figure 1.5a. Axial-Centrifugal-Flow Compressor.

1.12. ADVANTAGES OF TURBINE ENGINES

Keeping in mind the basic theory of turbine engines, compare the advantages and disadvantages of the turbine engine with the piston or reciprocating engine. The advantages are covered in the subparagraphs below, and disadvantages are discussed in paragraph 1.13.

a. Power-to-weight ratio. Turbine engines have a higher power-to-weight ratio than reciprocating engines. An example of this is the T55-L-111. It weighs approximately 650 pounds and delivers 3, 750 shaft horsepower. The power-to-weight ratio for this engine is 5.60 shp per pound, where the average reciprocating engine has a power-to-weight ratio of approximately .67 shp per pound.

b. Less maintenance. Maintenance per hour of operation is especially important in military operations. Turbine engines require less maintenance per flying hour than reciprocating engines generally do. As an aircraft maintenance officer, this advantage will appeal to you because of a greater aircraft availability and lower maintenance hour to flying hour ratio. The turbine engine also has fewer moving parts than a reciprocating engine; this is also an advantage over the reciprocating engine.

c. Less drag. Because of the design, the turbine engine has a smaller frontal area than the reciprocating engine. A reciprocating engine requires a large frontal area which causes a great deal of drag on the aircraft. Turbine engines are more streamlined in design, causing less drag. Figure 1.6 shows one of the two nacelles that contain reciprocating engines in the old CH-37 cargo helicopter. Figure 1.7 shows the smaller frontal area of the turbine engines that power the CH-47 Chinook helicopter. Because of this, the engine nacelles are more streamlined in design, causing less drag.

d. Cold weather starting. The turbine engine does not require any oil dilution or preheating of the engine before starting. Also, once started, the reciprocating engine takes a long time to warm up to operating temperatures, whereas the turbine engine starts readily and is up to operating temperature immediately.

e. Low oil consumption. The turbine engine, in general, has a lower rate of oil consumption than the reciprocating engine. The turbine engine does not require the oil reservoir capacity to be as large as the reciprocating engine's; because of this, a weight and economy factor is an additional advantage.

Figure 1.6. Reciprocating Engine Nacelles on CH-37.

Figure 1.7. Turbine Engine Nacelles on CH-47.

1.13. DISADVANTAGES OF TURBINE ENGINES

Just like everything else, along with the advantages or the good, we have to take the disadvantages or the bad. This also holds true with the turbine engine. The disadvantages of the turbine engine are discussed in the following subparagraphs.

a. Foreign object damage. One of the major problems faced by the turbine engine is foreign object damage (FOD). A turbine engine requires tremendous quantities of air. This air is sucked into the engine at extremely high velocities, and it will draw up anything that comes near the inlet area. The turbine engines used in Army aircraft are fitted with filters around the engine inlet to prevent foreign objects from entering the engine and damaging the compressor vanes. However, even with this precaution, FOD is still a menace to turbine engine operation, as shown in figure 1.8.

Figure 1.8. Compressor Foreign Object Damage.

b. High temperatures. In the combustion chamber, the temperature is raised to about 3, 500° F. in the hottest part of the flame. Because this temperature is above the melting point of most metals, proper cooling and flame dilution must be employed at all times to insure that the engine is not damaged.

c. Slow acceleration. The acceleration rate of a turbine engine is very slow in comparison with that of a reciprocating engine. The pilot must be aware of the time lag in the turbine engine acceleration between the instant when power is requested and when power is available.

d. High fuel consumption. Turbine engines are very uneconomical when it comes to the amount of fuel they consume. The Lycoming T53 turbine engine, for instance, uses approximately 1.5 gallons per minute of fuel. Compare it to a reciprocating engine of approximately the same horsepower which has a fuel consumption rate of 1 gallon per minute.

e. Cost. The initial cost of a turbine engine is very high when compared to the cost of a reciprocating engine. For example the T53-L-13B engine costs about $63,000, and the cost of a reciprocating engine of approximately the same horsepower is $20,000.

1.14. SUMMARY

The two turbine engines commonly in use today are the turbojet and turboshaft. The turbine has surpassed the piston engine in design efficiency. The advantages of the gas turbine are a high power-to-weight ratio, less maintenance, and low oil consumption. Because of the small frontal area, turbines have less aerodynamic drag. The disadvantages are foreign object damage to the compressor vanes, high operating temperatures, and high fuel consumption. The turbine also has a slower acceleration rate. Because of the high operating rpm, all rotating parts must be in perfect balance. The cost to manufacture a turbine is much higher than that of a reciprocating engine. Aircraft designers have always been limited by the powerplants available for use on aircraft of new design. Their constant plea has been for higher power, less weight, and a more compact design; the turbine engine has been the answer to some, if not all, of their pleas.

Section III. Major Engine Sections

1.15. GENERAL

Because of the many types of turbine engines, it is not possible to list all the major components and have the list apply to all engines. Several components are common to most turbine engines, and a knowledge of these will be helpful in developing a further understanding of aviation gas turbine engines. This section discusses the major engine sections individually.

1.16. ENGINE TERMINOLOGY

Engine terminology must be explained at this point to enable you to understand the terms used in discussing gas-turbine-engine operating theory explained in this course. Directional references are shown in figure 1.9. Table I shows engine symbols and abbreviations commonly used.

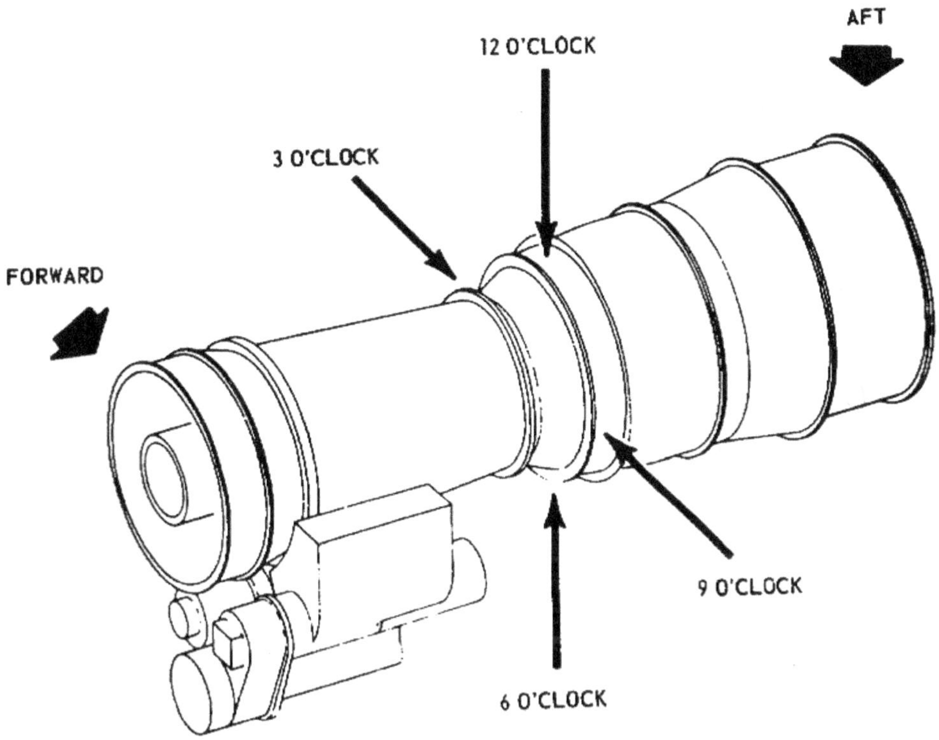

Figure 1.9. Directional References.

a. Directional references. Front or forward -- cold end of engine. Rear or aft -- hot end of engine. Right and left -- determined by viewing the engine from the rear. Bottom -- determined by the location of the combustor drain valve. Top -- directly opposite, or 180 degrees from the combustor drain valve.

These directional references hold true for most gas turbine engines. On some the power shaft is at the end where the exhaust gas is expelled. An engine of this design is the T73 installed on the CH-54 flying crane.

b. Engine station notation. The engine is divided into stations to designate temperature (T) or pressure (P) measuring locations. Figure 1.10 shows a T53-L-13, labeling the engine stations. Any time a number is placed after the letter T or P, it denotes a specific location in the engine.

Example: The symbol T_3, denotes the relative temperature at a specific location on the engine.

Table I. Commonly Used Gas Turbine Engine Symbols and Abbreviations

N	=	Engine rotational speed in rpm or percent of rpm
N_1	=	Speed of first rotating mass: i.e. gas producer (compressor) system
N_2	=	Speed of second rotating mass: i.e. power turbine
%	=	percent of engine rpm
P	=	Pressure
T	=	Temperature
shp	=	shaft horsepower
I.G.V.	=	Inlet guide vanes
V.I.G.V.	=	Variable inlet guide vanes
T.I.T.	=	Turbine inlet temperature
E.G.T.	=	exhaust gas temperature
O.A.T.	=	Outside air temperature

c. Engine speed notation. The rotational speed of the engine is represented by the capital letter N. The first rotating mass, the gas producer has the symbol N_1. Any time a number is placed after the letter N it denotes a specific system on the gas turbine engine.

1.17. ENGINE MODEL DESIGNATIONS

Letter designators are used to differentiate the jet-propulsion engines from reciprocating engines. A letter-number combination identifies each type of the various gas turbine engines. The J series designates true turbojet engines; the T designates turboprop or turboshaft engines; and the TF, turbofan engines.

One of these letters or letter combinations begins each engine model number, each part of which has a special significance. For example, in the engine model number T53-L13, the T means turboshaft; the 53 is simply the number assigned to this model by the Air Force when the engine was accepted or used experimentally. The Air Force/Army designation numbers are odd, while engines developed originally for the Navy get even numbers. The letter L is added by the manufacturer (in this case the Lycoming Division of

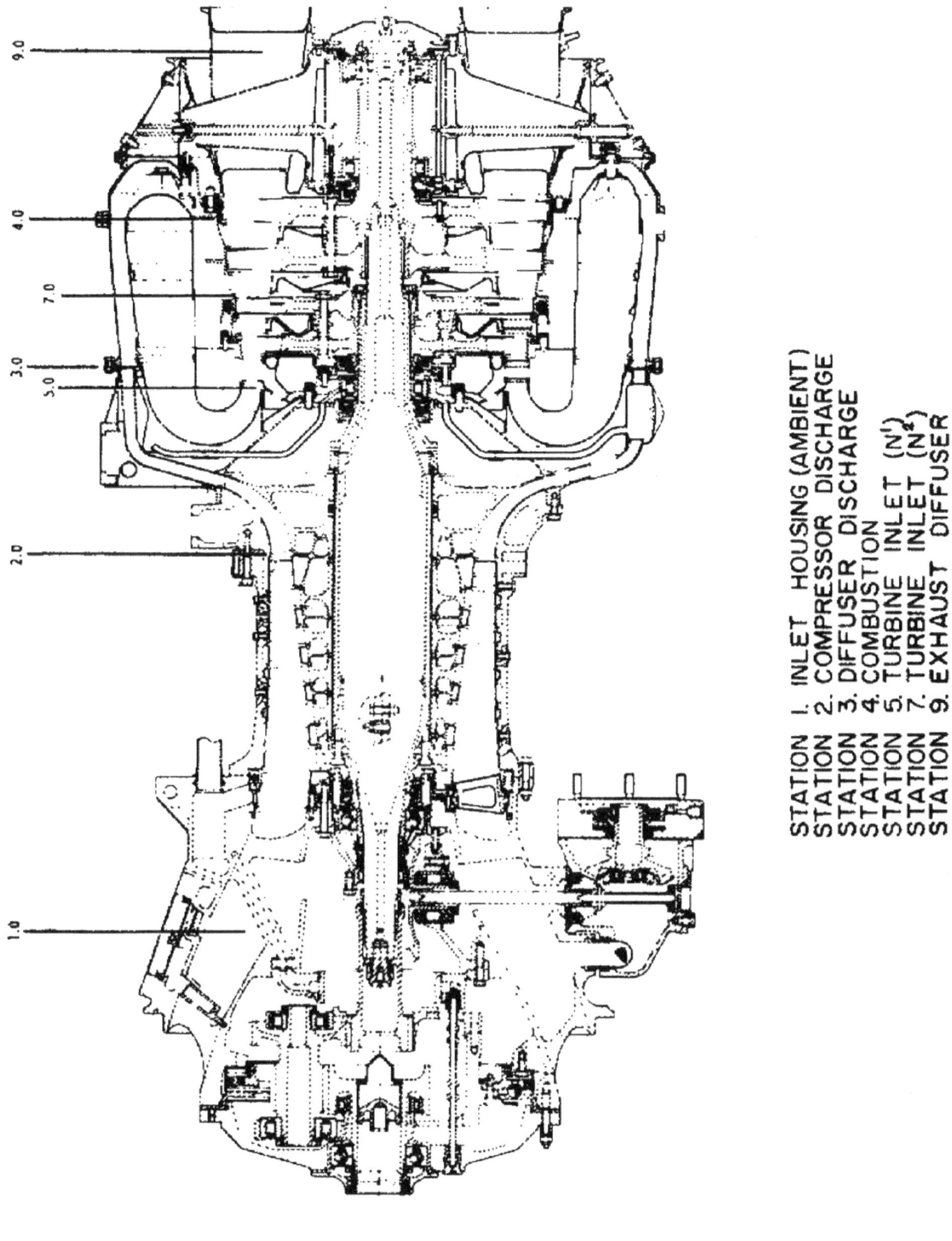

Figure 1.10. Engine Stations.

AVCO) and the -13 is referred to as the "dash number." These are also always odd numbers, if the engine was developed for the Air Force/Army. The dash number designates the particular version of this model engine. When a production model is improved by major modifications, the dash number is changed.

In the past the Army has been getting their engines through the Air Force, using Air Force designators. However, now the Army has its own series of number designators. The first engine procured under the new system was the T700-GE-700 engine. The next engine developed for the Army will be the T701-manufacturer's code-three-digit dash number. The engines that were in the Army inventory previous to this new designator system will keep their present designators. However, when these models are improved, they will get the new three-digit number for a dash number. For example the improved version of the T53-L-15 is the T53-L-701.

1.18. AIR INLET SECTION

The amount of air required by a gas turbine engine is approximately ten times that of a reciprocating engine. The air inlet is generally a large, smooth aluminum or magnesium duct which must be designed to conduct the air into the compressor with minimum turbulence and restriction. The air inlet section may have a variety of names according to the desire of the manufacturer. It may be called the front frame and accessory section, the air inlet assembly, the front bearing support and shroud assembly, or any other term descriptive of its function. Usually, the outer shell of the front frame is joined to the center portion by braces that are often called struts. The anti-icing system directs compressor discharge air into these struts. The temperature of this air prevents the formation of ice that might prove damaging to the engine. Anti-icing systems are discussed further in the chapter covering the engines they may be installed on. Figure 1.11 illustrates the variety of inlet duct designs of Army aircraft.

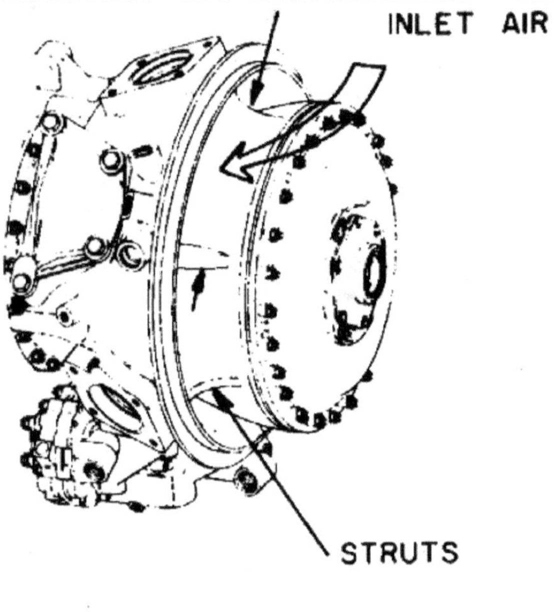

AIR INLET SECTION

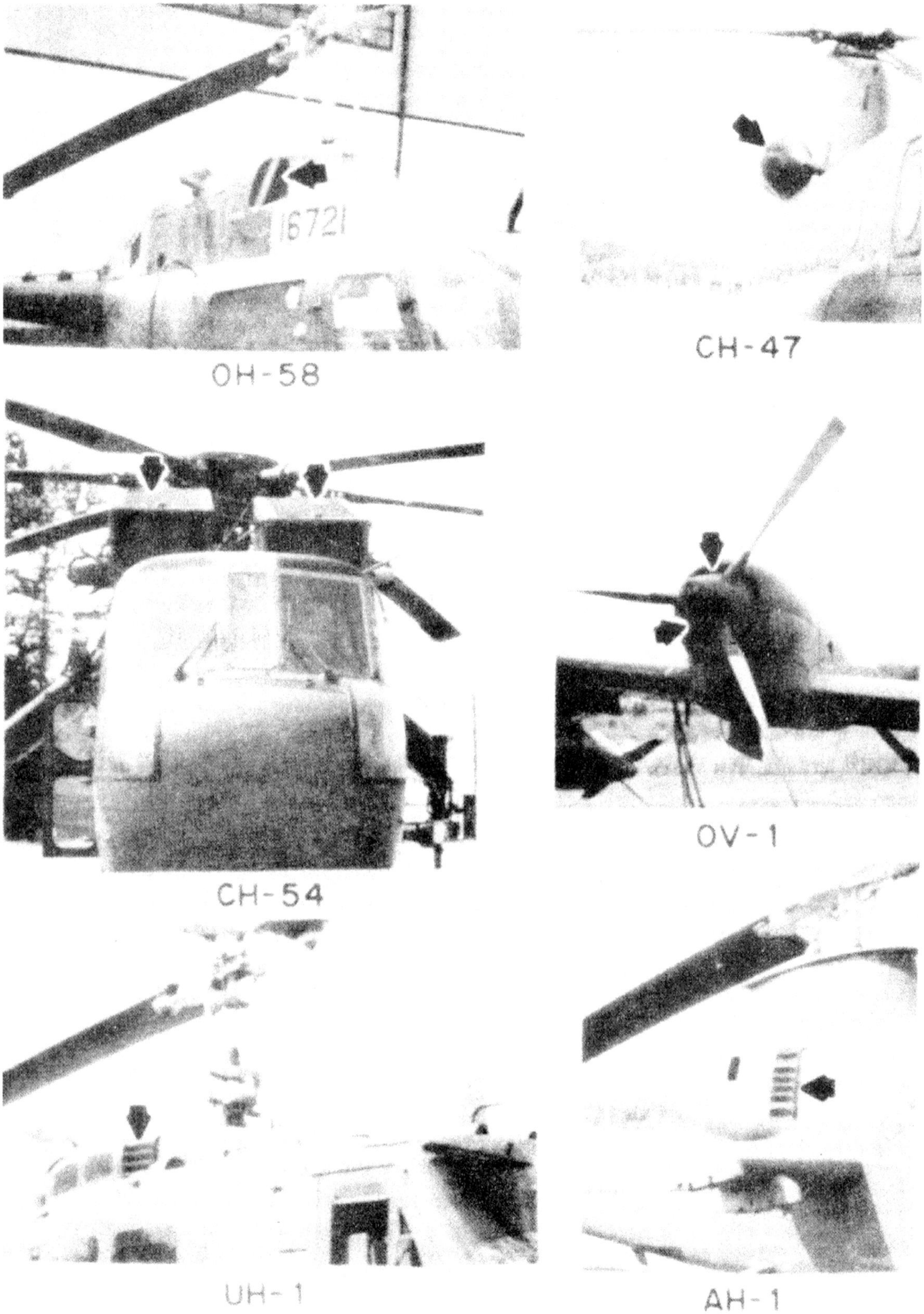

Figure 1.11. Inlet Duct Variations.

1.19. COMPRESSOR SECTION

The compressor is the section of the engine that produces an increase in air pressure. It is made up of rotating and stationary vane assemblies. The first stage compressor rotor blades accelerate the air rearward into the first stage vane assemblies. The first stage vane assemblies slow the air down and direct it into the second stage compressor rotor blades. The second stage compressor rotor blades accelerate the air rearward into the second stage vane assemblies, and so on through the compressor rotor blades and vanes until air enters the diffuser section. The highest total air velocity is at the inlet of the diffuser. As the air passes rearward through the diffuser, the velocity of the air decreases and the static pressure increases. The highest static pressure is at the diffuser outlet.

The compressor rotor may be thought of as an air pump. The volume of air pumped by the compressor rotor is basically proportional to the rotor rpm. However, air density, the weight of a given volume of air, also varies this proportional relationship. The weight per unit volume of air is affected by temperature, compressor air inlet pressure, humidity, and ram air pressure*. If compressor air inlet temperature is increased, air density is reduced. If compressor air inlet pressure is increased, air density is increased. If humidity increases, air density is decreased. Humidity, by comparison with temperature, and pressure changes, has a very small effect on density. With increased forward speed, ram air pressure increases and air temperature and pressure increase.

The following is an example of how air density affects compressor efficiency of the T62 gas turbine. At 100 percent N_1 rpm, the compressor rotor pumps approximately 40.9 cubic feet of air per second. At standard day static sea level conditions, 59° F outside air temperature and 29.92" Hg barometric pressure, with 0 percent relative humidity and 0 ram air, air density is .07651 pound per cubic foot. Under these conditions, 40.9 cubic feet per second times .07651 pound per cubic feet equals approximately 3.13 pounds per second air flow through the engine. If the air density at the compressor inlet is less than on a standard day, the weight of air flow per second through the engine is less than 3.13 pounds per second. If N_1 is less than 100 percent rpm on a standard day, the weight of air flow per second through the engine will be less than 3. 13 due to decreased volume flow at lower rpm. Because of this, N_1 rpm varies

*ram air pressure - free stream air pressure provided by the forward motion of the engine.

with the power output. If output power is increased, N_1 rpm will increase and vice versa. Thus, the weight of air pumped by the compressor rotor is determined by rpm and air density.

Compressor efficiency determines the power necessary to create the pressure rise of a given airflow, and it affects the temperature change which takes place in the combustion chamber. Therefore, the compressor is one of the most important components of the gas turbine engine because its efficient operation is the key to overall engine performance. The following subparagraphs discuss the three basic compressors used in gas turbine engines: the centrifugal-flow, the axial-flow, and axial-centrifugal-flow compressors. The axial-centrifugal-flow compressor is a combination of the other two and operates with characteristics of both.

a. Centrifugal-flow compressor. Figure 1.12 shows the basic components of a centrifugal-flow compressor: rotor, stator, and compressor manifold.

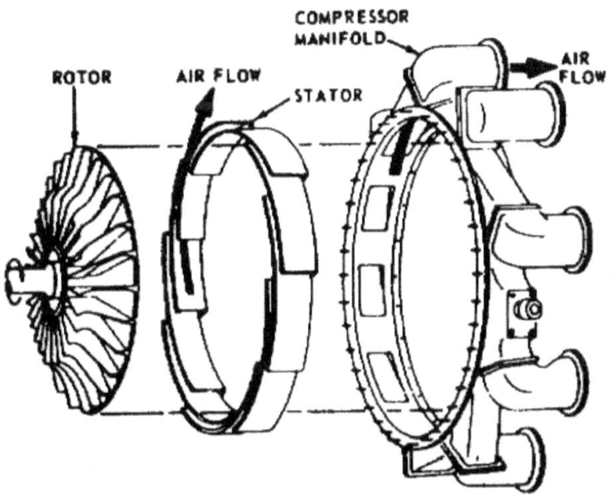

Figure 1.12. Typical Single-stage Centrifugal Compressor

As the impeller (rotor) revolves at high speed, air is drawn into the blades near the center. Centrifugal force accelerates this air and causes it to move outward from the axis of rotation toward the rim of the rotor where it is forced through the diffuser section at high velocity and high kinetic energy. The pressure rise is produced by reducing the velocity of the air in the diffuser, thereby converting velocity energy to pressure energy. The centrifugal compressor is capable of a relatively high compression ratio per stage. This compressor is not used on larger engines because of size and weight.

Because of the high tip speed problem in this design, the centrifugal compressor finds its greatest use on the smaller engines where simplicity, flexibility of operation, and ruggedness are the principal requirements rather than small frontal area and ability to handle high airflows and pressures with low loss of efficiency.

b. Axial-flow compressor. The air is compressed, as the name implies, in a direction parallel to the axis of the engine. The compressor is made of a series of rotating airfoils called rotor blades, and a stationary set of airfoils called stator vanes. A stage consists of two rows of blades, one rotating and one stationary. The entire compressor is made up of a series of alternating rotor and stator vane stages as shown in figure 1.13.

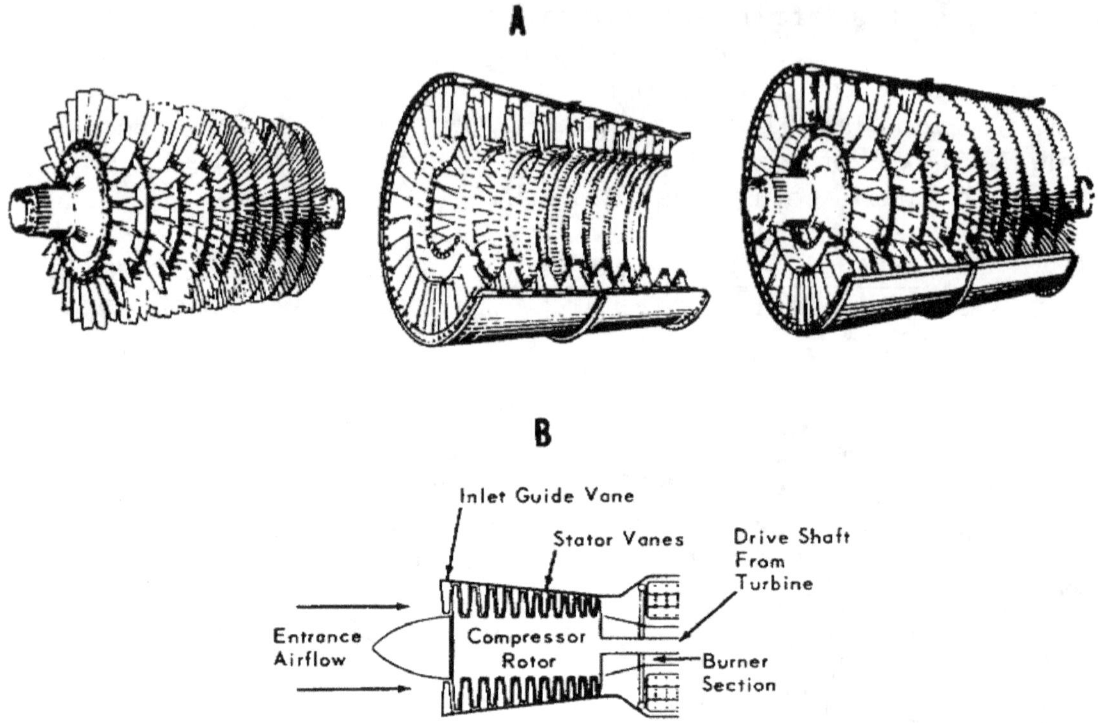

Figure 1.13. Axial-flow Compressor.

Axial flow compressors have the advantage of being capable of very high compression ratios with relatively high efficiencies; see figure 1.14. Because of the small frontal area created by this type of compressor, it is ideal for installation on high-speed aircraft. Unfortunately, the delicate blading and close tolerances, especially toward the rear of the compressor where the blades are smaller and more numerous per stage, make this compressor highly susceptible to foreign-object damage. Because of the close fits required

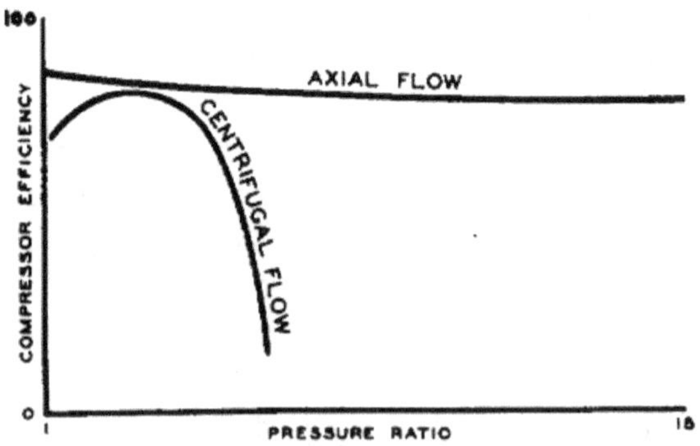

Figure 1.14. Compressor Efficiencies and Pressure Ratios.

for efficient air-pumping and higher compression ratios, this type of compressor is very complex and very expensive to manufacture. For these reasons the axial-flow design finds its greatest application where required efficiency and output override the considerations of cost, simplicity, and flexibility of operation. However, due to modern technology, the cost of the small axial-flow compressor, used in Army aircraft, is coming down.

c. Axial-centrifugal-flow compressor. The axial-centrifugal-flow compressor, also called the dual compressor, is a combination of the two types, using the same operating characteristics. Figure 1.15 shows the compressor used in the T53 turbine engine. Most of the gas turbine engines used in Army aircraft are of the dual compressor design. Usually it consists of a five- or seven-stage axial-flow compressor and one centrifugal-flow compressor. The dual compressors are mounted on the same shaft and turn in the same direction and at the same speed. The centrifugal compressor is mounted aft of the axial compressor. The axial compressor contains numerous air-foil-shaped blades and vanes that accomplish the task of moving the air mass into the combustor at an elevated pressure.

As the air is drawn into the engine, its direction of flow is changed by the inlet guide vanes. The angle of entry is established to ensure that the air flow onto the rotating compressor blades is within the stall-free (angle of attack) range. Air pressure or velocity is not changed as a result of this action. As the air passes from the trailing edge of the inlet guide vanes, its direction of flow is changed due to the rotational effect of the compressor. This change of airflow direction is similar to the action that takes place when a car is driven during a rain or snow storm. The rain or snow falling in a vertical direction strikes the windshield at an angle due to the horizontal velocity of the car.

In conjunction with the change of airflow direction, the velocity of the air is increased. Passing through the rotating compressor blades, the velocity is decreased, and a gain in pressure is obtained. When leaving the trailing edge of the compressor blades, the velocity of the air mass is again increased by the rotational effect of the compressor. The angle of entry on to the stationary stator vanes results from this rotational effect as it did on the airflow onto the compressor.

Passing through the stationary stator vanes the air velocity is again decreased resulting in an increase in pressure. The combined action of the rotor blades and stator vanes results in an

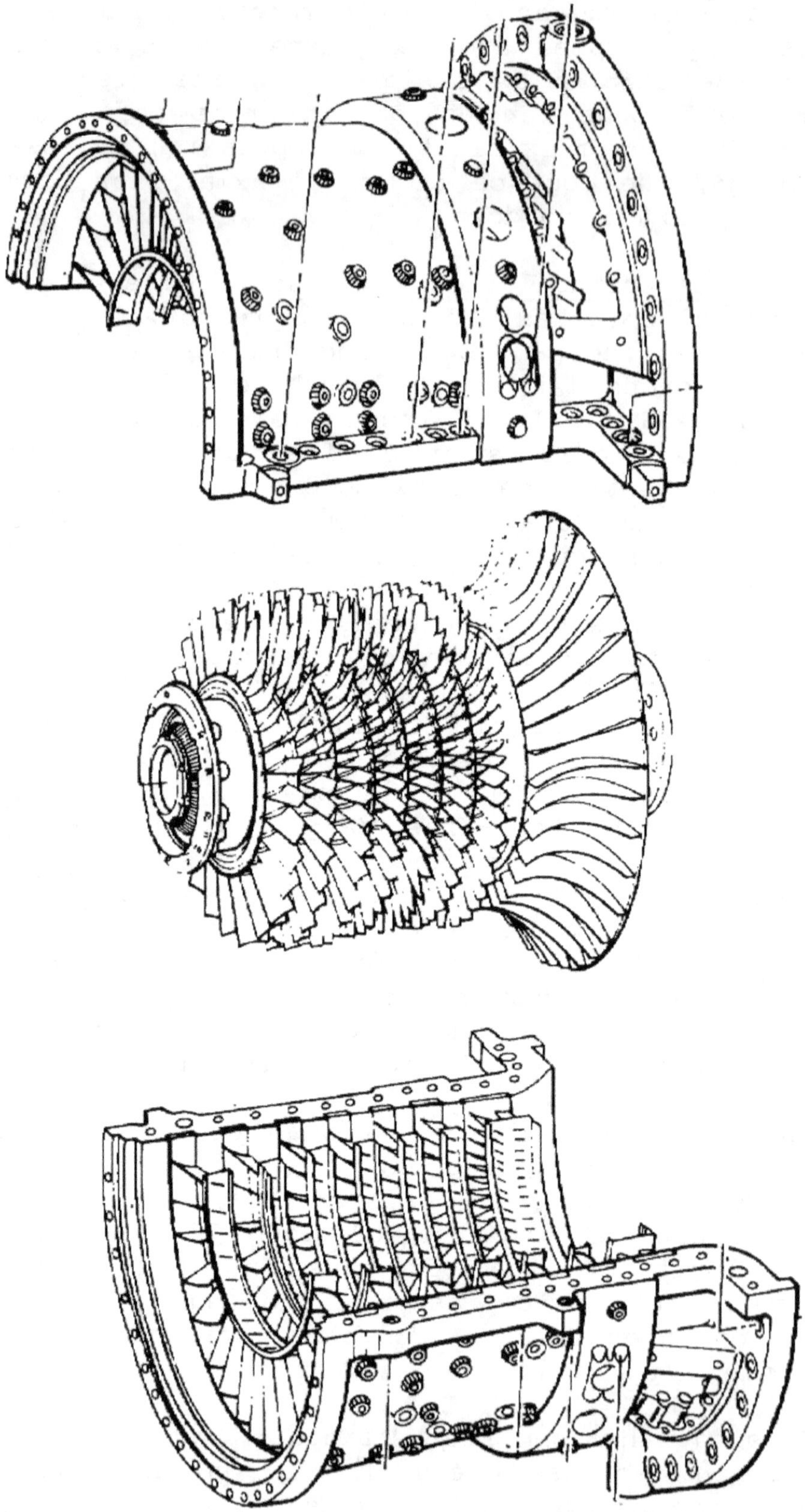

Figure 1.15. Axial-Centrifugal-Flow Compressor.

increase in air -pressure; combined they constitute one stage of compression. This action continues through all stages of the axial compressor. To retain this pressure buildup, the airflow is delivered, stage by stage, into a continually narrowing airflow path. After passing from the last set of stator vanes the air mass passes through exit guide vanes. These vanes direct the air onto the centrifugal impeller.

The centrifugal impeller increases the velocity of the air mass as it moves it in a radial direction. The axial-centrifugal-flow compressor is discussed further in chapter 4.

d. Compressor stall. Gas turbine engines are designed to avoid the pressure conditions that allow engine surge to develop, but the possibility of surge still exists in engines that are improperly adjusted or have been abused. Engine surge occurs any time the combustion chamber pressure exceeds that in the diffuser, and it is identified by a popping sound which is issued from the inlet. Because there is more than one cause for surge, the resultant sound can range from a single carburetor backfire pop to a machinegun sound.

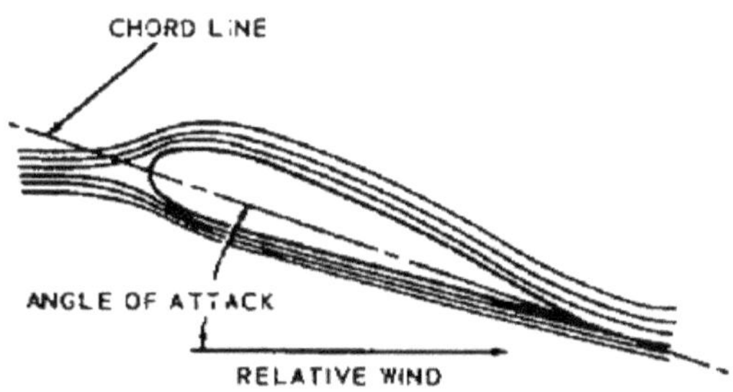

ANGLE OF ATTACK

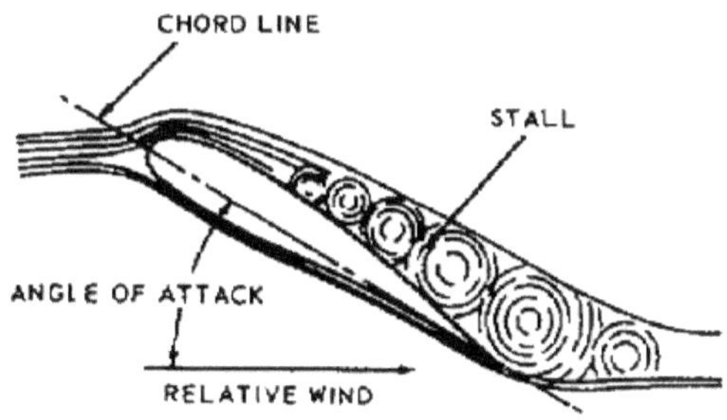

AIRFOIL ANGLES OF ATTACK VS LIFT

SCHEMATIC DIAGRAM ILLUSTRATING STALL

Engine surge is caused by a stall on the airfoil surfaces of the rotating blades or stationary vanes of the compressor. The stall can occur on individual blades or vanes or, simultaneously, on groups of them. To understand how this can induce engine surge, the causes and effects of stall on any airfoil must be examined.

All airfoils are designed to provide lift by producing a lower pressure on the convex (suction) side of the airfoil than on the concave (pressure) side. A characteristic of any airfoil is that lift increases with an increasing angle of attack, but only up to a critical angle. Beyond this critical angle of attack, lift falls off rapidly. This is due largely to the separation of the airflow from the suction surface of the airfoil, as shown in the sketch. This phenomenon is known as stall. All pilots are familiar with this condition and its consequences as it applies to the wing of an aircraft. The stall that takes place on the fixed or rotating blades of a compressor is the same as the stalling phenomenon of an aircraft wing.

1.20. COMPRESSOR CONSTRUCTION

Centrifugal-flow compressors are usually made of titanium. The diffuser is generally manufactured of a stainless steel alloy. A close fit is important between the compressor and its case to obtain maximum compressor efficiency. Correct rotor assembly balancing is essential for safe operation because of the high rpm. Balancing the rotor can be accomplished by removing metal from specified areas of the compressor or by using balancing weights installed in holes in the hub of the compressor. On some engines where the compressor and turbine wheel are balanced as a unit, special bolts and nuts having slight variations in weight are used.

Axial-flow compressors are constructed of many different materials, depending upon the load and temperature under which the unit must operate. The rotor blades are generally cast of stainless-steel alloy. Some manufacturers use mdybdenum coated titanium blades to dampen vibrations on some stages of rotor blades. The clearance between the rotor blades and the outer case is most important. Some companies coat the inner surface of the compressor case with a soft material that can be worn away by the blades as they expand because of the heat generated from compressing the air. This type of compressor uses the "wear-fit" method to form its own clearance between the compressor case and the rotor blade tip.

Methods of attaching the blade to the disk or hub vary between manufacturers, with the majority using some variation of the dove-tail method to hold the rotor blades to the disk. Various other methods are used to anchor the blades in place. Some blades do not have a tight fit in the disk, but rather are seated by centrifugal force during engine operation. By allowing the blades to move, vibrational stress is reduced during start and shutdown. Stator vanes, shown in figure 1.16, can be either solid or hollow construction, and are connected together at their tips by a shroud. This shrouding serves two purposes. First, it provides support, and second, it provides the necessary air seal between rotating and stationary parts. Most manufacturers use the split compressor cases, while some others favor a weldment, forming a continuous case. The advantages of the split case lie in the fact that the compressor and stator blades are readily available to inspection. The one-piece case offers simplicity and strength because it is one piece; in most instances, it is a principal structural part of the engine and is usually made of cast aluminum, magnesium, or steel. Figures 1.16 and 1.17 show shrouded compressor stators in both the split case and the one-piece case.

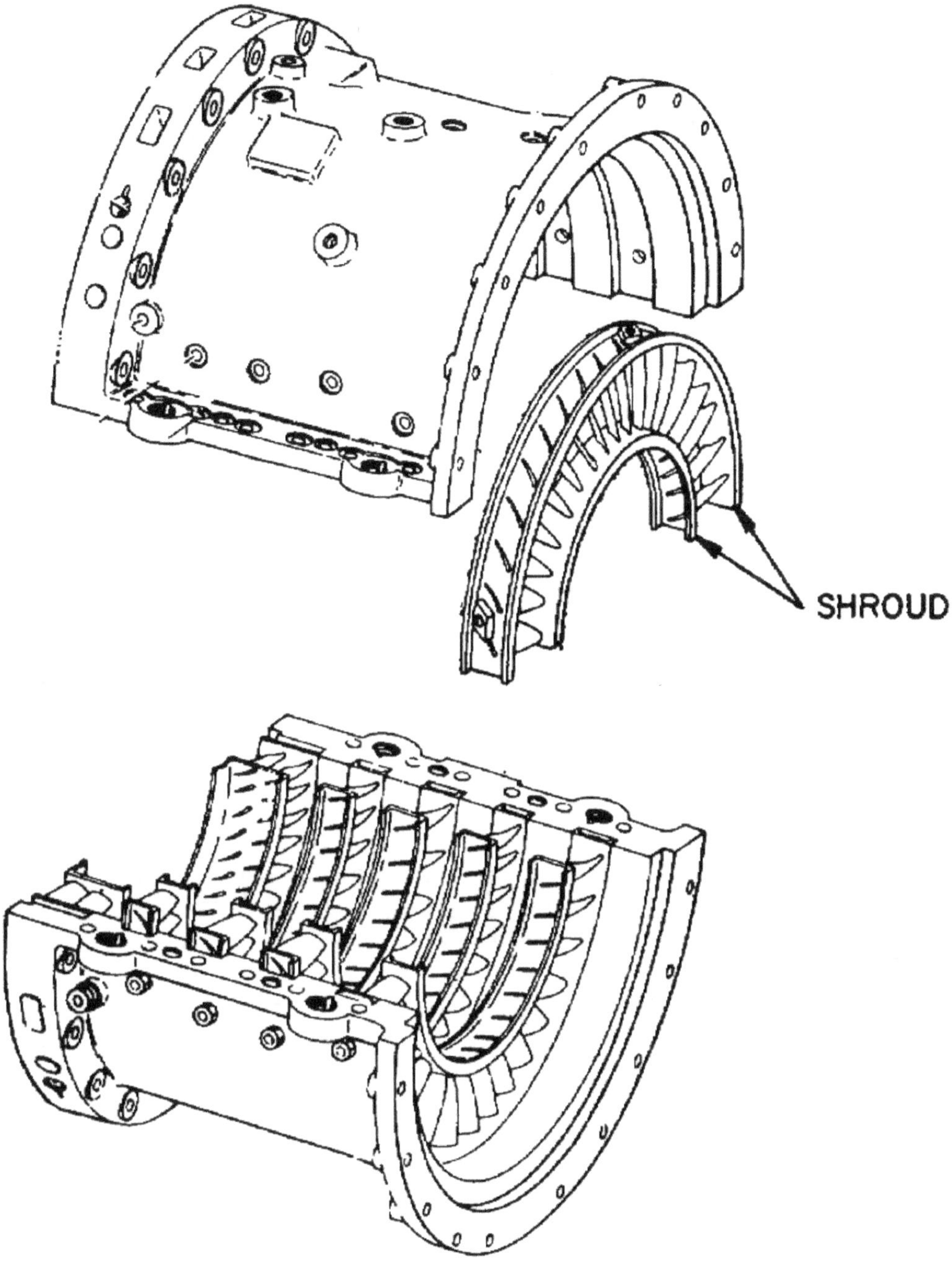

Figure 1.16. Shrouded Compressor Stators.

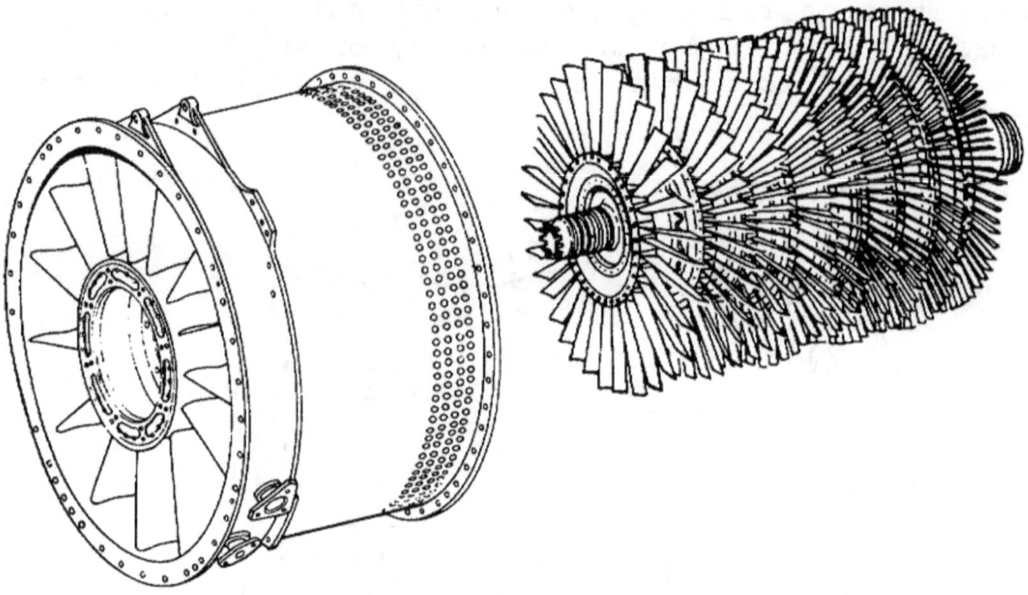

Figure 1.17. One-Piece Compressor Case.

1.21. COMBUSTION SECTION

Today, three basic combustion chambers are in use. They are the annular combustion chamber, the can type, and the combination of the two called the can-annular. Variations of these basic systems are used in a number of engines. The three systems are discussed individually in the following subparagraphs. The most commonly used gas turbine engine in Army aircraft is the annular reverse-Row type. The combustion section contains the combustion chambers, igniter plugs, and fuel nozzles or vaporizing tubes. It is designed to burn a fuel-air mixture and deliver the combusted gases to the turbine at a temperature which will not exceed the allowable limit at the turbine inlet.

Fuel is introduced at the front end of the burner in a highly atomized spray from the fuel nozzles. Combustion air flows in around the fuel nozzle and mixes with the fuel to form a correct fuel-air mixture. This is called primary air and represents approximately 25 percent of total air taken into the engine. The fuel-air mixture which is to be burned is a ratio of 15 parts of air to 1 part of fuel by weight. The remaining 75 percent of the air is used to form an air blanket around the burning gases and to lower the temperature. This temperature may reach as high as 3500° F. By using 75 percent of the air for cooling, the temperature operating range can be brought down to about half, so the turbine section will not be destroyed by excessive heat. The air used for burning is called primary air- and that for cooling is secondary air. The secondary air is controlled and directed by holes and louvers in the combustion chamber liner.

Igniter plugs function only during starting, being cut out of the circuit as soon as combustion is self-supporting. On engine shutdown, or, if the engine fails to start, the combustion chamber drain valve, a pressure-actuated valve, automatically drains any remaining unburned fuel from the combustion chamber. All combustion chambers contain the same basic elements: a casing or outer shell, a perforated inner liner or flame tube, fuel nozzles, and some means of initial ignition. The combustion chamber must be of light construction and is designed to burn fuel completely in a high velocity airstream. The combustion chamber liner is an extremely critical engine part because of the high temperatures of the flame. The liner is usually constructed of welded high-nickel steel. The most severe operating periods in combustion chambers are encountered in the engine idling and maximum rpm ranges. Sustained operation under these conditions must be avoided to prevent combustion chamber liner failure.

<u>a</u>. <u>The annular-type combustion chamber</u> shown in figure 1.18 is used in engines of the axial-centrifugal-flow compressor de-

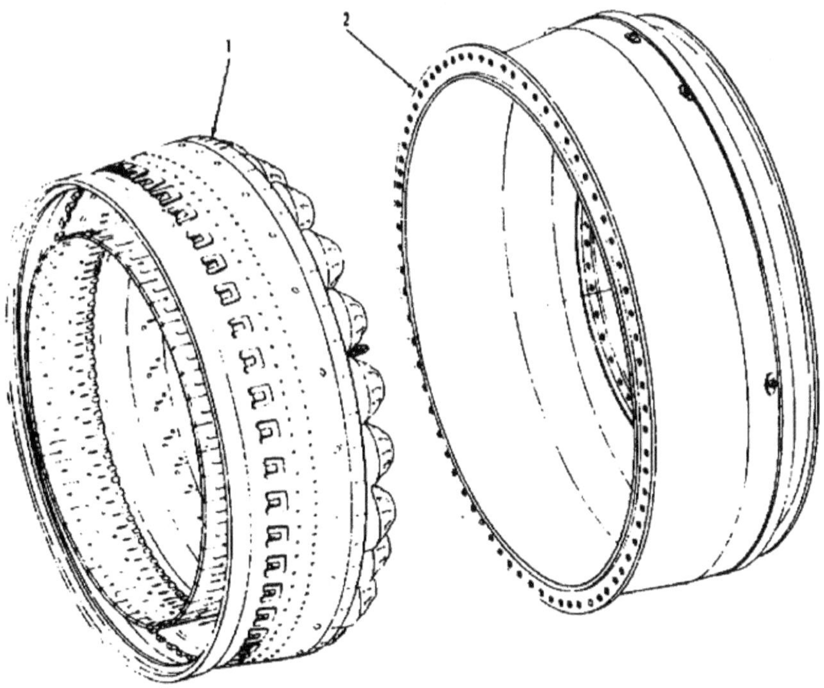

1. ANNULAR TYPE COMBUSTION CHAMBER LINER
2. COMBUSTION CHAMBER HOUSING ASSEMBLY

Figure <u>1.18</u>. <u>Annular-type Combustion Chamber</u>.

sign. The annular combustion chamber permits building an engine of a small and compact design. Instead of individual combustion chambers, the primary compressed air is introduced into an annular space formed by a chamber liner around the turbine assembly. A space is left between the outer liner wall and the combustion chamber housing to permit the flow of secondary cooling air from the compressor. Primary air is mixed with the fuel for combustion. Secondary (cooling) air reduces the temperature of the hot gases entering the turbine to the proper level by forming a blanket of cool air around these hot gases.

The annular combustion chamber offers the advantages of a larger combustion volume per unit of exposed area and material weight, a smaller exposed area resulting in lower pressure losses through the unit, and less weight and complete pressure equalization.

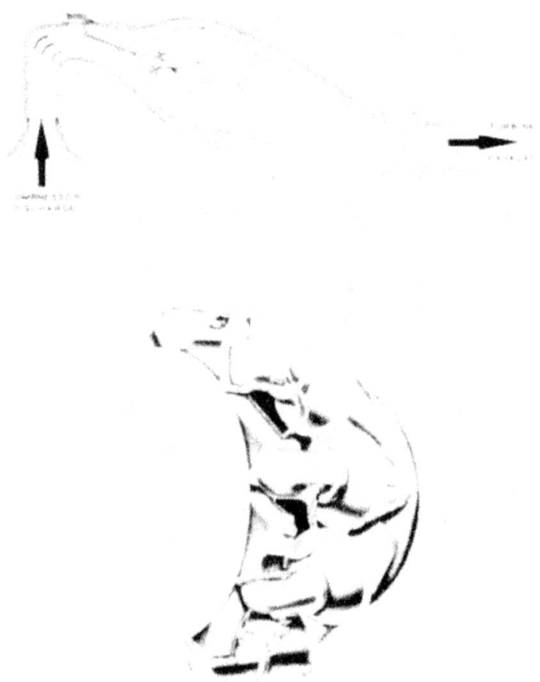

Figure 1.19. Can-type Combustion Chamber (Cutaway).

b. The can-type combustion chamber is one made up of individual combustion chambers. This type of combustion chamber is so arranged that air from the compressor enters each individual chamber through the adapter. Each individual chamber is composed of two cylindrical tubes, the combustion chamber liner and the outer combustion chamber, shown in figure 1.19. Combustion takes place within the liner. Airflow into the combustion area is controlled by small louvers located in the inner dome, and by round holes and elongated louvers along the length of the liner. Airflow into the combustion area is controlled by small louvers located in the inner dome, and by round holes elongated louvers along the length of the liner. Through these openings flows the air that is used in combustion and cooling. This air also prevents carbon deposits from forming on the inside of the liner. This is important, because carbon deposits can block critical air passages and disrupt airflow along the liner walls causing high metal temperatures and short burner life.

Ignition is accomplished during the starting cycle. The igniter plug is located in the combustion liner adjacent to the start fuel nozzle. The Army can-type engine employs a single can-type combustor.

c. Can-annular combustion chamber. This combustion chamber uses characteristics of both annular and can-type combustion chambers. The can-annular combustion chamber consists of an outer shell, with a number of individual cylindrical liners mounted about the engine axis as shown in figure 1.20. The combustion chambers are completely surrounded by the airflow that enters the liners through various holes and louvers. This air is mixed with fuel which has been sprayed under pressure from the fuel nozzles. The fuel-air mixture is ignited by igniter plugs, and the flame is then carried through the crossover tubes to the remaining liners. The inner casing assembly is both a support and a heat shield; also, oil lines run through it.

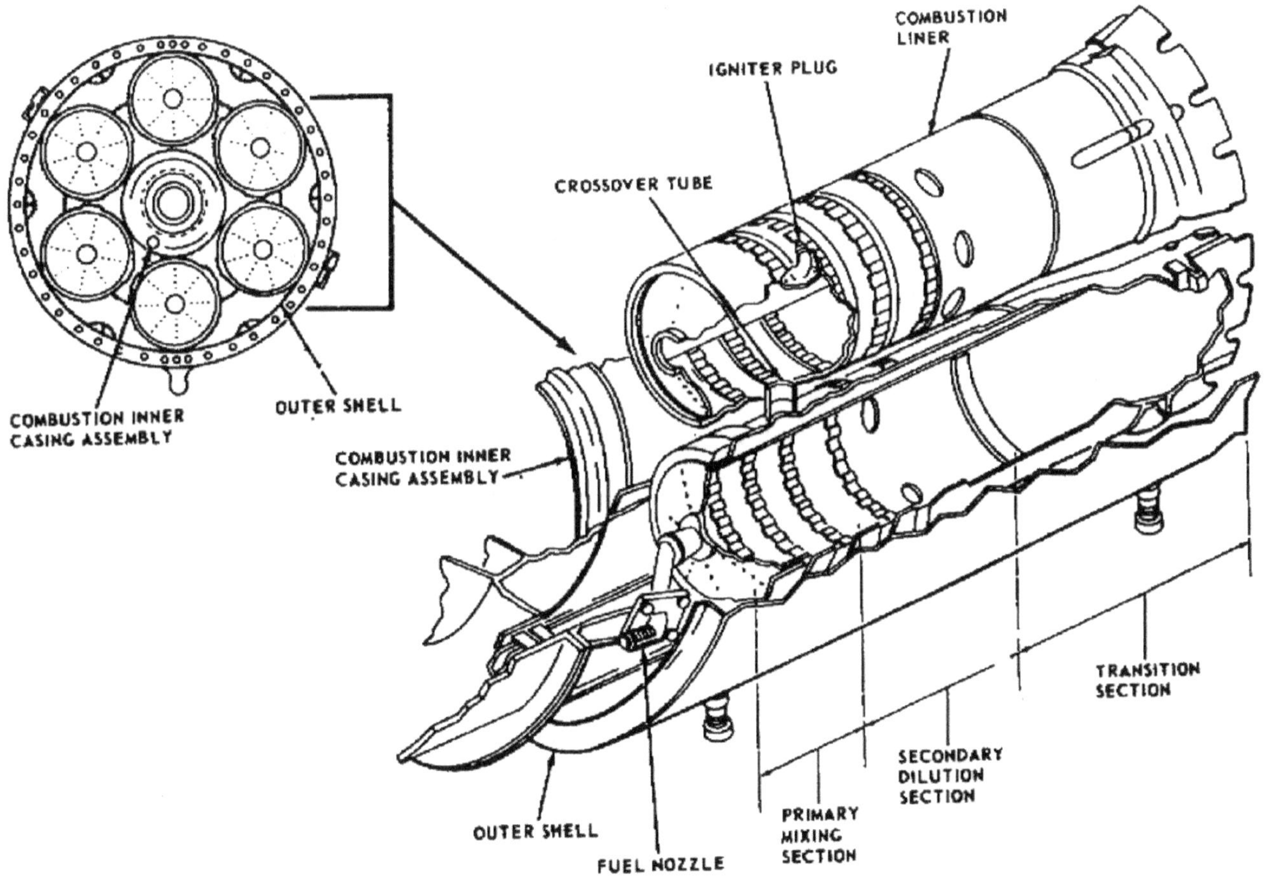

Figure 1.20. Can-Annular Combustion Chamber.

1.22. TURBINE SECTION

A portion of the kinetic energy of the expanding gases is extracted by the turbine section, and this energy is transformed into shaft horsepower which is used to drive the compressor and accessories. In turboprop and turboshaft engines, additional turbine rotors are designed to extract all of the energy possible from the remaining gases to drive a powershaft.

a. Types of turbines. Gas turbine manufacturers have concentrated on the axial-flow turbine shown in figure 1.21. This turbine is used in all gas-turbine-powered aircraft in the Army today. However, some manufacturers are building engines with a radial inflow turbine, illustrated in figure 1.22. The radial inflow turbine has the advantage of ruggedness and simplicity, and it is relatively inexpensive and easy to manufacture when compared to the axial-flow turbine. The radial flow turbine is similar in design and construction to the centrifugal-flow compressor described in paragraph 1.19a. Radial turbine wheels used for small

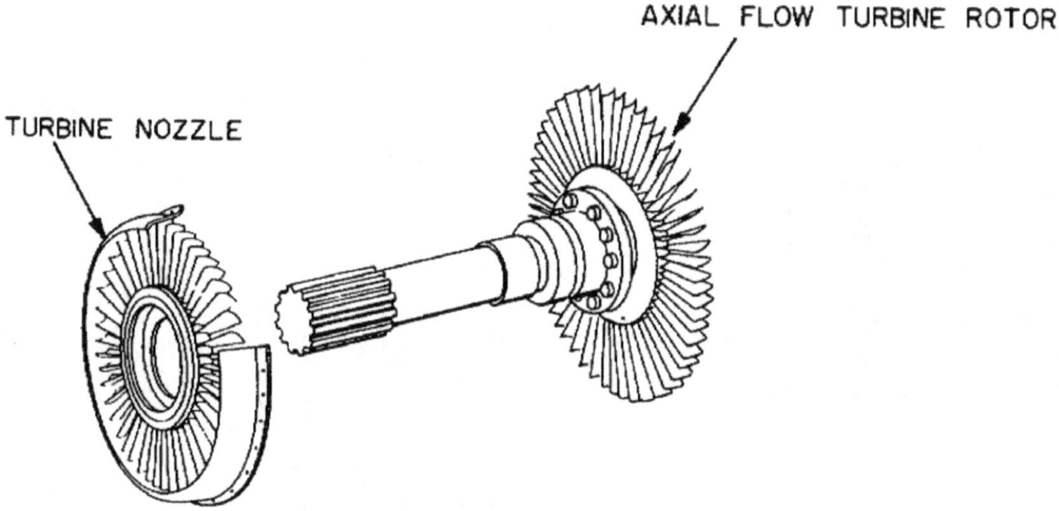

Figure 1.21. Axial-flow Turbine Rotor.

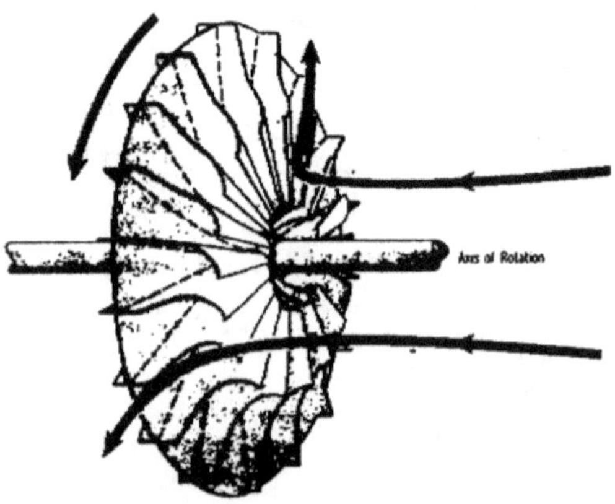

Figure 1.22. Radial Inflow Turbine.

engines are well suited for a higher range of specific speeds and work at relatively high efficiency.

The axial-flow turbine consists of two main elements, a set of stationary vanes followed by a turbine rotor. Axial-flow turbines may be of the single-rotor or multiple-rotor type. A stage consists of two main components: a turbine nozzle and a turbine rotor or wheel, as shown in figure 1.21. Turbine blades are of two basic types, the impulse and the reaction. Modern aircraft gas turbines use blades that have both impulse and reaction sections, as shown in figure 1.23.

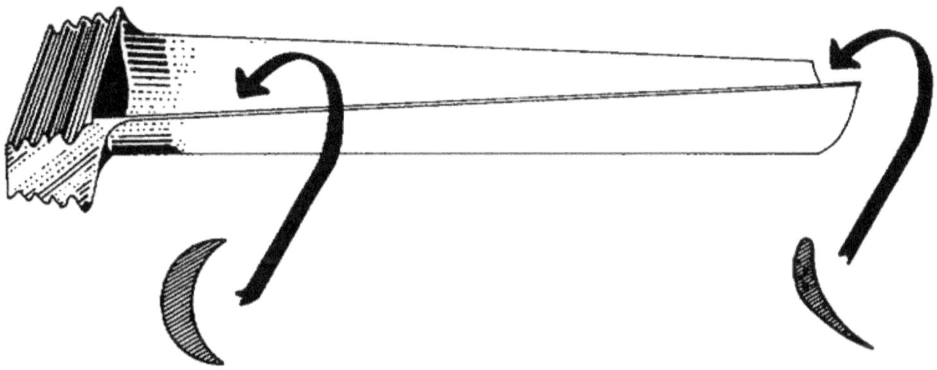

Figure 1.23. Impulse-Reaction Turbine Blade.

The stationary part of the turbine assembly consists of a row of contoured vanes set at a predetermined angle to form a series of small nozzles which direct the gases onto the blades of the turbine rotor. For this reason, the stationary vane assembly is usually called the turbine nozzle, and the vanes are called nozzle guide vanes.

b. Single-rotor turbine. Some gas turbine engines use a single-rotor turbine, with the power developed by one rotor. This arrangement is used on engines where low weight and compactness are necessary. A single-rotor, single-stage turbine engine is shown in figure 1.24, and a multiple-rotor, multiple-stage turbine engine is shown in figure 1.25.

c. Multiple-rotor turbine. In the multiple-rotor turbine the power is developed by two or more rotors. As a general rule, multiple-rotor turbines increase the total power generated in a unit of small diameter. Generally the turbines used in Army aircraft engines have multiple rotors. Figure 1.26 illustrates a multistage, multiple-rotor turbine assembly.

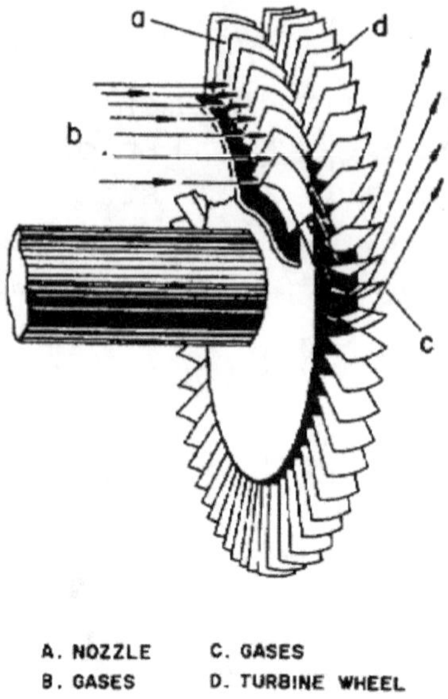

Figure 1.24. Single-rotor, Single-stage Turbine.

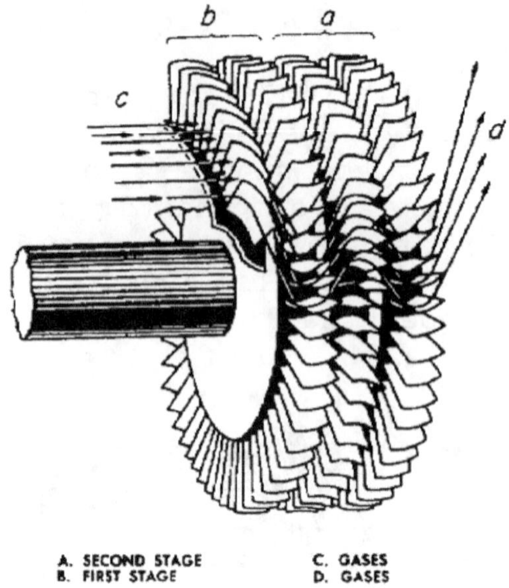

Figure 1.25. Multiple-rotor, Multiple-stage Turbine.

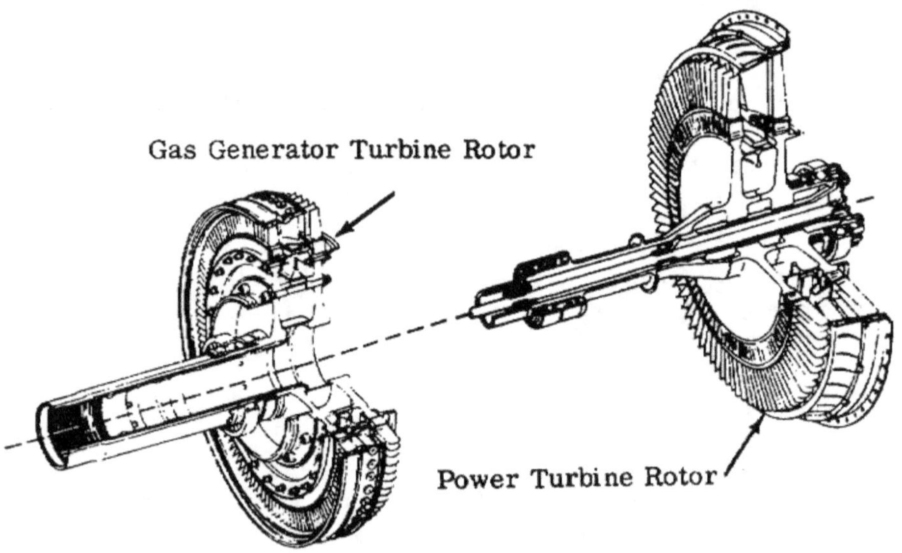

Figure 1.26. Multirotor - Multistage Turbine.

1.23. TURBINE CONSTRUCTION

The turbine rotor is one of the most highly stressed parts in the engine. It operates at a temperature of approximately 1,700° F. Because of the high rotational speeds, over 40,000 rpm for the smaller engines, the turbine rotor is under severe centrifugal loads. Consequently, the turbine disk is made of specially alloyed steel, usually containing large percentages of chromium, nickel, and cobalt. The turbine rotor assembly is made of two main parts, the disk and blades.

Nozzle vanes may be either cast or forged. Some vanes are made hollow to allow cooling air to flow through them. All nozzle assemblies are made of very high-strength steel that withstands the direct impact of the hot gases flowing from the combustion chamber.

The turbine blades are attached to the disk by using the "fir tree" design, shown in figure 1.27, to allow for expansion between the disk and the blade while holding the blade firmly to the disk against centrifugal loads. The blade is kept from moving axially either by rivets or special locking devices. Turbine rotors are of the open-tip type as shown in figure 1.27, or the shroud type as shown in figure 1.28.

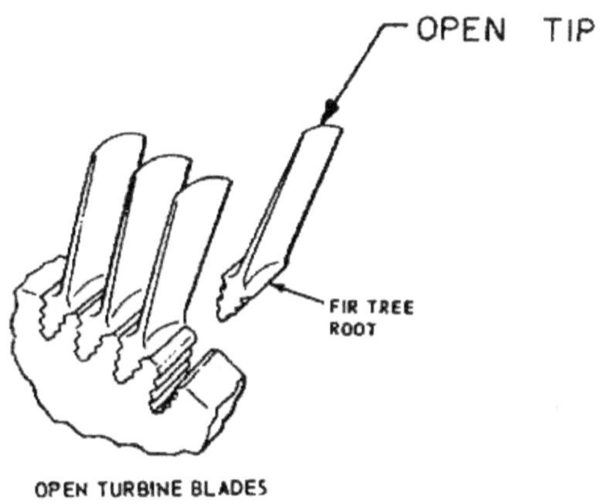

Figure 1.27. Turbine Wheel Open Tip.

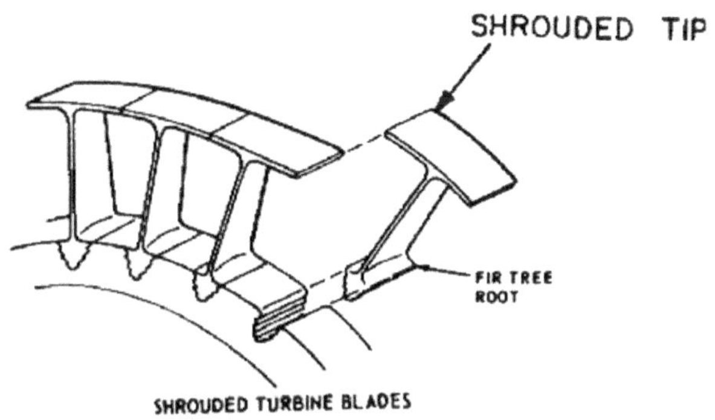

Figure 1.28. Turbine Blade "Fir Tree Root" Shroud.

The shroud acts to prevent gas losses over the blade tip and excessive blade vibrations. Distortion under severe loads tends to twist the blade toward low pitch, and the shroud helps to reduce this tendency. The shrouded blade has an aerodynamic advantage in that thinner blades can be used with the support of the shroud. Shrouding, however, requires that the turbine run cooler or at reduced rpm because of the extra mass at the tip.

Blades are forged or cast from alloy steel and machined and carefully inspected before being certified for use. Manufacturers stamp a "moment weight" number on the blade to retain rotor balance when replacement is necessary. Turbine blade maintenance and replacement are covered in chapter 3.

1.24. EXHAUST SECTION

The hot gases are exhausted overboard through the exhaust diffuser section. Internally, this section supports the power turbine and aft portion of the powershaft. The exhaust diffuser is composed of an inner and outer housing, separated by hollow struts across the exhaust passage. The inner housing is capped by either a tailcone or a cover plate which provides a chamber for cooling the powershaft bearing. A typical exhaust diffuser section is shown in figure 1.29.

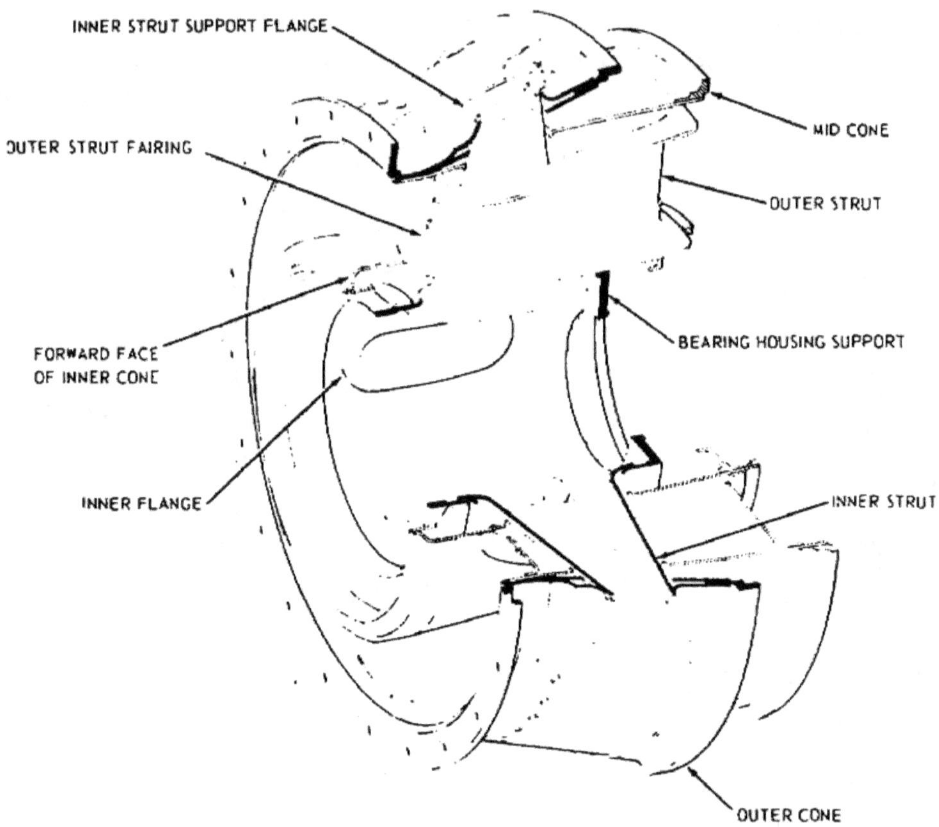

Figure 1.29. Exhaust Diffuser Section.

Turboshaft engines used in helicopters do not develop thrust by use of the exhaust duct. If thrust were developed by the engine exhaust gas, it would be impossible to maintain a stationary hover; therefore, helicopters use divergent ducts. These ducts reduce gas velocity and dissipate any thrust remaining in the exhaust gases. On fixed wing aircraft, the exhaust duct may be the convergent type, which accelerates the remaining gases to produce thrust which adds additional shaft horsepower to the engine rating. The combined thrust and shaft horsepower is called equivalent shaft horsepower (ESHP).

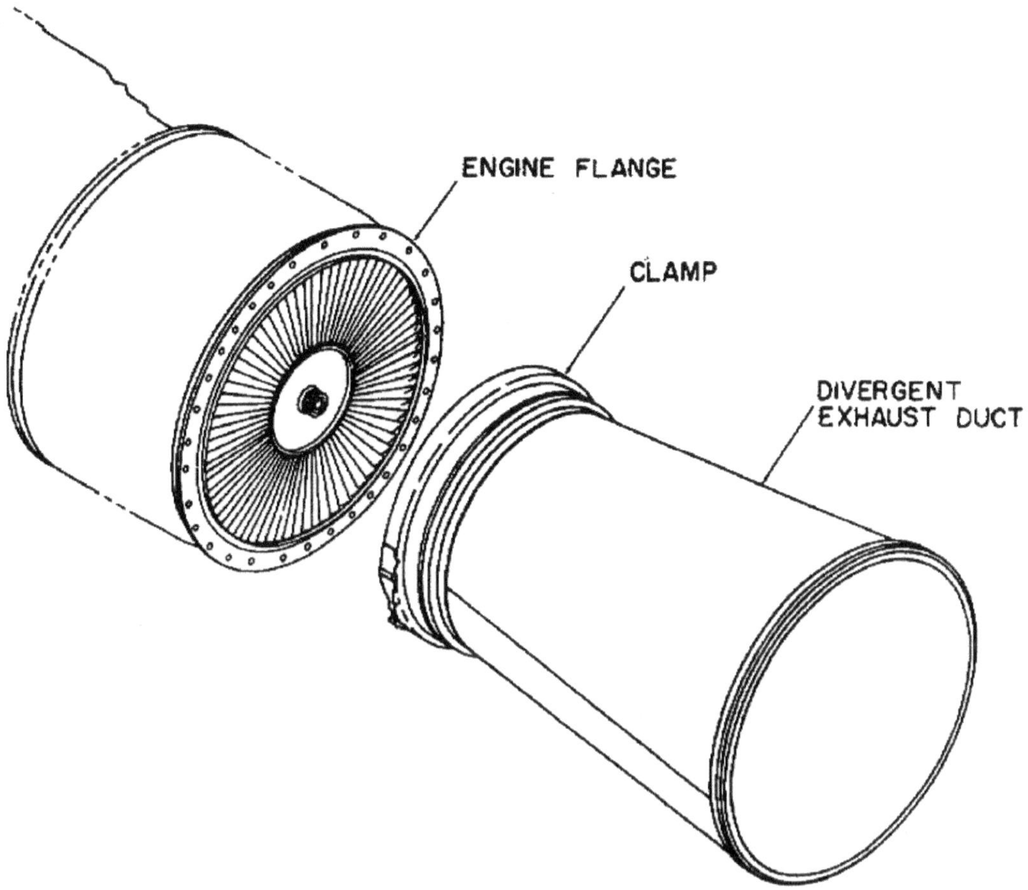

Figure 1.30. Divergent Exhaust Duct.

1.25. SUMMARY

The gas turbine engine has five major sections: inlet, compressor, combustion, turbine, and exhaust. Engine terminology includes directional references, engine stations, and model designations.

Gas turbine engine construction is not limited to one type of compressor. The compressor may be either centrifugal or axial or a combination of the two. Compressors are made in single or multiple stage assemblies.

Three basic types of combustion chambers are in use: the annular, can, or a combination of the two called can-annular or cannular.

Gas turbine engines may use either an axial-flow turbine or a radial-inflow turbine. The turbine section may have a single- or multiple-stage turbine. The hot exhaust gases are exhausted overboard through the exhaust section. Exhaust ducts used on helicopters are divergent. The ducts used on fixed-wing aircraft may be of the convergent type.

Chapter 2

SYSTEMS AND ACCESSORIES

2.1. INTRODUCTION

This chapter introduces the fundamental systems and accessories of the gas turbine engine. Each one of these systems must be present to have an operating turbine engine. Section I describes the fuel system and related components that are necessary for proper fuel metering to the engine.

The second section discusses the theory and components of the lubricating system. Oil is the lifeblood of any engine. If the oil supply to the bearings should cease, within a matter of seconds the lubricating films would break down and cause scoring, seizing, and burning of the vital moving parts.

The third section tells of the ignition system used in the gas turbine engines and of various cockpit instruments used to measure engine performance.

Section I. Fuel Systems and Components

2.2. GENERAL

The fuel system consists of the fuel control, speed governors, fuel pumps, starting fuel nozzles, main fuel system flow divider, main fuel manifold, and vaporizing tubes or nozzles. Fuel is conducted between these components by flexible or rigid lines. The fuel system must supply clean, accurately metered fuel to the combustion chambers. All fuel systems have basically the same components; how these specific units do their jobs differs radically from one engine to another. Some systems incorporate features that are not necessary to the metering of fuel, such as fuel and oil heat exchangers, use of fuel pressure to operate variable inlet guide vanes, and compressor bleed mechanisms. It is the purpose of this section to illustrate typical fuel systems so that the reader may obtain some idea of the route of fuel and location of the components that make up the system. Figure 2.1 shows a typical schematic of a gas turbine engine fuel system.

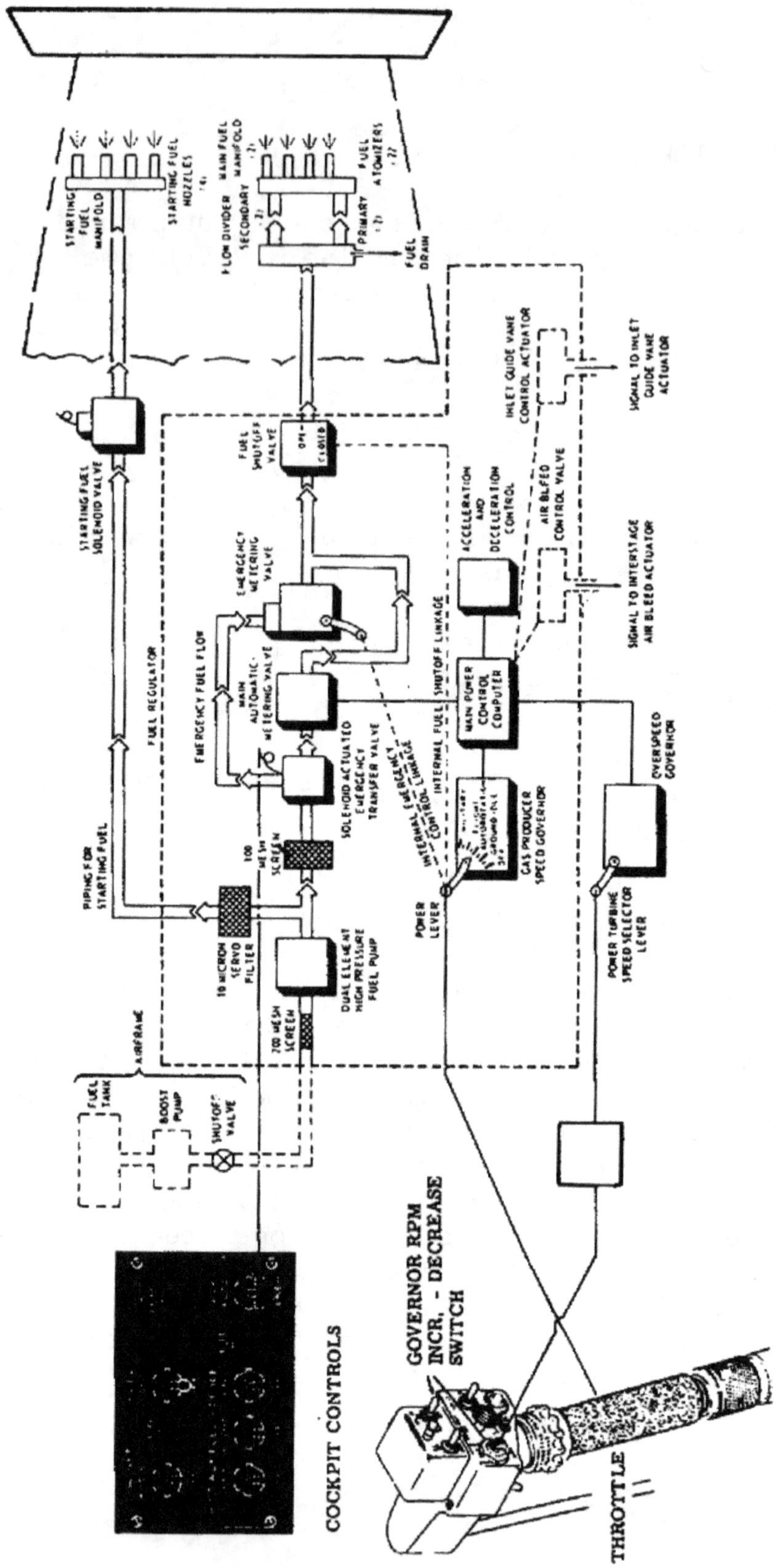

Figure 2.1. Fuel System Schematic Diagram.

2.3. FUEL CONTROLS

The principles and operation of fuel controls used on current engines are discussed in this paragraph. Depending upon the type of engine and the performance expected of it, fuel controls may range from simple valves to automatic computing controls containing hundreds of intricate parts.

Strictly speaking, a pilot of a gas-turbine-powered aircraft does not directly control his engine. His command over the engine corresponds to that of the captain of a ship who obtains engine response by relaying orders to an engineer below deck who, in turn, moves the throttle of the engine. But before he moves the throttle, he monitors certain operating pressures, temperatures, and rpm that are not apparent to the captain. The engineering officer then refers to a chart and computes a fuel flow or throttle change which will not allow the engine to exceed its operating limitations. If you think of the pilot as the captain of the ship, then think of the automatic controls as the engineer. They, too, monitor operating pressures, temperatures, and rpm, and make the necessary fuel and throttle adjustments.

Fuel controls can be divided into two basic groups: hydromechanical, and electronic. There are as many variations in controls as there are engines. Although each type of fuel control has its particular advantage, most controls in use today are hydromechanical. Some fuel controls are extremely complex devices composed of speed governors, servo systems, valves, metering systems, and sensing pickups.

This section limits discussion mainly to fuel control theory of the hydromechanical type. A schematic of one is shown in figure 2.2. A fuel control in the simplest form consists of a plain metering valve to regulate fuel flow to the engine. A hydromechanical fuel control consists of the following main components, but it is not limited to only these.

1. Pump to pressurize fuel.

2. Governors to control rpm.

3. Relief valves to protect the control.

4. Manual control systems (emergency control system).

5. Fuel shutoff valve.

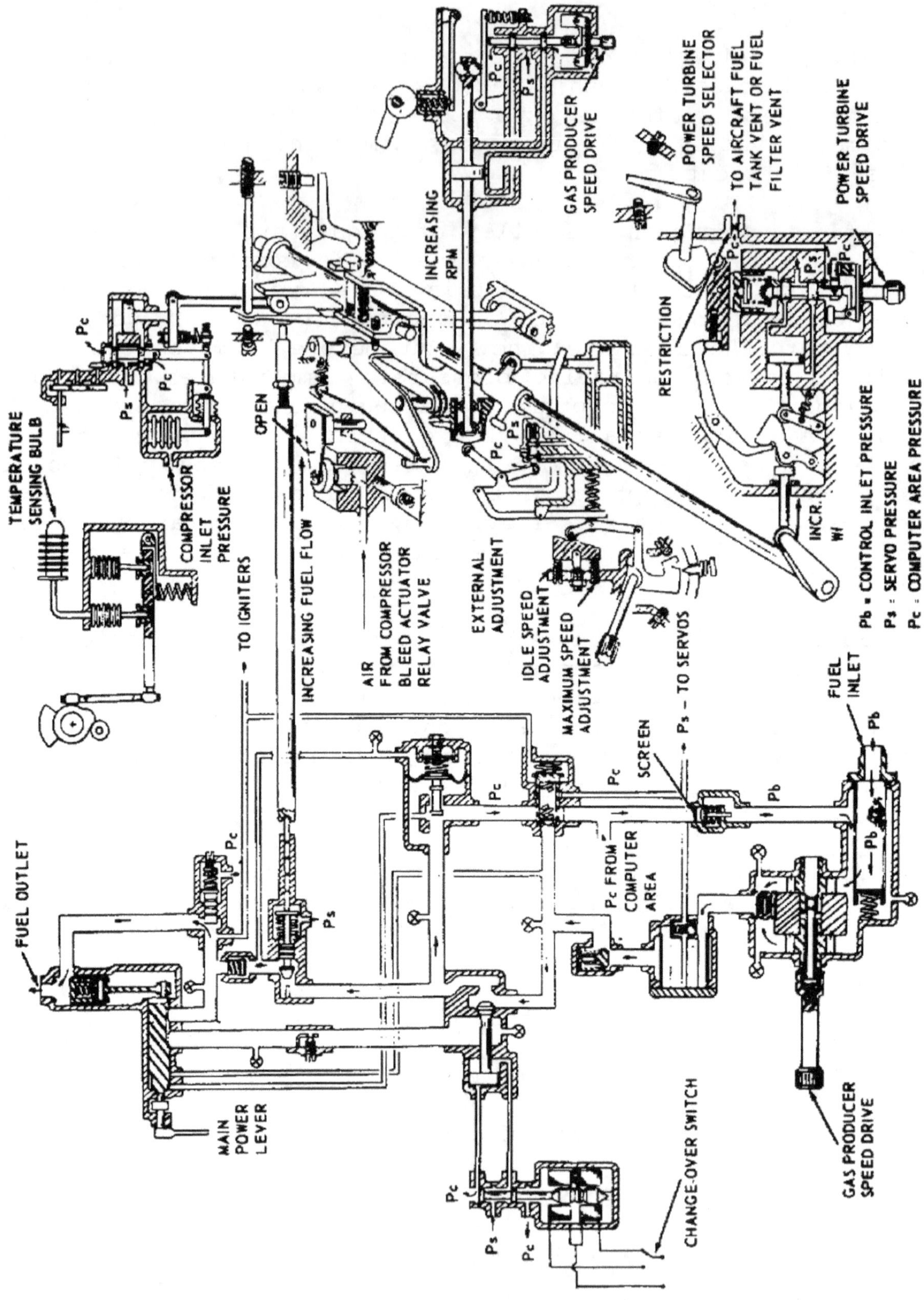

Figure 2.2. Hydromechanical Fuel Control Schematic.

Most modern fuel control units meter the flow of fuel by keeping the pressure drop or difference across the metering valve a constant value, while varying the orifice of the metering valve. Another way to control fuel is to keep the valve orifice a constant size and vary the pressure acting upon the fluid. The operation of a gas turbine requires that a number of variable conditions be given careful thought to provide for safe, efficient operation. Among these are engine rpm, acceleration, exhaust gas temperature (egt), compressor inlet temperature, compressor discharge pressure, and throttle or power control setting. All these conditions affect or are affected by fuel flow, which is increased only to the point where the limiting temperature is reached. As the engine accelerates and airflow through the engine increases, more fuel is added. If turbine inlet temperature were the only engine limitation, a temperature pickup sensing this temperature could be used. However, it is also necessary to avoid the operating range that would cause a compressor surge and stall. Because more than one factor limits engine operation, it is necessary to schedule the accelerating fuel in accordance with a combination of these factors. Because turbine engine compressors are susceptible to surges and stalls, a control with a longer acceleration time is used than is needed for a reciprocating engine. This acceleration time is known as a "lag," and the pilot must be aware of the time it takes the engine to accelerate and give him the power change he requires. Compressor discharge pressure or burner pressure is commonly used as the variable for these controls, since they vary both with engine speed and inlet air temperature. By evaluating these variable conditions, a fair indication of the amount of fuel which can be burned without exceeding engine limitations is obtained.

Two fuel control systems are discussed in the following subparagraphs.

a. Automatic control system. The amount of fuel required to run the engine at rated rpm varies with the inlet air temperature and pressure. For example, it requires less fuel to run the engine on a hot day than on a cold day. To relieve the pilot of the necessity of resetting the power lever to compensate for changes in outside air temperature and pressure, a speed governor is used. A simple speed governor consists of flyweights balanced by a spring. When the engine is running unloaded, at rated speed, the metering valve is open only far enough to supply the small amount of fuel required. If a load is applied to the engine, the speed decreases. This decrease in rpm causes the flyweights to move in under the force of the spring tension and the fuel valve to open wider and admit more fuel. With the additional fuel, the engine picks up speed again, and, as the rated speed is reached, the flyweights move the fuel valve in the closing direction until the proper steady-state fuel flow is reached.

b. Manual (emergency) control system. When the governor control switch in the cockpit is moved from the automatic position to the manual (emergency), a valve is actuated in the fuel control, and fuel is redirected to the manual system metering valve. The throttle in a helicopter is of the motorcycle twist-grip type. When the governor is in the automatic position the throttle is rolled full open and left there, with the fuel control making all fuel-flow changes automatically. If the automatic fuel control fails, the pilot switches to the emergency mode and takes manual control of the throttle, which is mechanically linked to the manual metering valve. The manual throttle control has no compensation for altitude or temperature, and it has no protection against an engine overspeed.

Keep in mind that so far the discussion has been on principles of operation, and any specific fuel control may differ.

2.4. FUEL PUMP

Main fuel pressure pumps for gas turbine engines generally have one or two gear-type, positive-displacement, high-pressure elements. Each of these elements discharges fuel through a check valve to a common discharge port. Thus, if one element fails, the remaining element continues to supply sufficient fuel for engine operation. On some engines, the fuel pump is built in to the fuel control. However, on other engines the fuel pump may be a separate component.

2.5. STARTING-FUEL SYSTEM

Fuel flows through an external line from the fuel control to the starting-fuel solenoid. During the starting sequence, the pilot actuates the start-fuel solenoid switch in the cockpit. The solenoid actuates the valve to the open position, then fuel flows through an external line to the start-fuel manifold. The start-fuel nozzles are attached to the manifold; the number of nozzles varies according to engine design. The nozzles introduce atomized fuel in the combustion chamber during the starting sequence. After the engine has attained a specified speed, the main fuel starts to flow automatically. After the engine is running on the main fuel system, the start fuel system is shut off. A starting fuel system is shown in Figure 2.3.

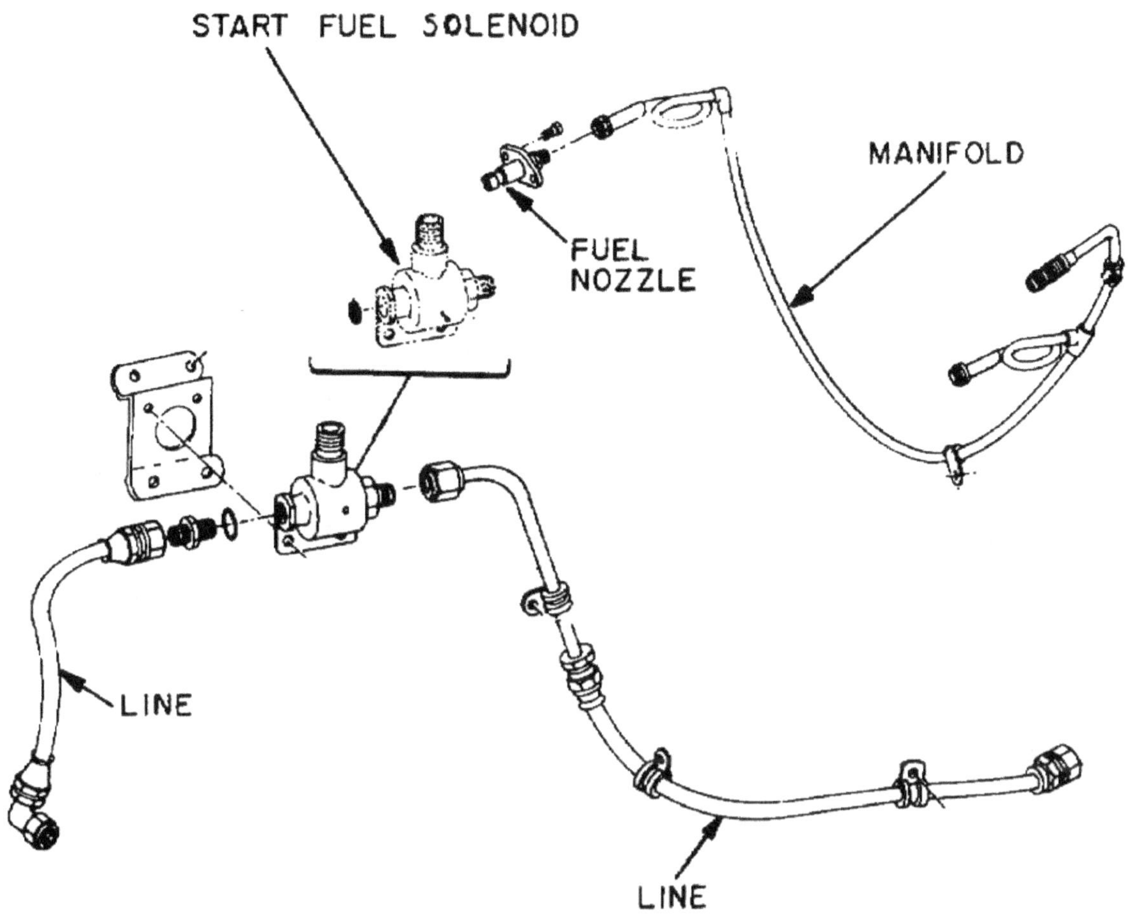

Figure 2.3. Starting Fuel System.

2.6. MAIN FUEL SYSTEM

Main fuel is delivered from the fuel control to the main fuel manifold assembly by external lines. The main fuel manifold delivers fuel to the fuel nozzles, which may be of the single or dual orifice injector type, designed to introduce the fuel into the combustion chamber. Some earlier engines use fuel vaporizer tubes in place of the more efficient fuel nozzles.

2.7. FUEL NOZZLES

On most gas turbine engines, fuel is introduced into the combustion chamber through a fuel nozzle that creates a highly atomized and accurately shaped spray of fuel suitable for rapid mixing and combustion. Most engines use either the simplex or the duplex nozzle. The exception to this is the Lycoming T53-L-11 engine which uses vaporizer tubes in place of fuel nozzles. Each type of nozzle is discussed in the following subparagraphs.

a. Simplex nozzle. Figure 2.4 illustrates a typical simplex nozzle; as its name implies, it is simpler in design than the duplex nozzle. Its big disadvantage lies in the fact that a single orifice cannot provide a satisfactory spray pattern with the changes in fuel pressure.

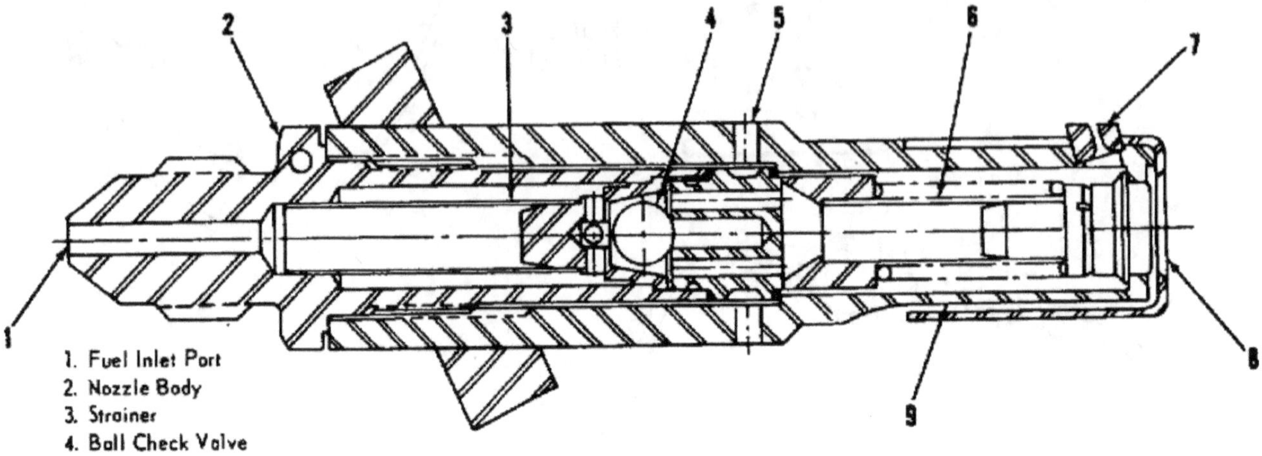

1. Fuel Inlet Port
2. Nozzle Body
3. Strainer
4. Ball Check Valve
5. Air Inlet Port
6. Strainer
7. Hot Streak Fuel Outlet Port
8. Fuel Outlet
9. Air Inlet

Figure 2.4. Simplex Fuel Nozzle.

b. Duplex nozzle. Because the fuel-flow divider and the duplex nozzle work hand in hand, the description of these units is combined. The chief advantage of the duplex nozzle is its ability to provide good fuel atomization and proper spray pattern at all fuel pressures. For the duplex nozzle to work, there must be a fuel-flow divider to separate the fuel into low (primary) and high (secondary) pressure supplies. Single-entry duplex nozzles have an internal flow divider and require only a single fuel manifold, while, as shown in figure 2.5, dual-entry fuel nozzles require a double fuel manifold. The flow divider, whether self-contained in each nozzle, or installed separately with the manifold, is usually a spring-loaded valve set to open at a specific fuel pressure. When the pressure is below this value, the flow divider directs fuel to the primary manifold. Pressures above this value cause the valve to open and fuel is allowed to flow in both manifolds. A fuel flow divider is shown in figure 2.6.

In addition, an air shroud surrounding the nozzle, as shown in figure 2.7, cools the nozzle tip and improves combustion by retarding the accumulation of carbon deposits on the face. The shroud also helps to contain the flame in the center of the liner.

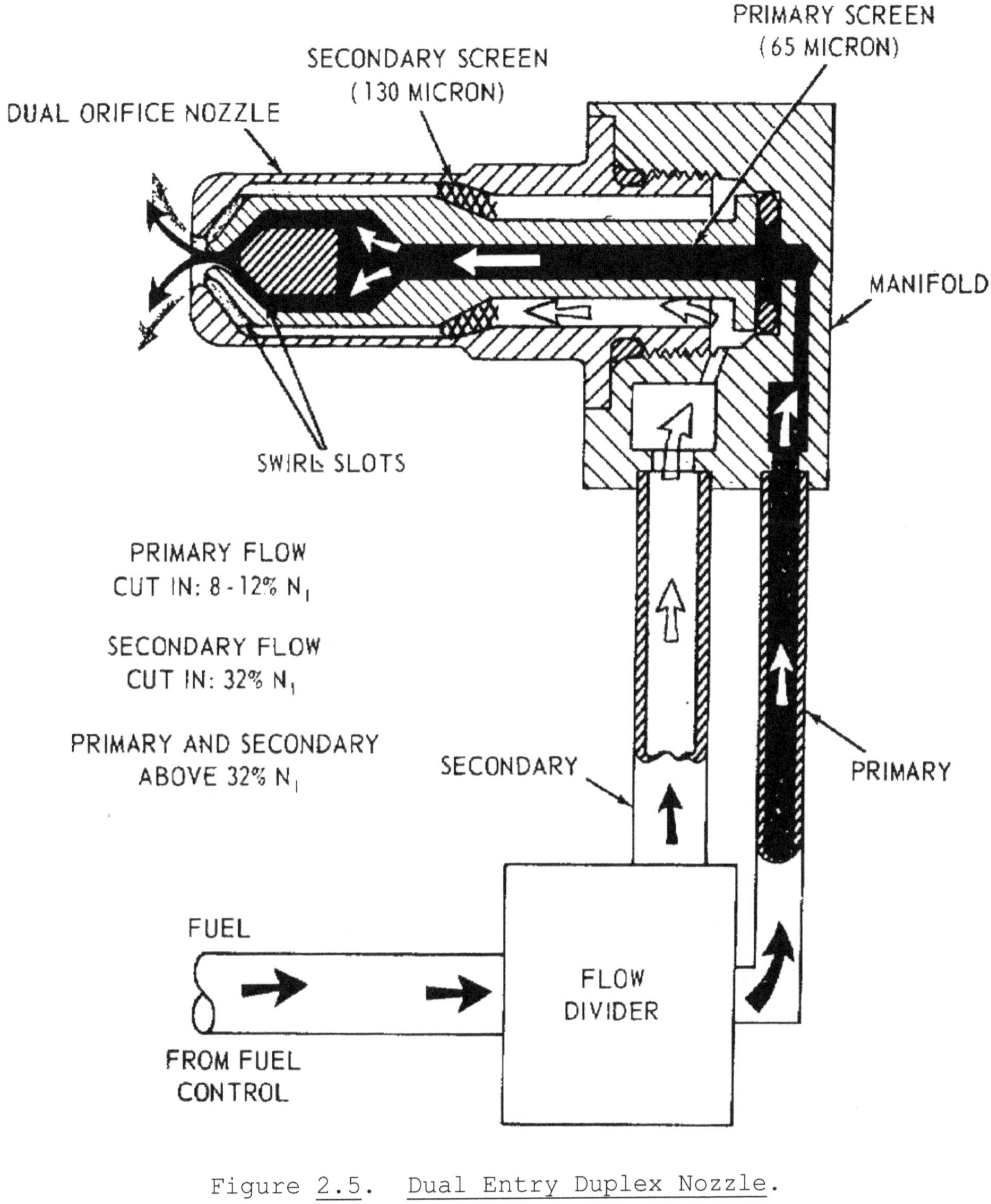

Figure 2.5. Dual Entry Duplex Nozzle.

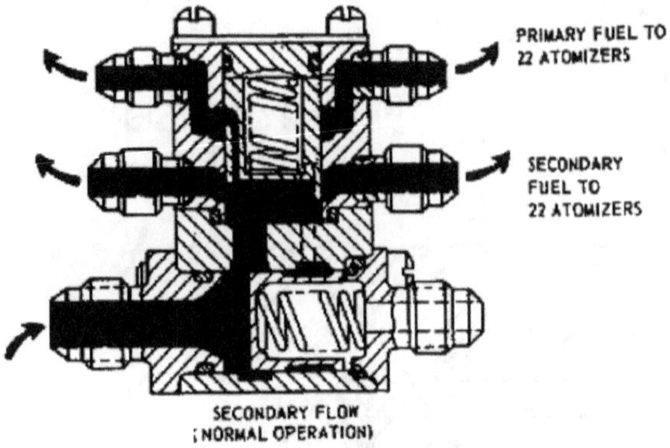

Figure 2.6. Fuel Flow Divider.

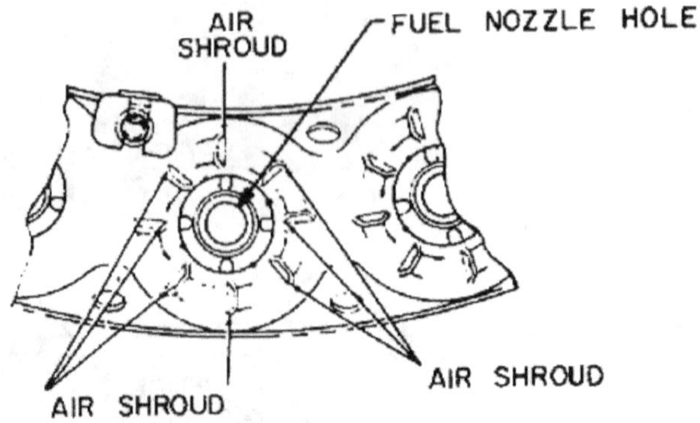

Figure 2.7. Air Shroud.

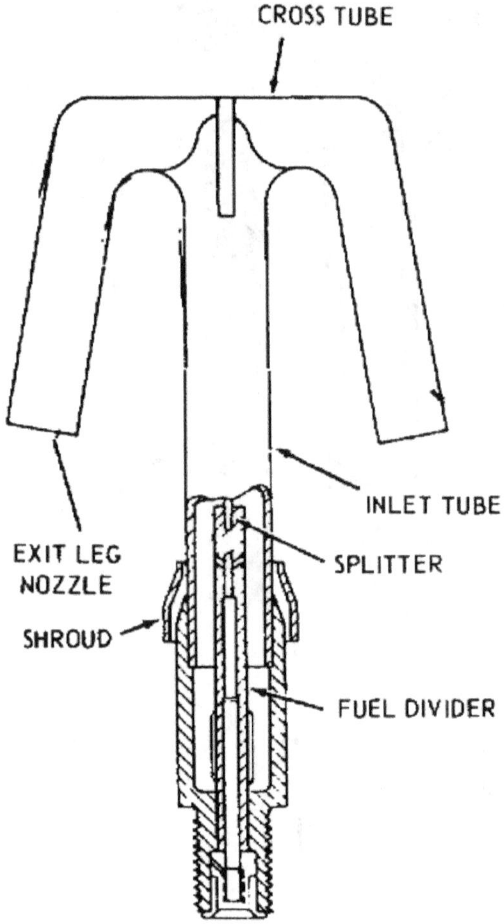

Figure 2.8. Vaporizing Tube.

A word of caution; extreme care must be taken when cleaning or handling the nozzles, since even the acid on the fingers may corrode and produce a spray pattern which is out of tolerance.

c. Vaporizing tube. Engines such as the Lycoming T53-L-11 use vaporizing tubes instead of injector nozzles. The vaporizing tube is a T-shaped, ceramic-coated pipe, whose exit faces upstream to the airflow. Figure 2.8 shows a vaporizing tube that is used on the T53-L-11.

2.8. FUEL FILTERS

Gas turbine engines may have several fuel filters installed at various points throughout the systems, one fuel filter before the fuel pump and one on the high-pressure side after the pump. In most cases the filter includes a relief valve set to open at a specified differential pressure (PSID) between inlet and outlet pressure. This gives the fuel a bypass if the filter becomes clogged from contamination.

More than one kind of filter is used on turbine engines. A paper cartridge filter is usually used on the low-pressure side of the pump. It uses a replaceable paper element, shown in figure 2.9, capable of filtering out particles larger than 100 microns, or about the diameter of human hair.

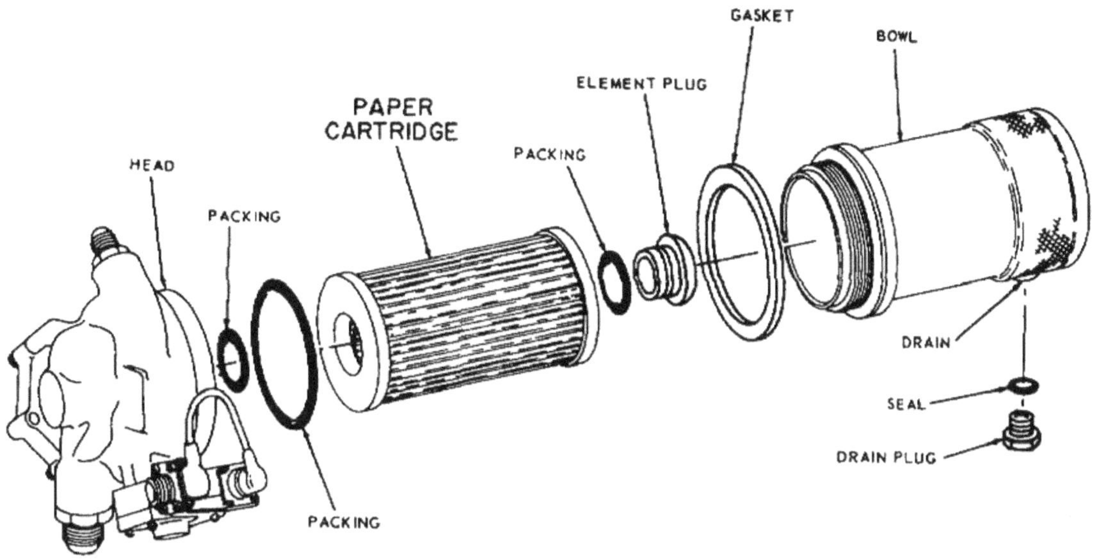

Figure 2.9. Paper Cartridge Fuel Filter.

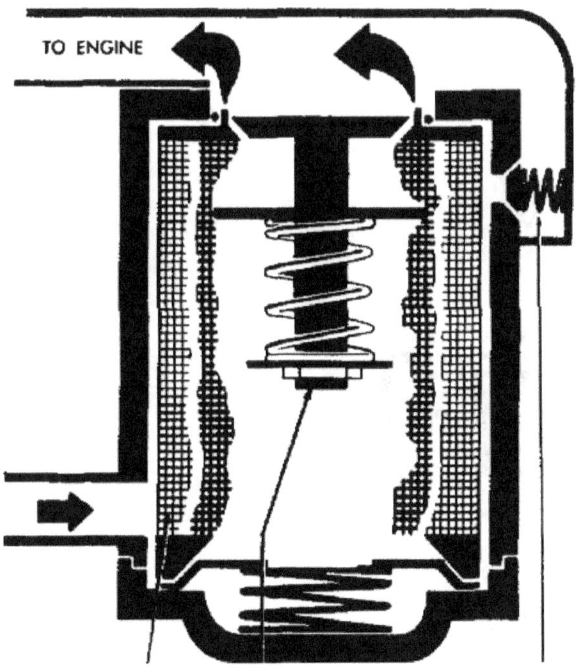

Figure 2.10. Cylindrical Screen Filter.

A cylindrical screen filter is generally used where the fuel pressure is low. The filter is constructed of stainless steel wire mesh cloth and is capable of filtering out particles larger than 40 microns. Such a filter, shown in figure 2.10, may be cleaned, preferably ultrasonically, and reused.

In addition to the main line filters, other filtering elements may be located in the fuel tanks, fuel control, fuel nozzles, and just about any other place fuel is routed.

2.9. PRESSURIZING AND DRAIN DUMP VALVES

Until sufficient pressure is attained in the fuel control to compute the fuel flow schedules, flow to the main fuel nozzle is prevented by the pressurizing and drain dump valve. This valve also drains the fuel manifold at engine shutdown to prevent post-shutdown fires, and it traps fuel in the upstream portion of the system to keep the fuel control primed to permit faster starts.

All manufacturers install a combustion chamber drain valve in the combustion section. During normal engine operation this valve is closed. The drain valve is located at the lowest part of the combustion chamber. When the combustion pressure in the chamber drops below a specified minimum, usually a few pounds per square inch, this valve opens and drains any fuel remaining after a false or aborted start. The fuel drained from this valve is dumped overboard.

2.10. FUEL OIL-COOLER

Some turbine engines use a fuel oil-cooler or heat exchanger to cool the lubricating oil. This unit is discussed under the lubrication system because its prime function is to help cool the oil. It consists of a cylindrical oil chamber surrounded by a jacket through which the fuel passes. Heat from the oil is transferred to the fuel via conduction*. Figure 2.11 shows a typical fuel oil-cooler.

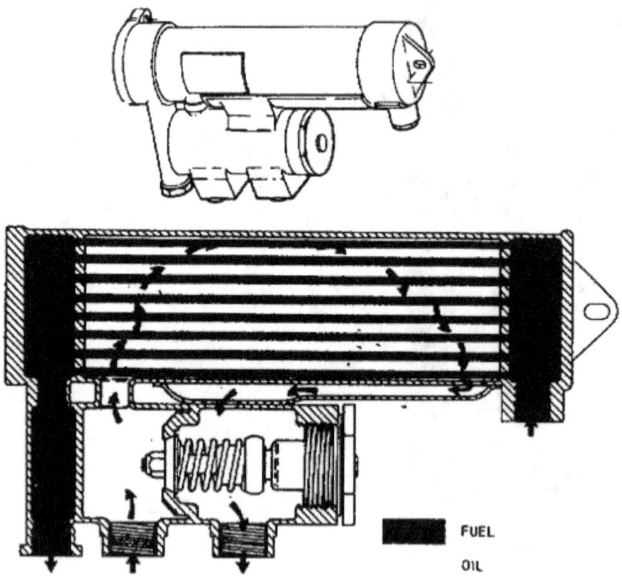

Figure 2.11. Fuel Oil-Cooler.

*See glossary.

2.11. SUMMARY

The fuel system must supply clean, accurately metered fuel to the combustion chamber. Most turbine engine fuel systems have the same components: fuel control, pressure pumps, fuel flow divider, manifold, and atomizers. There are two types of fuel controls: hydromechanical and electronic. Engine-driven fuel pumps are high-pressure, positive-displacement, gear type pumps, and the fuel nozzles are either simplex or duplex. However, some engines use vaporizer tubes in place of fuel nozzles. Some gas turbine engines use a fuel oil-cooler to cool the oil.

Section II. Lubrication Systems

2.12. GENERAL

During the first few years of gas turbine experience, lightweight, petroleum-base oil was suitable for gas turbines as well as other types of engines. Most of the early engines used lubricating oil conforming to MIL-O-6081A, Grade 1010. Engines requiring an extremely light oil were operated on MIL-O-3519, Grade 1005. These were conventional petroleum oils of high quality and light weight which met the requirements of all the older engines.

Because of the continuous demand for greater power, gas turbine engines have been designed to operate at higher temperatures and pressure ratios. Some gas turbine engine oil temperatures encountered are considerably above the flash point of the petroleum oils. Because of this, a high temperature lubricant had to be developed. The oil used in all Army gas turbine engines is MIL-L-23699, or MIL-L-7808. These are synthetic lubricants which have wide operating ranges and load carrying capabilities. The MIL-L-7808 is used in engines operating below -25° F. OAT, and MIL-L-23699 is used when temperatures are above -25° F. This section discusses the various components that make up a typical lubricating system.

2.13. LUBRICATING SYSTEMS

Lubricating systems for modern gas-turbine engines are relatively simple in design and operation, but their function is of vital importance. The principal purposes of the lubricating system are to clean, reduce friction, and to cool the bearing surfaces. The main units of the typical system are the reservoir or oil tank, the pressure pump, scavenger pumps, filters, oil cooler, and spray oil jets. A schematic illustration of a gas turbine engine oil system is shown in figure 2.12.

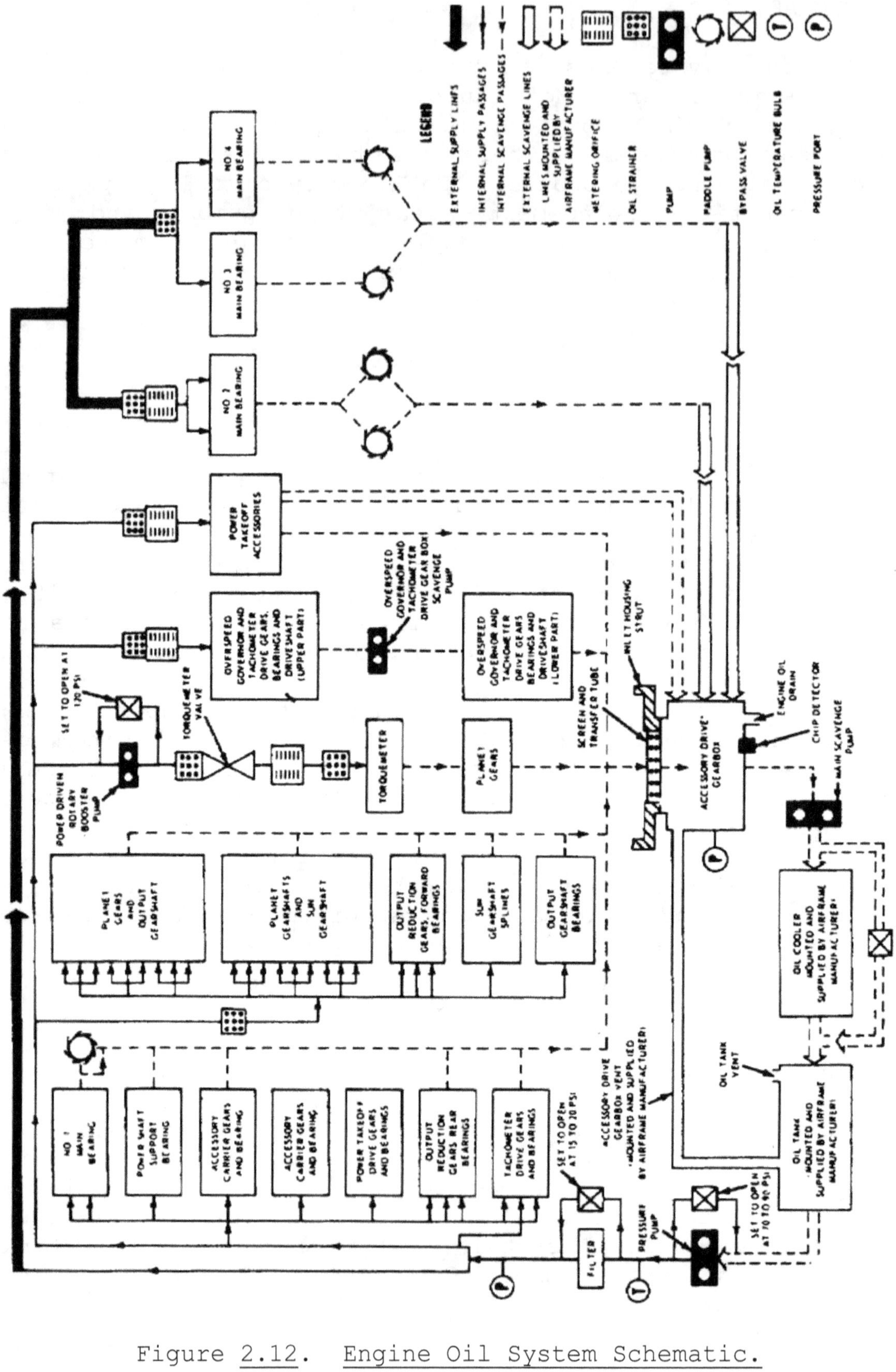

Figure 2.12. Engine Oil System Schematic.

2.14. OIL TANKS

Most gas turbine engines are of the dry-sump type, meaning the on is stored separately from the engine, or the tank may be attached to a structural part of the engine. Usually constructed of welded aluminum or steel, it can contain a venting system, a deaerator (baffles) to separate air from the oil. Some systems use an oil level transmitter to indicate quantity, where others have a dipstick or visual sight gage.

2.15. PRESSURE PUMPS

Oil pumps for turbine engines are usually of the positive-displacement gear type, with a relief valve to prevent excessive pressure. A modified gear-type pump is called the "gerotor pump."

The gear-type pump consists of a driving and driven gear. The pump is driven from the engine accessory section and causes the oil to pass around the outside of the gears in pockets formed by the gear teeth and the pump casing. The pressure developed is proportional to engine rpm up to the point where the pressure relief valve opens and limits the pressure output of the pump.

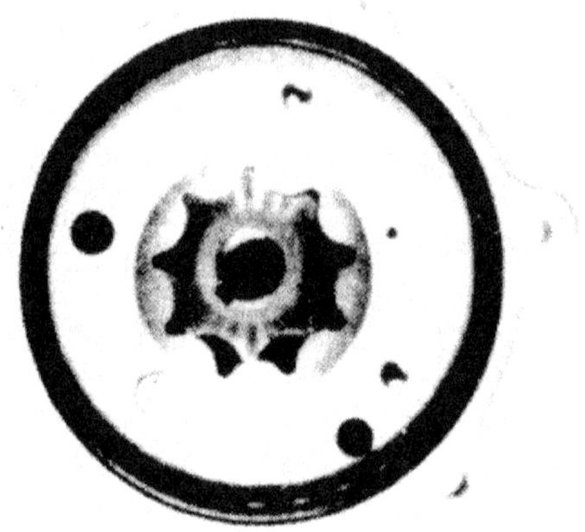

Figure 2.13. Gerotor Booster Pump.

The gerotor pump has two moving parts, an inner toothed element meshing with an outer toothed element. The inner element has one less tooth than the outer, and the missing tooth provides a chamber to move the fluid from the intake to the discharge port. Both elements are mounted eccentrically to each other, the inner one mounted on the shaft and the outer one meshed with it. Figure 2.13 is a picture of the gerotor pump, showing both inner and outer toothed elements.

2.16. SCAVENGE PUMPS

Although much larger in total capacity, scavenge pumps are usually constructed in the same manner as pressure pumps. Engines are generally provided with several scavenge pumps to drain oil from various parts of the engine. Often such a pump shares the same housing as the pressure pump. These pumps are used to draw the oil from the sumps at the bearings, accessory gearbox housings, and other drainage points and return the oil back to the tank.

2.17. FILTERS

Three basic oil filters or strainers are made: cartridge, screen-disk, and screen. These filters are the same design as the filters used in the fuel system, as covered in paragraph 2.8. The main objective of a filter is to remove all foreign particles from the lubricant without creating excessive back pressure against the pumps. Filters are usually provided with bypass valves to permit the flow of oil in case the filter becomes clogged.

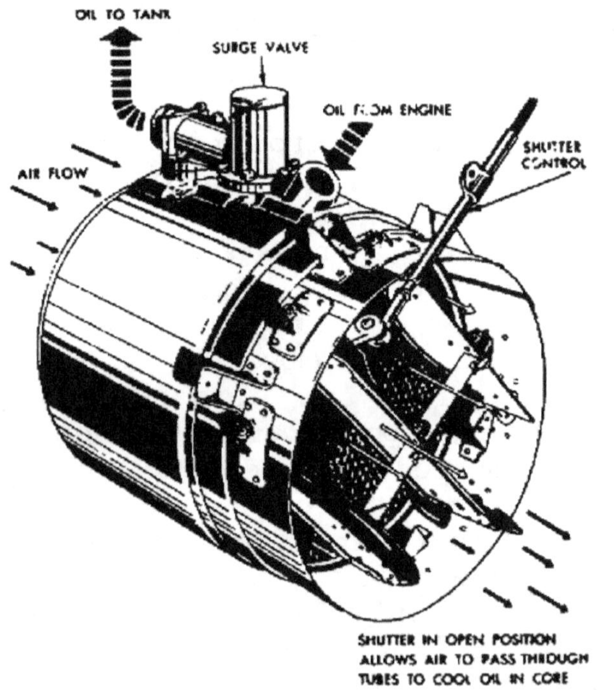

Figure 2.14. Air Oil Cooler.

2.18. OIL COOLER

Oil coolers for aviation gas-turbine engines are either simple oil radiators with air cooling or the kind that uses fuel as the cooling medium. The latter type of unit is used on the Lycoming T55 engine. The fuel oil-cooling unit is a heat exchanger which transfers the heat in the oil to the fuel flowing to the fuel nozzles. Since the fuel flow through the cooler is much greater than the oil flow, the fuel is able to absorb a considerable amount of heat from the oil, thereby reducing the size and weight of the cooler. The fuel oil-cooler is shown in figure 2.11 on page 57. An air cooler is shown in figure 2.14.

2.19. SPRAY OIL JETS

The lubrication method most generally used is known as a calibrated system, where oil is specifically controlled by a calibrated orifice which provides the proper oil flow at all engine operating speeds. The oil is supplied from the oil pressure pump through tubing and internal passageways to the spray jets, where the oil is sprayed on the bearing surfaces.

2.20. SUMMARY

Gas turbine engine oil systems perform three major functions. They clean and reduce friction, and they cool and dissipate heat. They also clean the engine interior through the use of oil filters and strainers. Because much of the aircraft powerplant consists of moving parts, lubricants are needed to overcome friction caused by one metal surface sliding or rolling over another. Friction causes heating of parts, excessive wearing, and useless expenditure of horsepower. Lubricating systems used in gas turbine engines have oil tanks, pressure pumps, scavenger pumps, filter, oil coolers, and spray oil jets. The system most widely used on turbine engines is the dry sump lubrication system which uses a separate or external oil tank, located near the engine.

The two kinds of pumps are pressure pumps and scavenge pumps, the first to put oil into the system, and the second to collect oil from the system. Filters remove foreign matter from the oil, and either a fuel oil-cooler or an air cooler takes the heat out of it. Oil is sprayed on the bearing surface by spray jets.

Section III. Ignition Systems and Engine Instrumentation

2.21. GENERAL

Gas turbine ignition systems fall into three general types: first, the induction type, that produces high tension voltage by conventional induction coils; second, the capacitor type that causes ignition by means of high energy and very high temperature sparks produced by a condenser discharge; and a third type of ignition system, not widely adopted, that uses a glow plug.

Most ignition systems used on Army aircraft are of the high-energy capacitor type. This system has been accepted for gas turbine engines because it produces high voltage and an exceptionally hot spark, and the high voltage covers a large area.

The tachometer is one of the cockpit instruments described briefly in this section. Others are indicating systems for torque, engine oil pressure, engine oil temperature, exhaust gas temperature, and fuel pressure.

2.22. IGNITION UNIT

Usually, gas turbine engines are equipped with two or more igniter plugs; however, the smaller engines like the T63 have only one igniter plug, sometimes called the spark plug. Igniter plugs serve a purpose similar to the spark plug in a reciprocating engine, although operation of the ignition system and the igniter plugs is necessary only for a short period during the engine starting cycle. On many installations, ignition is initiated simultaneously with the starter. The ignition cycle takes place several times per second and continues to operate as long as the ignition switch is on.

The term "high energy" is used in the section to describe the capacitor type of ignition system. However, the amount of energy produced is very small. The intense spark is obtained by expending a small amount of electric energy in a very short time. Energy is the capacity for doing work. It can be expressed as the product of the electrical power and time. Gas turbine ignition systems are rated in joules. The joule is also an expression of electric energy, being equal to the amount of energy expended in one second by an electric current of one ampere through a resistance of one ohm. All other factors being equal, the temperature of the spark is determined by the power level reached. A high-temperature spark can result from increasing the energy level, or by shortening the duration of the spark. Increasing the energy level requires a heavier, more bulky ignition unit, since the energy delivered to the spark plug is only about 30 to 40 percent of the total energy stored in the capacitor. Also the higher the current flow, the higher the erosion rate on the igniter plug electrodes. Furthermore, much of the spark would be wasted, because ignition takes place in a matter of microseconds. In a capacitor discharge ignition system, most of the total energy available to the igniter plugs is dissipated in 10 to 100 microseconds, with up to 80, 000 watts with a spark duration of 50 microseconds. Figure 2.15 shows a wiring schematic of a typical ignition unit.

WARNING: When working around the ignition unit of the engine, disconnect the input lead to the ignition exciter unit. Remove the igniter plugs from the combustion chamber and ground them to the engine. You do this to dissipate any charge that might be left in the exciter unit.

Some ignition exciter units contain a very small amount of radioactive material (cesium-barium 137) and normally require no handling precautions. However, severely damaged units that have been broken open must be handled with forceps or gloves and disposed of in accordance with AR 755-15.

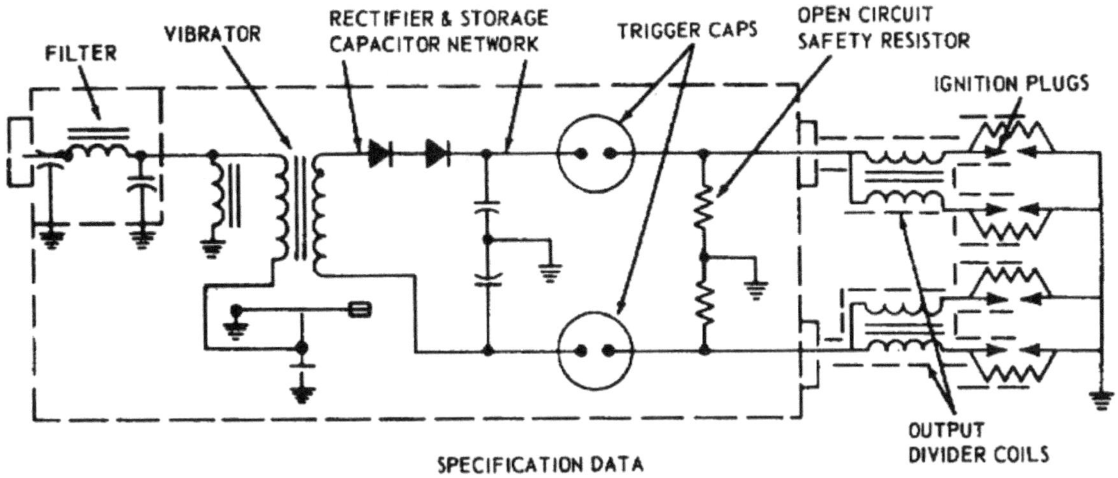

SPECIFICATION DATA

INPUT VOLTAGE	14-29 VDC
INPUT CURRENT	3.0 AMPS MAX
NUMBER OF PLUGS FIRED	4
OPERATING ALTITUDE	50,000 FEET
AMBIENT OPERATING TEMPERATURE RANGE	-65° TO +250°F
SPARK RATE	1.5/SEC MIN AT 14 VDC
STORED ENERGY	4 JOULES NOMINAL
DUTY CYCLE	2 MIN ON, 3 MIN OFF 2 MIN ON, 23 MIN OFF PER MIL-E-8595
OUTPUT VOLTAGE	2.5 KV NOMINAL

Figure 2.15. Wiring Schematic of Typical Ignition Unit.

2.23. IGNITERS

Gas turbine igniters come in many sizes and shapes depending upon the duty they will be subjected to. The electrodes of the plugs used with high-energy ignition systems must be able to accommodate a current of much higher energy than the electrodes of conventional spark plugs are capable of handling. Although the high-energy current causes more rapid igniter-electrode erosion than that encountered in reciprocating-engine spark plugs, this is not a major disadvantage, because of the relatively short time that the ignition system is in operation. Most igniter plugs used in turbine engines are of the annular-gap type, shown in figure 2.16.

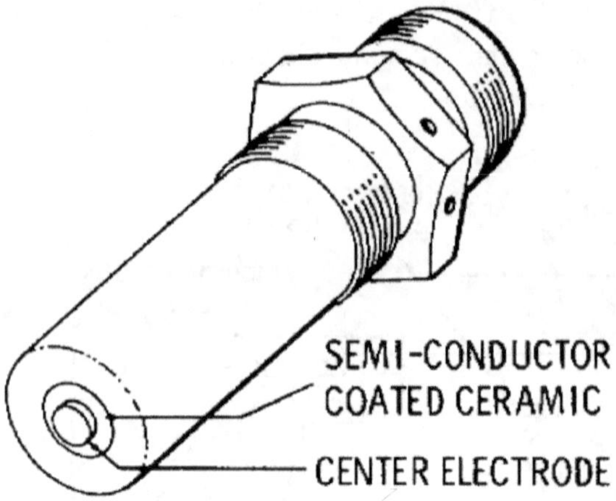

Figure 2.16. Annular Gap Igniter Plug.

The annular-gap igniter plug protrudes slightly into the combustion chamber liner to provide an effective spark. Another type of igniter is the constrained-gap plug which does not closely follow the face of the plug; instead it tends to jump in an arc which carries it beyond the face of the chamber liner. Because the constrained-gap plug does not have to protrude into the liner, the electrode operates at a cooler temperature than that of the annular-gap plug.

2.24. INTERNAL COOLING SYSTEM

The intense heat generated when combustion takes place means that all internal combustion engines must be cooled by some means. Air-cooled reciprocating engines are cooled by air passing over fins attached to the cylinders. Liquid-cooled engines, as in an automobile, use a liquid coolant that passes through jackets surrounding the cylinders. In a reciprocating engine, combustion takes place only during every fourth stroke of a four-cycle engine. However, in a gas turbine engine, where the burning process is continuous, nearly all the cooling air must pass through the inside of the engine. If only enough air were admitted to the engine to provide combustion, internal temperatures would increase to more than 4,000° F. Because of this, the amount of air admitted to the engine is in excess of the amount required for combustion only; indeed, about 75 percent of the air is used for cooling and 25 percent for combustion. This large surplus of air (secondary air) cools the hot expanding gases just before they enter the turbines. In some engines, internal air is bled from the engine compressor section and is vented through passages to the bearings and other parts of the engine. This air is then vented into the exhaust stream.

2.25. ENGINE INSTRUMENTATION

Engine performance is monitored by instruments mounted on the instrument panel in the cockpit.

a. Tachometer system. The tachometer gives the pilot a continuous indication of engine rpm. A variety of systems or a combination of systems may be used on gas turbine engines. Gas producer or gas generator tachometers, turbine and rotor tachometers, and N_1 and N_2 tachometers are some of the tachometer systems used. The system may consist of dual indicators, registering rpm for multiengine aircraft, registering engine and rotor rpm for rotary-wing aircraft, or engine and propeller rpm for fixed-wing aircraft. A typical tachometer indicator is driven by a tachometer-generator. The generator supplies power at a frequency proportional to the driven speed which drives the synchronous motors in the indicator.

b. Torquemeter indicating system. Sometimes called a torque pressure indicating system, the typical torquemeter indicating system is a pressure indicator for continuous readings of engine output-shaft torque. It is powered by an electrical transmitter mounted on the engine inlet section.

c. Engine oil pressure indicating system. A typical engine oil pressure indicating system gives continuous readings of engine oil pump pressure in psi to the indicator, by means of an electrical transmitter mounted on the engine inlet section. The transmitter is connected to the 28-volt ac electrical system, and by a hose to a pressure tap on the engine oil filter housing.

d. Engine oil temperature indicating system. In a typical engine oil temperature indicating system, the indicator is electrically connected to the 28-volt dc system. An electrical resistance type thermobulb installed in the engine oil pump housing measures temperatures of the oil entering that unit. The temperature readings are transmitted to the indicator in degrees centigrade.

e. Exhaust gas temperature indicating system. The indicator in a typical exhaust gas temperature indicating system operates on electrical potential from an engine thermocouple harness with probes mounted in the aft section of the engine exhaust diffuser. The thermocouple is a device which converts heat into electricity. The exhaust gas temperature indicator (thermocouple thermometer indicator) is actually a sensitive millivoltmeter, calibrated in degrees centigrade. Its D'Arsonval movement is activated by an electrical force generated by its related thermocouple. The indicator circuit is entirely independent of any other electrical power source, and includes a coil resistor which provides a means of systems calibration.

f. Fuel pressure indicating system. A typical fuel pressure indicating system gives continuous readings of fuel pressure(psi) in the main fuel supply line from the boost pumps in the tanks, by means of an electrical transmitter. The transmitter is connected to a tap on the valve manifold where all the fuel supply lines join to deliver fuel to the engine through the fuel control inlet hose. Electricity is supplied to the transmitter by the 28-volt ac system.

2.26. SUMMARY

The three types of ignition systems used on turbine engines are induction, capacitor discharge, and glow plug. The most common ignition system used on Army aircraft is the capacitor discharge. The induction and capacitor systems use a spark producing plug to ignite the fuel air mixture.

Because of the high operating temperatures of turbine engines, an internal cooling system is used. Cooling air forms a blanket of air around the combustion chamber.

Instrumentation consists of tachometers, torquemeters, and pressure and temperature gages for monitoring engine performance.

Chapter 3

TESTING, INSPECTION, MAINTENANCE, AND STORAGE PROCEDURES

3.1. INTRODUCTION

The information in this chapter is important to you because of its general applicability to gas turbine engines. The information covers the procedures used in testing, inspecting, maintaining, and storing gas turbine engines. Specific procedures used for a particular engine must be those given in the technical manual (TM) covering that engine

3.2. THE TEST CELL

Before any engine is shipped to the user, the manufacturer or overhaul facility has test-run the engine in a test cell to ensure quality control. The test cell building is usually constructed of concrete and contains both the control room and engine room, although in some test cells only the control room is enclosed. A typical test cell and a control room are shown in figure 3.1. If an engine fails during a test run or does not perform to the standards set by the manufacturer, that engine and a specified number of previous engines are disassembled to check for faults.

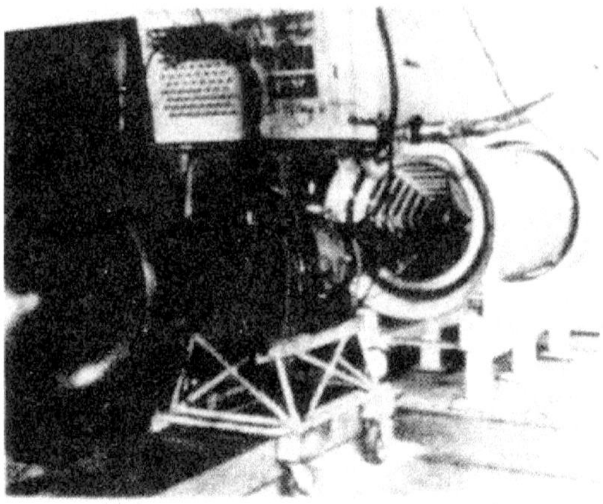

ENGINE RUNNING IN TEST CELL

TEST CELL CONTROL ROOM

Figure 3.1. Test Cell Control Room.

Ground operation, or testing of an engine may also be performed in the mobile engine test unit (METU). The mobile trailer unit contains engine testing equipment similar to that available in an engine test cell. The mobile unit increases engine availability by eliminating most of the need to return engines to an overhaul depot.

Test-cell instrumentation usually includes temperature and pressure gages to monitor engine performance. The engine is run in the test cell with the same demands placed upon it as if it were installed in an aircraft. The performance of any engine is considerably influenced by changes in ambient temperature and pressure, because of the way these conditions affect the weight of the air entering the engine. To compare the performance of similar engines on different days, under different atmospheric conditions, a given engine's performance must be corrected to the standard day condition of 29.92 inches of mercury and 59° F.

During the initial run after assembly, or after extensive maintenance or overhaul, engine statistics are recorded on a test log. This log sheet remains with the engine historical records until such time as another data sheet is completed.

During testing, any problem that would limit the engine's performance, such as exhaust gas temperature, torque, fuel flow, or maximum speed, is corrected. In addition, oil temperature, bearing scavenge-oil temperature, seal leakage, and oil consumption must be within established limits. These tests are usually performed under other than standard day conditions, and data will then be computed to a standard day rating by using the charts and tables in the engine manual. This new information is entered on the engine test log sheet as shown in figure 3.2 and becomes a permanent part of the engine records.

3.3. VIBRATION EQUIPMENT

High-frequency vibrations must be detected and eliminated because they can cause mechanical failure and extensive engine damage. This paragraph discusses the cause of vibrations and the equipment to analyze vibrations. The main source of vibration in the gas turbine engine is the imbalance of rotating parts. Imbalance is caused by an uneven distribution of weight and is measured in inch-grams or inch-ounces. An inch-gram is one gram of unbalanced weight concentrated one inch from the center of a rotating part. When an unbalanced part is rotated, a force is generated. This force is a product of the amount of imbalance and rotating speed.

To analyze this force, a vibration transducer is used; this is a miniature generator. When attached to a vibrating object it generates an electrical signal that is proportional to the force being analyzed. The signal is sent to a meter that amplifies the signal so it can be conveniently read. The meter has four input channels that independently accommodate a signal from a transducer. The meter,

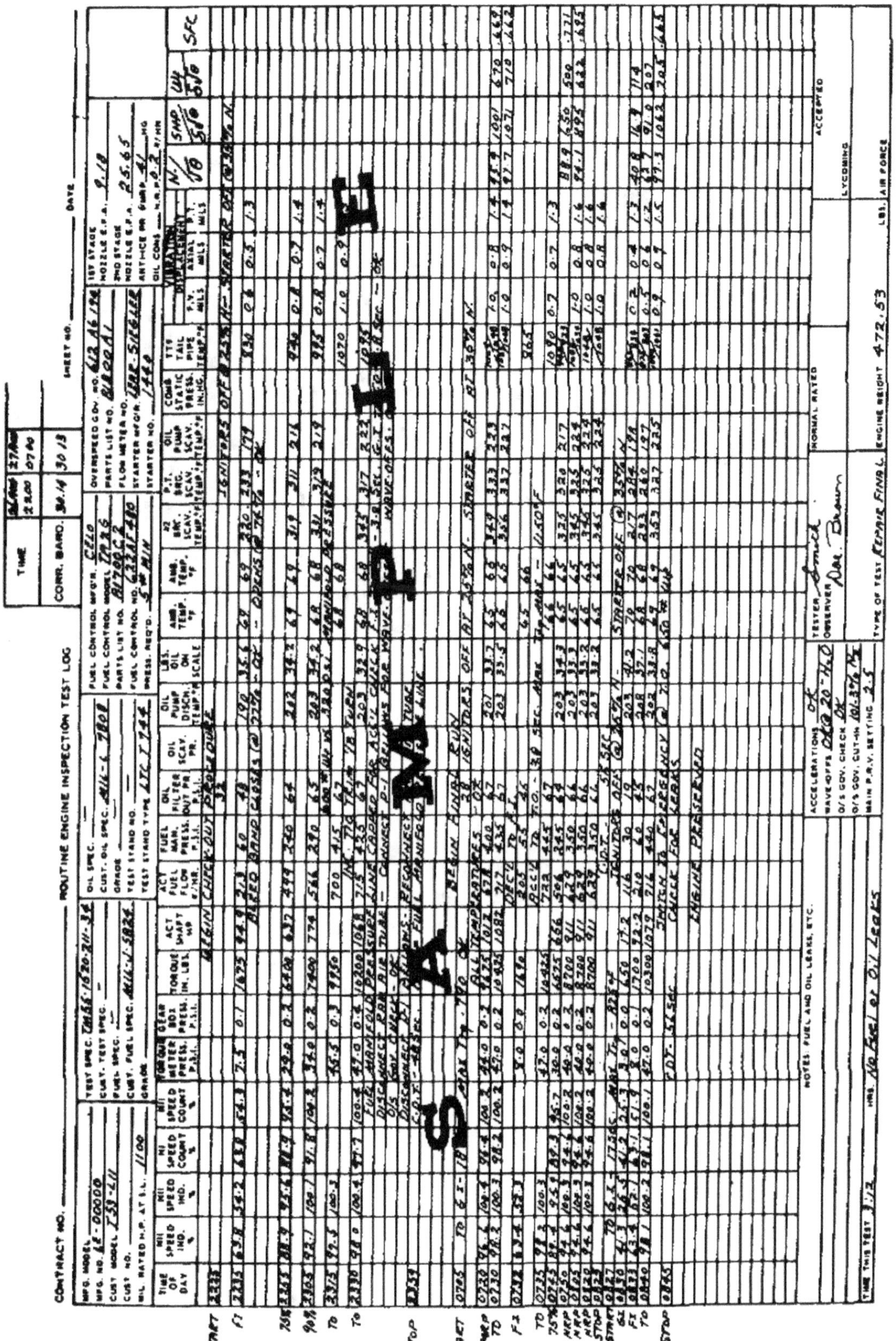

Figure 3.2. Engine Test Log Sheet.

shown in figure 3.3, also has connections for an oscilloscope or an oscillograph for closer examination. Filters are available to eliminate low frequency vibrations for a clearer picture of the higher, more damaging vibration signals being studied.

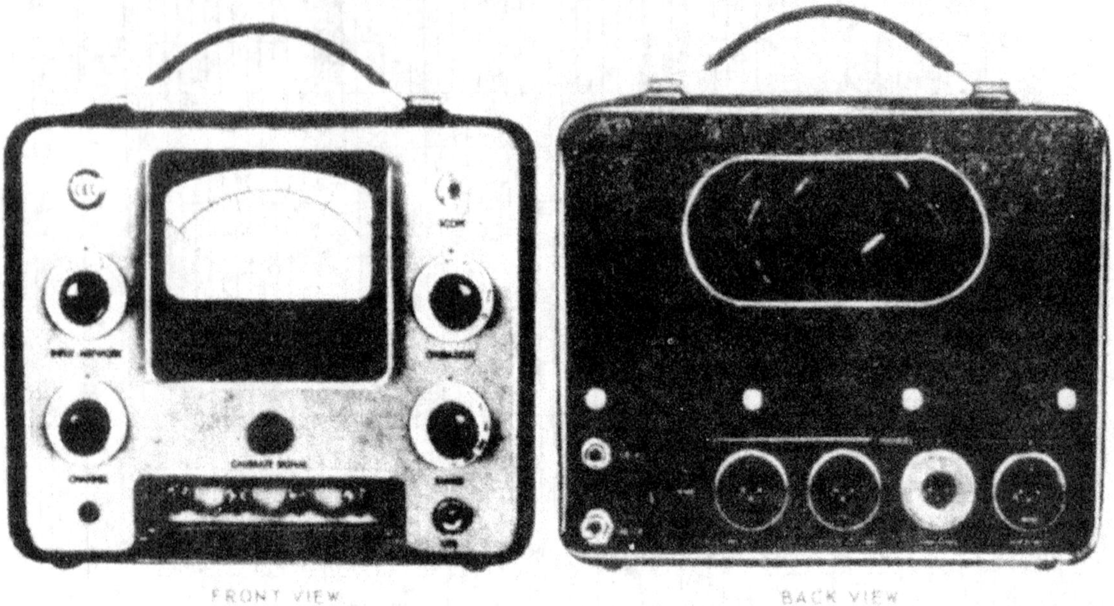

Figure 3.3. Vibration Meter.

Gas turbine vibration sources fall in two general classifications: forced vibration, and externally excited vibration.

Forced vibration is due to unbalanced rotating parts. The uneven weight distribution that causes the imbalance may be due to manufacturing methods, or improper assembly of components without regard for the balance in relation to other components. Other engine imbalances may be the result of a bent shaft, or a distortion caused by temperature.

Externally excited vibrations are caused by means other than an imbalance of rotating engine components. They may be caused by associated accessories, such as loose engine mounts or clamps, improperly mounted accessories, engine-driven transmissions, or airframe structure members.

An engine vibration test is preferred after major repair, removal, or replacement of any rotating part, or when excessive engine vibrations are suspected. Vibration pickups, attached to adapters mounted on the engine, transmit electrical impulses through cables to a vibration meter. The vibration meter indicates the total amount of engine movement in mils, one mil being 1/1000 of an inch. Meter indications are recorded on an Engine Vibration Test Data Sheet, shown in figure 3.4. The recorded figures are compared with the figures shown in parentheses on the data sheet for maximum permissible engine vibration. If these maximum figures are exceeded, the cause of the excessive vibration must be found and corrected before the engine can be accepted for flight.

3.4. JETCAL ANALYZER

To check the exhaust gas temperature (ECT) system when periodic maintenance inspections are required, or to troubleshoot the system if abnormally high or low temperatures are noted, the jetcal analyzer shown in figure 3.5 is used.

When checking the engine exhaust gas temperature (EGT)system, the jetcal is used to heat the thermocouple to the desired temperature without running the engine. The portable jetcal analyzer is equipped with a handle and two rubber wheels for easy movement. The jetcal operates on 95 to 135 volts, 50 to 400 cycles ac power supply.

3.5. SCHEDULED AND SPECIAL INSPECTIONS

Gas turbine engines are inspected at regular intervals -- scheduled. The inspection requirements are stated in a required order to assure that defects are discovered and corrected before malfunctioning or serious trouble results.

A special inspection is required whenever any of the operating limitations have been exceeded. Table II is a list of some of the conditions when a special inspection is required.

3.6. ARMY SPECTROMETRIC OIL ANALYSIS PROGRAM (ASOAP)

In the ASOAP program samples of used oil containing microscopic metal particles are sent periodically to an oil analysis laboratory. There the oil and its metal particles are burned by an electric or gas flame. The wave length of the light emitted from the burning oil and metal particles is measured to determine the kind and quality of metal in the oil. The identification gives advance warning of excessive wear on particular engine parts, thereby aiding in preventing inflight engine failures.

T53-L-9/9A/11 ENGINE VIBRATION TEST DATA SHEET

Engine Serial No. ______		(TSN) ______	
Last Overhaul Location ______		(TSLO) ______	
Aircraft Serial No. ______		A/C Time ______	
Test Location ______		Date ______	
Tested By ______		Observer ______	
Engine Passed Test – Accepted	☐	Engine Failed Test	☐

n_{II} Speed in rpm (± 50 rpm)	n_I Speed in Percent (± 0.5%)	No. 1 Pickup S/N ______ (70 cps Filter)	No. 1 Pickup S/N ______ (200 cps Filter)	No. 2 Pickup S/N ______ (200 cps Filter)	No. 3 Pickup S/N ______ (200 cps Filter)	REMARKS
6200	90	(2.5)	(2.5)	(2.5)	(3.7)	
5400	75	(3.5)	(3.5)	(3.5)	(3.7)	
6400	85	(2.5)	(2.5)	(2.5)	(3.7)	
6400	90	(2.5)	(2.5)	(2.5)	(3.7)	
6400	95	(2.5)	(2.5)	(2.5)	(3.7)	
6400	MAX.	(2.5)	(2.5)	(2.5)	(3.7)	
6600	90	(2.5)	(2.5)	(2.5)	(3.7)	

Figure 3.4. Engine Vibration Test Data Sheet.

3.7. ENGINE MAINTENANCE PRECAUTIONS

Personnel performing maintenance on gas turbine engines must observe the precautions stated in the applicable engine manual. Disregarding these warnings and precautionary measures can result in serious injury, illness, or death. The following subparagraphs discuss some of the precautions that must be taken while performing engine maintenance.

Figure 3.5. Jetcal Analyzer.

a. Use of lubricating oil. Prolonged contact with lubricating oil may cause a skin rash. Skin and clothing that come in contact with lubricating oil must be thoroughly washed immediately. Saturated clothing should be removed without delay. Areas in which lubricating oil is used must be ventilated to keep mist and fumes to a minimum. Because lubricating oil can soften some paint, oil spilled on painted surfaces must be promptly and thoroughly washed off.

b. Cadmium plated tools. Be sure tools used on engine are not cadmium plated. The cadmium plating on tools chips off, and oil contaminated with cadmium chips can cause magnesium parts to deteriorate.

c. Handling of parts. When handling combustion chamber internal parts that have been exposed to fuels containing tetraethyl lead compounds, be sure that the poisonous lead-oxide residue is not

TABLE II
SPECIAL INSPECTION TABLE

Condition
Exhaust gas temperature (egt) limits exceeded
Engine speed limits exceeded
Engine torque limits exceeded
Sudden engine stoppage experienced
Engine dropped during handling
Excessive G-load experienced
Impending bypass indicator button extended
Oil pump failure
Metal chip accumulations in chip detector, screens, or strainers
Spectrometric analysis indicates oil contamination
Engine compressor stall suspected
Foreign object ingestion suspected
Grass blockage inspection
Oil overtemperature inspection -- Engine models T53-L-5/9A/11/13/13A/13B
Excessive engine vibration experienced

inhaled or taken into the body through cuts or other external openings. If accidental exposure occurs, flush the affected area thoroughly with clear water and obtain immediate medical attention. Gloves and a face mask should be worn at all times when handling parts contaminated by lead oxide. hi addition bearings must be handled with special care. Gloves must be worn to prevent skin oil and acid from etching the bearing surface.

d. Marking on high-temperature materials. Using marking materials such as a common lead pencil on metals subject to high temperatures can cause the metal parts to crack. Approved marking materials are specified in the applicable engine manual. Only these marking materials are authorized for use.

e. Performing maintenance while engine is operating. Maintenance personnel must use caution when performing maintenance on operating engines. Because of the high temperature and velocity of the exhaust gases, personnel must stay clear of exhaust areas. Turbine intake areas are also a hazard. Large jet engines have been known to suck men into the engine. The smaller turbine engines in Army aircraft are capable of picking up small objects that are close to the intake. Anyone working around turbine engines should remove headgear and loose articles such as pens and pencils from shirt pockets. Figure 3.6 shows the exhaust-blast area of an OV-1 aircraft, to be avoided when the engine is running.

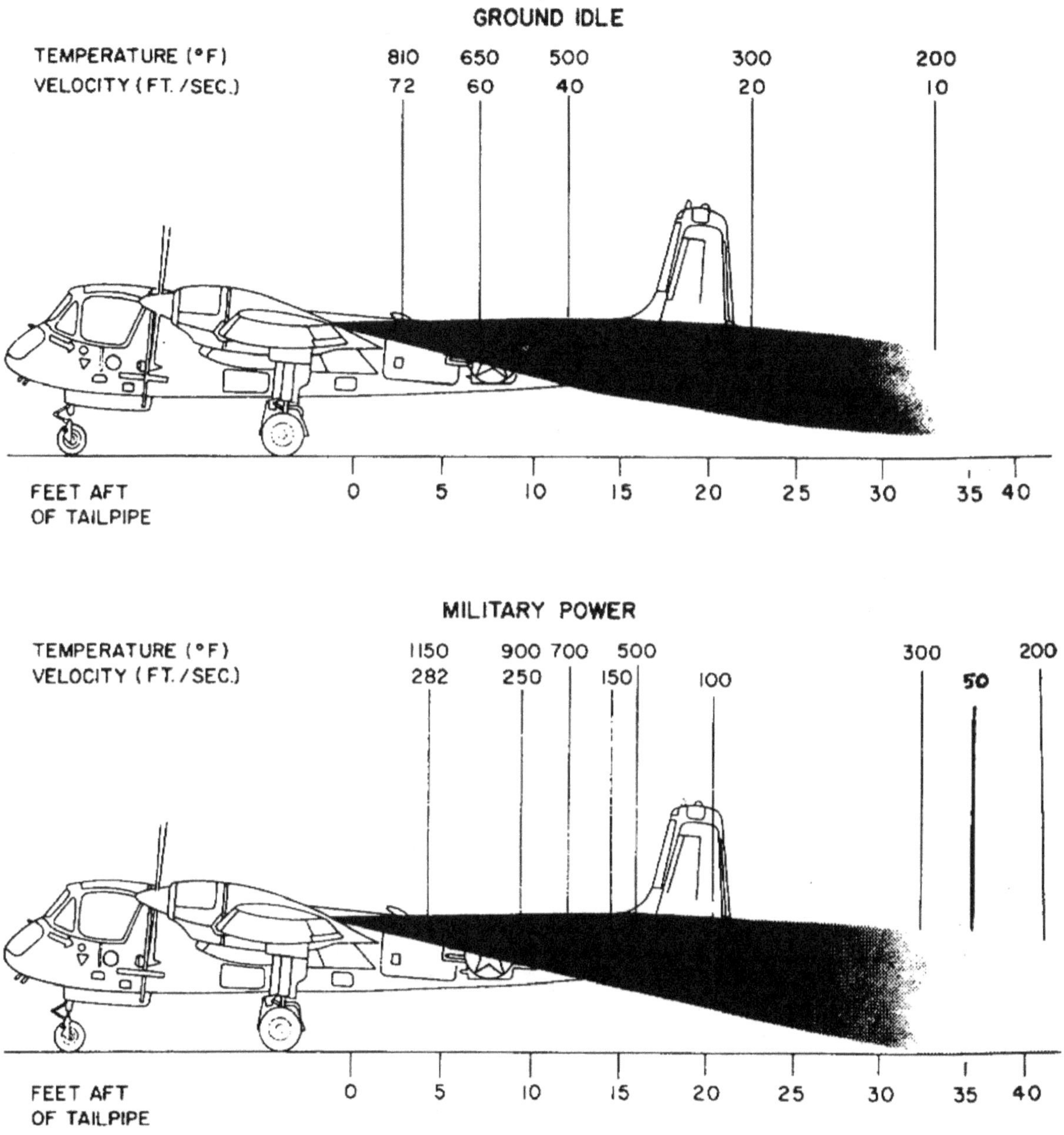

NOTE
THE BLAST AREA DOES NOT INCLUDE PROPELLER WAKE WHICH IS HIGHER IN VELOCITY BUT LOWERS THE EXHAUST GAS TEMPERATURE

Figure 3.6. Exhaust - Blast Area.

3.8. MAINTENANCE PROCEDURES

It is important to see that the engine compartment is kept as clean as possible because the high-velocity airflow through the engine will draw any foreign objects into the compressor. All loose parts, such as safety wire, cotter pin clippings, and nuts and bolts should be removed immediately. Tubing and lines should be checked for security, nicks, chafing, dents, and leaks.

Inspection and maintenance of gas-turbine engines are somewhat easier than those of reciprocating engines because the gas turbines stay cleaner. Besides, the first several stages of most compressors can be inspected for FOD by using a strong light. Also, the last two turbine stages are readily opened for inspection of heat damage.

The oil system is checked on the daily inspection for proper oil level. However, when adding oil, different types should not be mixed. In the past the Army has used MIL-L-7808 lubricating oil in turbine engines. Because of the higher operating temperatures encountered in the current gas turbine engines, a new oil has been developed. Military Specification No. MIL-L-23699 uses a new synthetic base and new additive combination to cope with the more severe operating conditions and higher temperature ratings. When changing from MIL-L-7808 to MIL-L-23699 lubricating oil, check the engine TM for proper procedures.

3.9. CLEANING ENGINE ASSEMBLY

The exterior of the engine, and its attached components, can be cleaned with a suitable cleaning solvent, such as P-D-680. If the solvent is sprayed on the engine with compressed air, care must be taken to avoid forcing dirt, solvent, or moisture into engine openings and electrical connections. The primary purpose of cleaning is to remove contaminants that might conceal minor cracks and defects which if not detected could eventually lead to failure. Under normal circumstances, engine components are cleaned only as necessary to perform required inspection and repair. After using alternate or emergency fuels, cleaning internal hot-end parts may be required to remove lead oxide deposits. These deposits, if not removed, are detrimental to engine life and performance. The choice of any particular cleaning agent or process depends upon the engine part to be cleaned and the contaminants to be removed.

Take particular care in selecting a cleaning method to ensure that anodizing or dichromating is not removed from the surfaces. Do not use caustics on aluminum, magnesium, ceramic-coated, aluminized, painted, nitrated, or carbonized parts. In most cases the engine manual prescribes the approved cleaning procedure to be used. Most engine parts may be cleaned by using the following methods.

a. Vapor degreasing. Used only on unpainted metal parts or aluminum-painted steel parts, vapor degreasing using heated trichloroethylene, type II, or perchloroethylene, specification No. O-T-634, removes oil, grease, and sludge. The hot vapor condenses on metal surfaces, liquefies, and carries away the oil, grease, and sludge. Parts may be flushed while held in the vapor. To prevent corrosion, the parts should not be removed from solvent vapors until they have reached the temperature of the vapor.

b. Solvent immersion. In another cleaning method, the parts are immersed in Carbon Removing Compound MIL-C-19853, to remove carbon, gum, grease, and other surface contaminants. This method is used on steel and stainless steel parts. Parts with painted finishes should not be cleaned by this method, because the carbon cleaning compound attacks the paint.

c. Vapor blasting. An abrasive method used to clean combustor parts, vapor blasting must not be used on ceramic, magnesium, painted, or aluminum surfaces. Be sure that metal is not removed during cleaning and that cooling slots, holes, ridges, and overlap areas do not become clogged with blasting grit.

d. Dry-cleaning solvent. All metal parts may be cleaned with dry-cleaning solvent, P-D-680 Type I. This method is suitable for removing heavy oil and grease deposits from most parts, including flexible hoses and carbon seals. Dry-cleaning solvent leaves an oily film that protects steel parts from corrosion for a short time.

3.10. CLEANING COMPRESSOR ROTOR BLADES

When a particular engine's performance decreases to or below the point specified in the applicable TM, and the EGT increases steadily during normal operation, the compressor rotor blades need cleaning. Compressor rotor blades should also be cleaned whenever the engine has been operating in areas where the air is salt laden, or when the engine has been subjected to contamination with fire extinguishing agent residue (chlorobromomethane and soda ash). Cleaning can be accomplished while the engine is installed in the aircraft.

Before cleaning any engine the applicable engine technical manual must be consulted for the proper procedures to follow. On some engines, temperature and pressure lines must be disconnected and capped to prevent solvent and water from entering.

The following is the preferred method for cleaning the compressor on the T53-L-13. Refer to figure 3.7 as you read the following steps in the cleaning method.

a. Remove airframe air intake components as necessary.

b. Remove the inlet air-temperature sensing element from the inlet housing.

c. Disconnect the pressure line to bleed band actuator and cap the diffuser fitting with AN929-6 cap assembly.

d. Block off the customer bleed air supply at the customer airbleed port in the airbleed adapter assembly.

e. While the engine is cold, rotate it with the starter and spray one quart of dry-cleaning solvent (P-D-680 Type 1) evenly through all sections of the inlet housing. Make sure both sides of the inlet guide vanes are covered with solvent.

f. Stop motoring the engine and let it stand for at least one hour to permit the dry cleaning solvent to loosen dirt.

g. Clean the inlet guide vanes with a small, round fiber brush with a long handle.

h. Start the engine and operate it at flight idle.

i. Spray CLEAN fresh water evenly into all sections of the inlet housing at the rate of two gallons per minute for approximately two minutes. To avoid freezing at ambient temperatures below 35° F (1.5° C), use anti-detonating injection fluid or a mixture containing 40 percent methanol and 60 percent water in lieu of water.

j. Allow the engine to run for 2 to 5 minutes to dry out; then shut the engine down.

k. Inspect the inlet guide vanes and compressor for cleanliness.

l. Repeat the cleaning procedure if necessary.

Figure 3.7. Compressor Blade Cleaning.

m. Reconnect the lines for normal operation.

n. Clean the temperature sensing element with dry-cleaning solvent and reinstall it.

3.11. OVERHAUL AND REPAIR

The time between overhauls (TBO) varies considerably between different engines. The TBO is established by the Army and the engine manufacturer who take into account the kind of operation and use expected for the engine, also the environment it will be operating in. As a specific model engine builds up operating time, it is inspected for signs of wear and impending failure of parts. If the engine is wearing well, the TBO is extended. The large improvement of TBO has been accomplished mainly through improvements in engine design, metallurgy, manufacturing, overhaul, inspection, and maintenance procedures. The use to which the engine is put is especially important in determining the TBO. For example, if the mission the aircraft is designed for calls for frequent starts and stops, or for power changes as in a helicopter, the resultant rapid temperature changes will shorten the allowed TBO. The following paragraphs cover disassembly, assembly, and repair procedures.

_a_. _Disassembly_. Engine disassembly can be accomplished on a vertical or horizontal disassembly stand as shown in figure 3.8. Some engines can be disassembled either horizontally or vertically, while others have to be done in only one position. After the engine is disassembled, the major components and section assembly are mounted on individual stands. To disassemble an engine, instructions in the TM must be followed, and a large number of special tools is required. A set of these tools may cost as much as the engine.

_b_. _Assembly_. Engine assembly also follows instruction in the TM; it is done on the same stand as disassembly. During assembly, care must be taken to prevent dirt and other foreign materials from entering the engine. The procedures and use of special tools as outlined in the maintenance manual must be followed to minimize possible injury to the mechanic and damage to the engine.

_c_. _Repair_. All engine parts must be repaired using methods approved by the engine TM. Figure 3.9 shows an illustration of typical repair limits for compressor rotor blades on the Lycoming T53 series engines.

3.12. MAINTENANCE ALLOCATION CHART

Maintenance function assignments are determined by the maintenance allocation chart found in the aircraft -20 technical manual. The maintenance allocation chart assigns functions to the lowest capable maintenance level based on past experience, and the skills, tools,

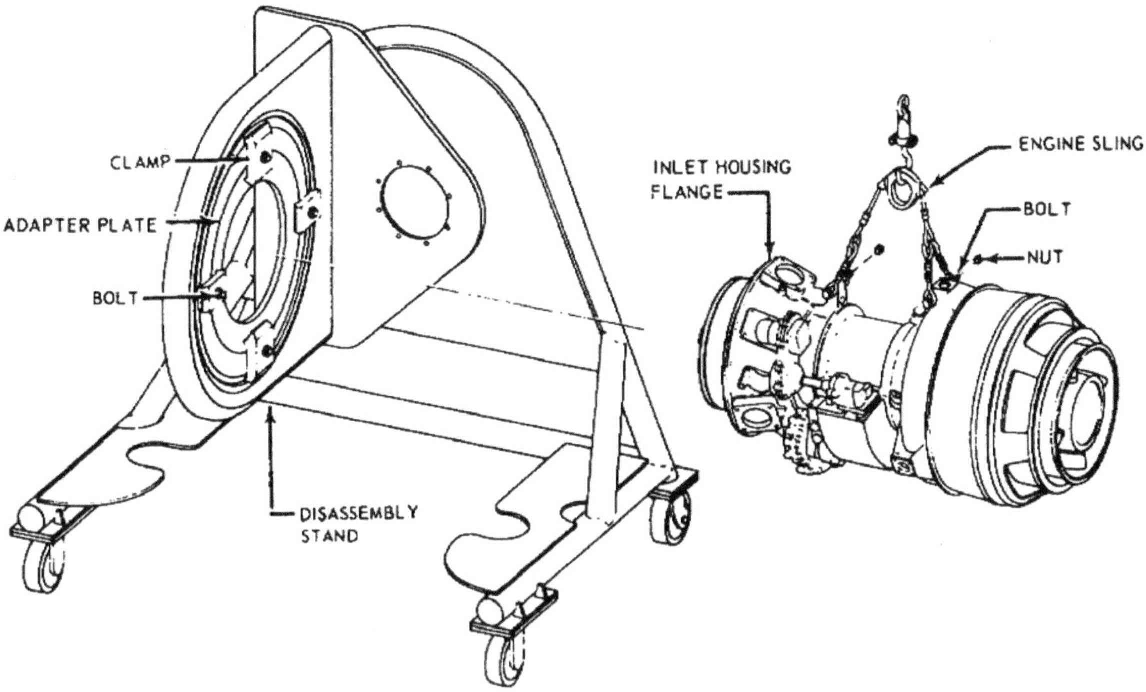

Figure 3.8. Engine Disassembly Stand.

and time available. Maintenance that cannot be performed at the assigned level may be reassigned to the next higher level. Generally, there is no deviation from the assigned level of maintenance. However, in cases of operational necessity, higher level function is assigned to a lower level by the maintenance officer of the level to which the function is originally assigned. Figure 3.10 shows an example of a maintenance allocation chart for UH-1D and H series helicopters. The symbols "O, F, H and D" represent: organizational maintenance (O), direct support maintenance (F), general support maintenance (H), and depot maintenance (D). When one of these symbols is placed on the allocation chart, it indicates the lowest level of maintenance responsible for performing the particular maintenance function. Maintenance levels higher than the level symbolized on the chart are authorized to perform the indicated maintenance.

The terms used in block (3) of the maintenance allocation chart are explained in table HI for convenience in reviewing and for future reference.

3.13. STORAGE AND PRESERVATION

The degree of preservation is determined by the anticipated length of time an engine is expected to be inactive. The three categories of storage are:

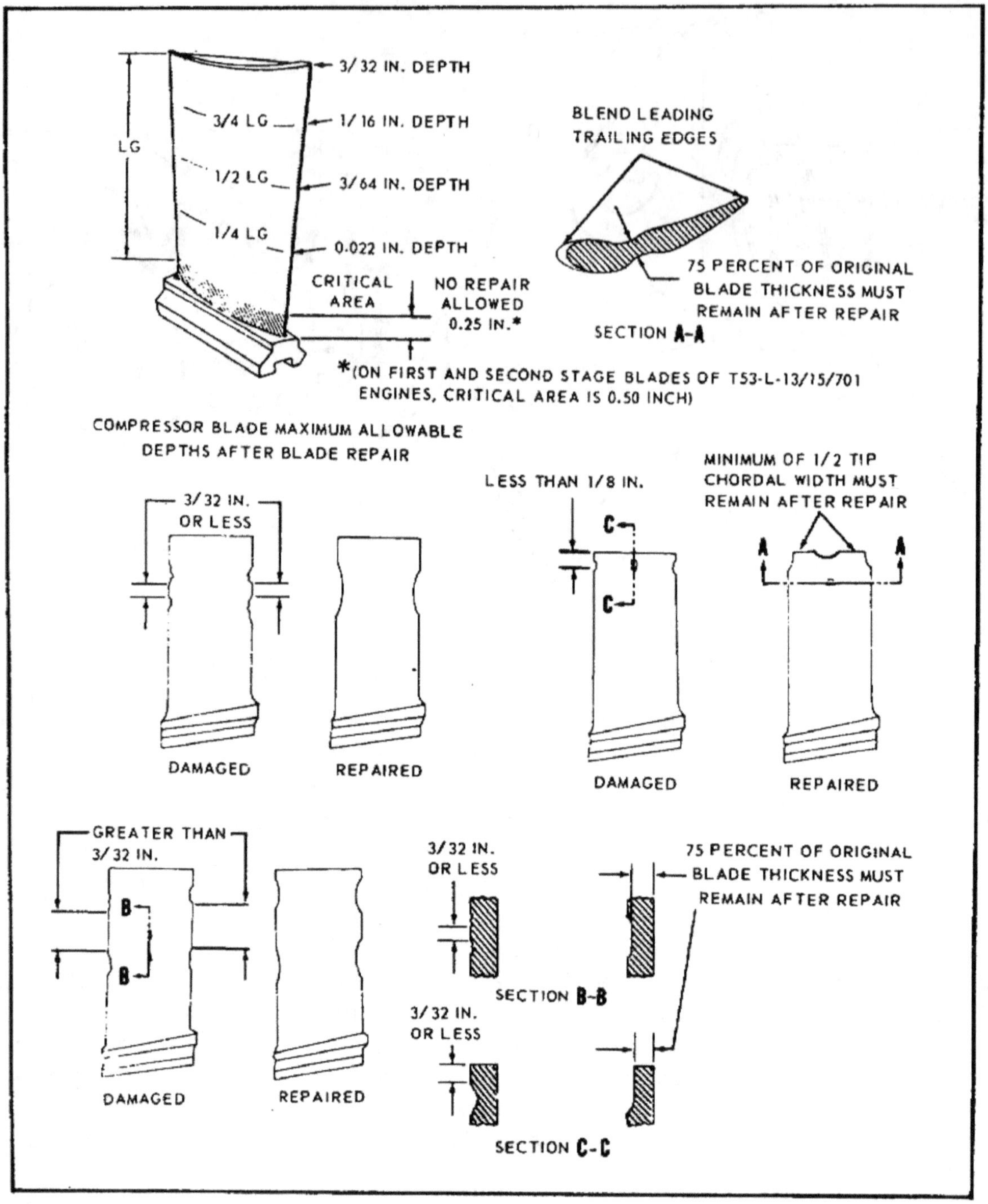

Figure 3.9. Compressor Rotor Blade Damage Before and After Repair.

a. Flyable storage. An engine that will not be operated for a period of at least 72 hours, nor more than 14 days, must be preserved and maintained with all components and systems in an operable condition. On the third day, the engine must be run-up or motored with the starter. If the engine is only motored on the third day, it must be run up on the seventh.

MAINTENANCE ALLOCATION CHART FOR YUH-1D, UH-1D and UH-1H														
(AR 310-3)														
(1)	(2)	(3)											(4)	(5)
		MAINTENANCE FUNCTION												
GROUP NO	FUNCTIONAL GROUP	INSPECT	TEST	SERVICE	ADJUST	ALIGN	CALIBRATE	INSTALL	REPLACE	REPAIR	OVERHAUL	REBUILD	TOOLS AND EQUIPMENT	REMARKS
03	ENGINE AND RELATED SYSTEMS													
	Engine as Complete Assembly	O	O			O			%%	O	D			Turbine Engine Test Stand at H_5
	Fuel and Oil Hoses	Q							O					
	Electrical Harness & Ignition Leads	O							O					
	Ignition Exciter	O							O					
	Igniter Plugs	O							O					
	Fuel Control Assembly	O		O	O				O		D			
	Electrical Solenoid Valve	O							O					
	Overspeed Governor Assembly	O							O		D			
	Fuel System Filter and Strainers	O		O					O					
	Starting Fuel Solenoid Valve	O							O					
	Main Fuel Manifolds	O							O	F				
	Starting Fuel Manifold	O							O					
	Starting Fuel Nozzles	O							O	H				
	Oil Pump	O			O				O		H			
	Bearing Housing Oil Strainers	O		O					O					
	Main Oil Filter	O		O					O	O				
	Engine Oil Tank	O		O					O	F				
	Engine Oil Coolers	O		O					O	F				
	Engine Oil Cooler Blower	O		O					O	F				

Figure 3.10. Maintenance Allocation Chart.

TABLE III
DEFINITIONS OF MAINTENANCE TERMS

a. Inspect. To determine the serviceability of an item by comparing its physical, mechanical, and electrical characteristics with established standards.

b. Test. To verify serviceability and to detect electrical or mechanical failure by using test equipment.

c. Service. To clean, to preserve, to charge, and to add fuel, lubricants, cooling agents, and air.

d. Adjust. To ratify to the extent necessary to bring into proper operating range.

e. Align. To adjust specified variable elements of an item to bring it to optimum performance.

f. Calibrate. To determine the corrections to be made in the readings of instruments or test equipment used in precise measurement. This consists of comparing two instruments, one of which is a certified standard of known accuracy, to detect and adjust any discrepancy in the accuracy of the instrument or test equipment being compared with the certified standard.

g. Install. To set up for use in an operational environment such as an emplacement, site, or vehicle.

h. Replace. To replace unserviceable assemblies, subassemblies, or parts.

i. Repair. To restore an item to serviceable condition through correction of a specific failure or unserviceable condition. This includes, but is not limited to, inspection, cleaning, preserving, adjusting, replacing, welding, riveting, and strengthening.

j. Overhaul. To restore an item to a completely serviceable condition as prescribed by maintenance serviceability standards prepared and published for the specific item to be overhauled.

k. Rebuild. To restore an item to a standard as near as possible to the original or new condition in appearance, performance, and life expectancy. This is accomplished through the maintenance technique of complete disassembly of the item, inspection of all parts or components, repair or replacement of worn or unserviceable elements (items) using original manufacturing tolerance and specifications, and subsequent reassembly of the item.

b. Temporary storage. An engine that will not be operated for over 14 days, but less than 45 days, must be placed in temporary storage. Engines normally falling in this category are those undergoing minor repair or modification, awaiting assignment or disposition, being held in operational reserve, or any other condition which requires idleness for a period not to exceed 45 days.

c. Extended storage. An engine that will be inactive for more than 45 days, but not exceeding 180 days, must be preserved and maintained in extended storage. Usually, this includes those engines undergoing major repair or modification, those declared surplus and awaiting final disposition, or any other circumstance that would warrant idleness for 45 to 180 days.

NOTE

Permanent storage is a depot level function.

ENGINE PRESERVATION - GENERAL. All preservation procedures require that any accumulation of dirt be removed from the engine with dry cleaning solvent. Under usual conditions, it will not be necessary to clean the entire external surface of the engine. If necessary, perspiration residues can be removed from close tolerance bare metal surfaces by wiping with a clean cloth dampened in fingerprint remover before cleaning with solvent.

CAUTION

To prevent oil contamination, never mix synthetic-base oils with mineral-base oils. Synthetic-base lubricating oil is required for the engine. Only a synthetic-base corrosion preventive oil can be used to spray the compressor for corrosion prevention.

3.14. TURBINE ENGINE TROUBLESHOOTING

Engine malfunctions can be recognized and diagnosed by comparing actual engine instrument reading with normal readings. To aid maintenance personnel in engine troubleshooting, the engine technical manual has troubleshooting charts to analyze, isolate, and correct engine malfunctions. Proper utilization of the troubleshooting charts will save time, provide a logical method of isolating the causes of malfunctions, and eliminate the unnecessary replacement of parts. Figure 3.11 explains how to use the troubleshooting charts.

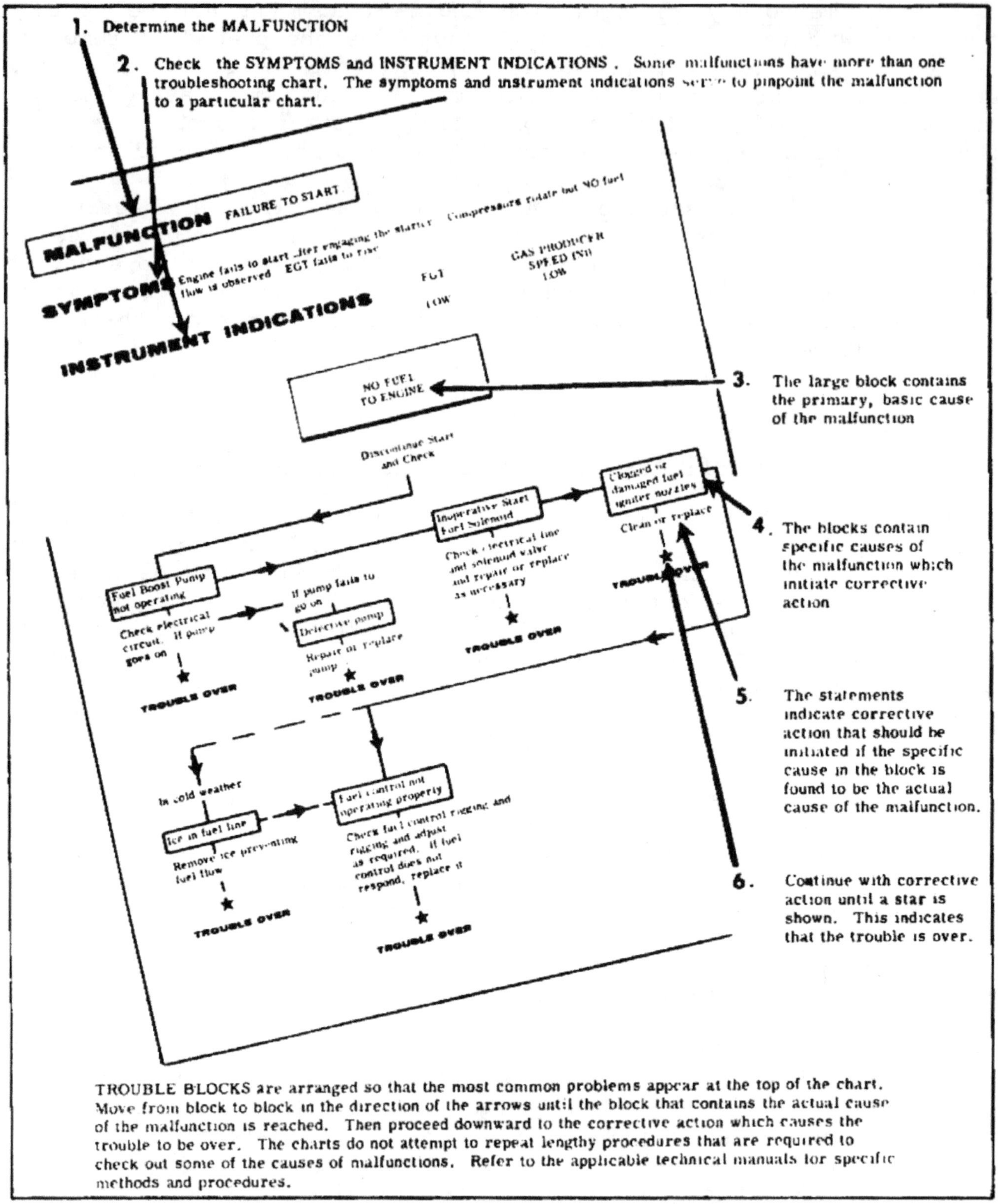

Figure 3.11. How to Use Troubleshooting Charts.

3.15. SUMMARY

Gas turbine engines are run in test cells to ensure quality control before they are shipped to the user for installation in aircraft. The test cell is equipped with instruments to monitor engine performance. Engine vibration tests can be performed with the engine in the test cell or installed in the aircraft. Vibration tests are required after any maintenance on rotating parts or when excessive engine vibration is suspected. A jetcal analyzer is used to check the accuracy of the egt system and to calibrate it.

There are two kinds of engine inspections, scheduled and special. Scheduled inspections are required whenever any of the operating limits have been exceeded.

Under the Army Spectrometric Oil Analysis Program (ASOAP) oil samples are analyzed for metal content, to prevent in-flight engine failures.

Personnel performing maintenance on gas turbine engines should observe the precautions stated in the engine manual to avoid serious personnel injury or engine damage. All engine cleaning, both internal and external, should be performed in accordance with the appropriate engine manual. In most cases the engine manual prescribes the approved cleaning procedure to be used. Most engine parts may be cleaned by the vapor degreasing, solvent immersion, or vapor blasting methods.

The TBO of a gas turbine engine depends upon such things as operating environment, mission to be performed, and how will the engine wear as flight time is built up.

Maintenance function assignments are determined by the maintenance allocation chart found in the aircraft -20 manual. Three categories of engine storage are used. The decision as to which category of storage is to be used depends upon the length of time the engine will be inactive.

Chapter 4

LYCOMING T53

4.1. INTRODUCTION

The two sections of this chapter discuss, in detail, the Lycoming T53 series gas turbine engine used in Army aircraft. Section I gives a general description of the T53, describes the engine's five sections, explains engine operation, compares models and specifications, and describes the engine's airflow path. The second section covers major engine assemblies and systems.

Basically, all models of the T53 engine are of the same design. The major difference on models later than the T53-L-11 is that they have two gas producer turbines (N_1) and two free-power turbines (N_2) instead of the single stage turbine used on the L- 11 and earlier models. The engine models described in this chapter are primarily the T53-L-13, and T53-L-701. However, the description and information given is applicable to all models except where noted.

Section I. Operational Description of the T53 Gas Turbine Engine

4.2. GENERAL

The information in this section is important to you because it describes the engine's airflow path through the inlet, compressor, diffuser, combustion, and exhaust sections, and explains the operational relationship of these sections. In addition differences between models and specifications are compared. Except for the paragraph comparing models, this section's coverage is limited to the T53-L-13 and -701.

4.3. GENERAL DESCRIPTION

The T53 series gas turbine engine is an annular reverse-flow, free-power turbine powerplant developed for fixed- and rotary-wing aircraft. As shown in figure 4.1, the engine consists of inlet, compressor, diffuser, combustion, and exhaust sections. All these are designed to include an annular or circular flow path for the air or hot gases, and they are structurally dependent on one another. These sections support all internal rotating systems and have attaching capabilities for engine accessories.

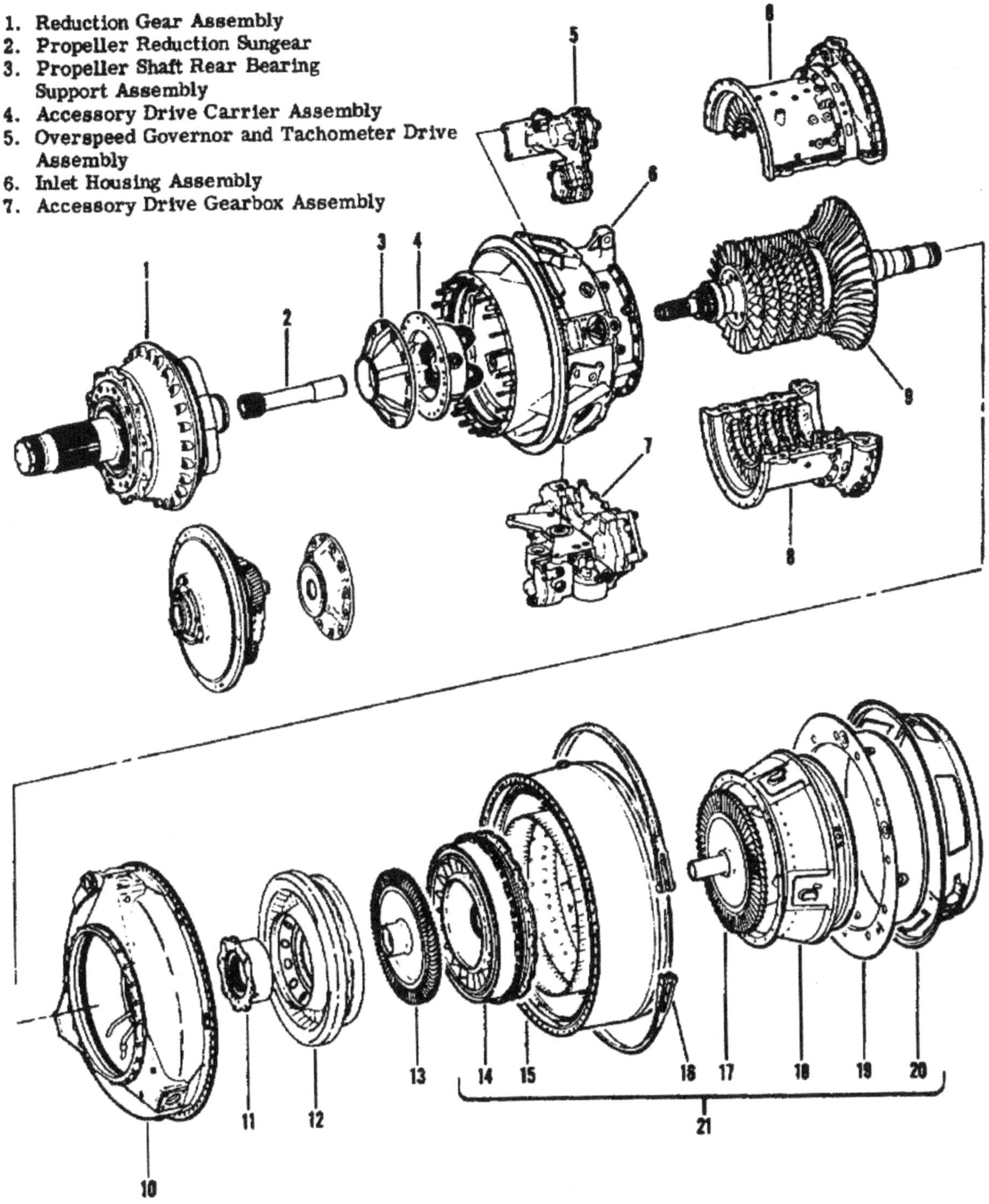

Figure 4.1. T53 Engine (Exploded View).

4.4. OPERATIONAL DESCRIPTION

The discussion in this paragraph describes briefly the airflow through the engines and the operation of the Lycoming T53. The capital letters in parentheses such as (A), (B), (C) and so on in the discussion correspond to similar letters in figure 4.2 and refer you to that particular portion of the engine diagram.

a. Air flow. Atmospheric air (A) is drawn into the annular air passageway of the inlet housing (B) and passes rearward across the variable inlet guide vanes (C). The vanes direct the air into the engine compressor section. The air passageway in the compressor section contains Eve rotating axial compressor stages with five sets of stationary stator vanes, a set of exit guide vanes (D), and one centrifugal compressor (E). As the air passes through this section, each rotating axial compressor stage increases the pressure. The exit guide vanes guide the air onto the centrifugal compressor which further accelerates the air as it passes radially into the diffuser housing air passageway (F). Vanes in the diffuser air passageway convert the high velocity of the air into pressure and also change the radial airflow to a rearward flowing direction.

At this point the air enters the combustor section, passing around and into the annular combustion area (G) through slots, louvers, holes, and scoops fabricated in the combustion liner. On entering the combustion area, flow direction is reversed while both air velocity and pressure drop. The air, at the same time, performs the multiple functions of cooling the combustor liner; mixing with fuel, and burning, sustaining, and maintaining the high heat combustion within a confined area; and absorbing the heat of combustion so as to lower the heat to a usable temperature. Combustion is made possible by introducing fuel into the combustion area through 22 atomizers. The atomized fuel mixes with the air, burns, and produces temperatures as high as 3,500 degrees F.

As previously stated, this exceedingly hot gas is cooled as it flows forward in the combustion area to the deflector, which reverses the hot gas flow. Now flowing rearward, the gas is directed across the two-stage gas producer nozzle turbine system. The first stage nozzle (H) directs the high energy gas onto the first stage turbine (I), across the second stage nozzle (J) onto the second stage turbine (K). The power system also uses the two-stage nozzle turbine concept. Therefore, on leaving the second stage gas producer turbine, the gas, still possessing a high work potential, flows across the third stage nozzle (L) onto the third stage turbine (M), across the fourth stage nozzle (N) onto the fourth stage turbine (O). On passing from the fourth stage turbine, the gas is exhausted into the atmosphere through the exhaust diffuser passageway (P).

Wexford
2008

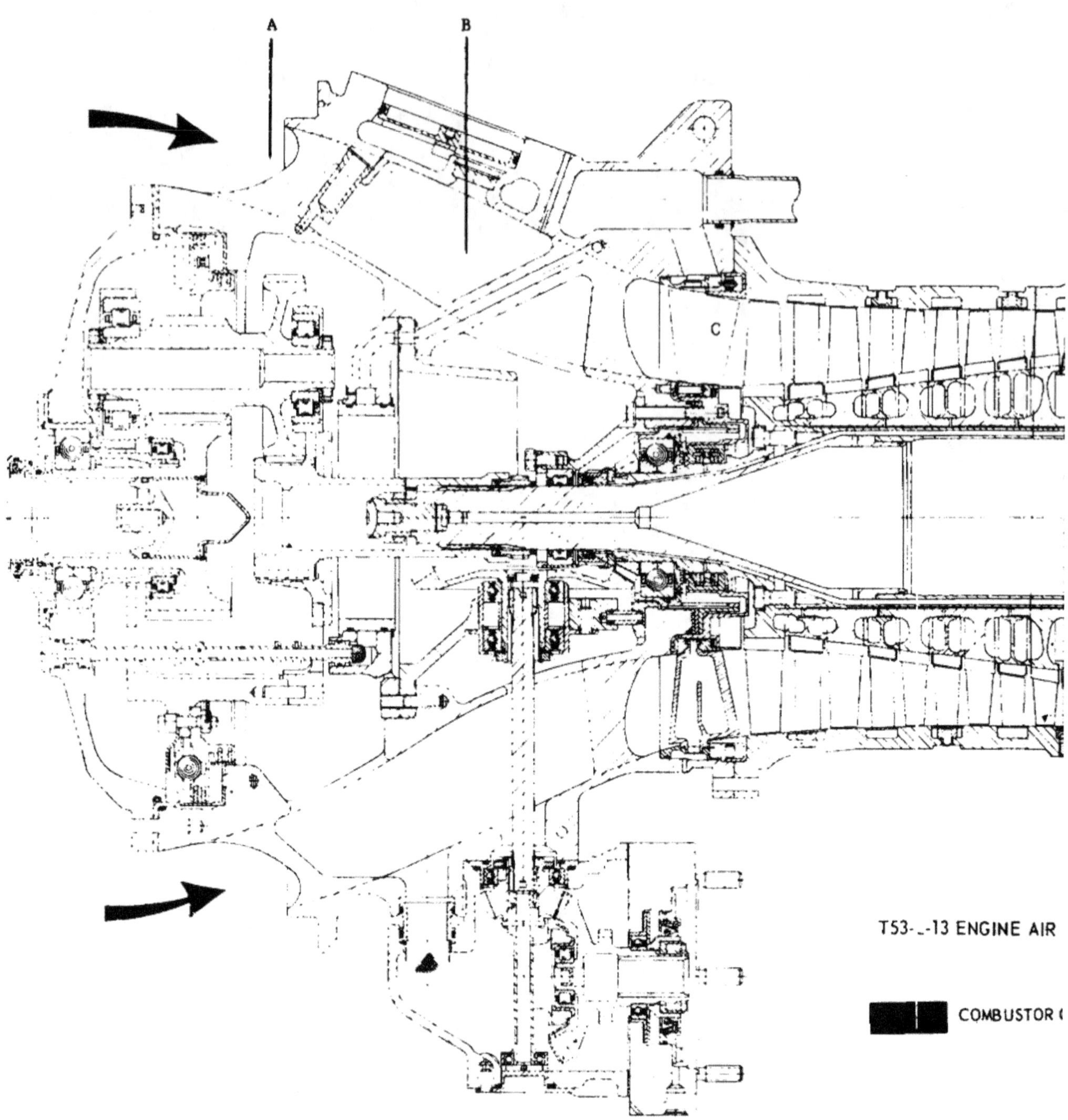

(continued on next page)

Figure 4.2. T53-L-13 Engine Airflow.

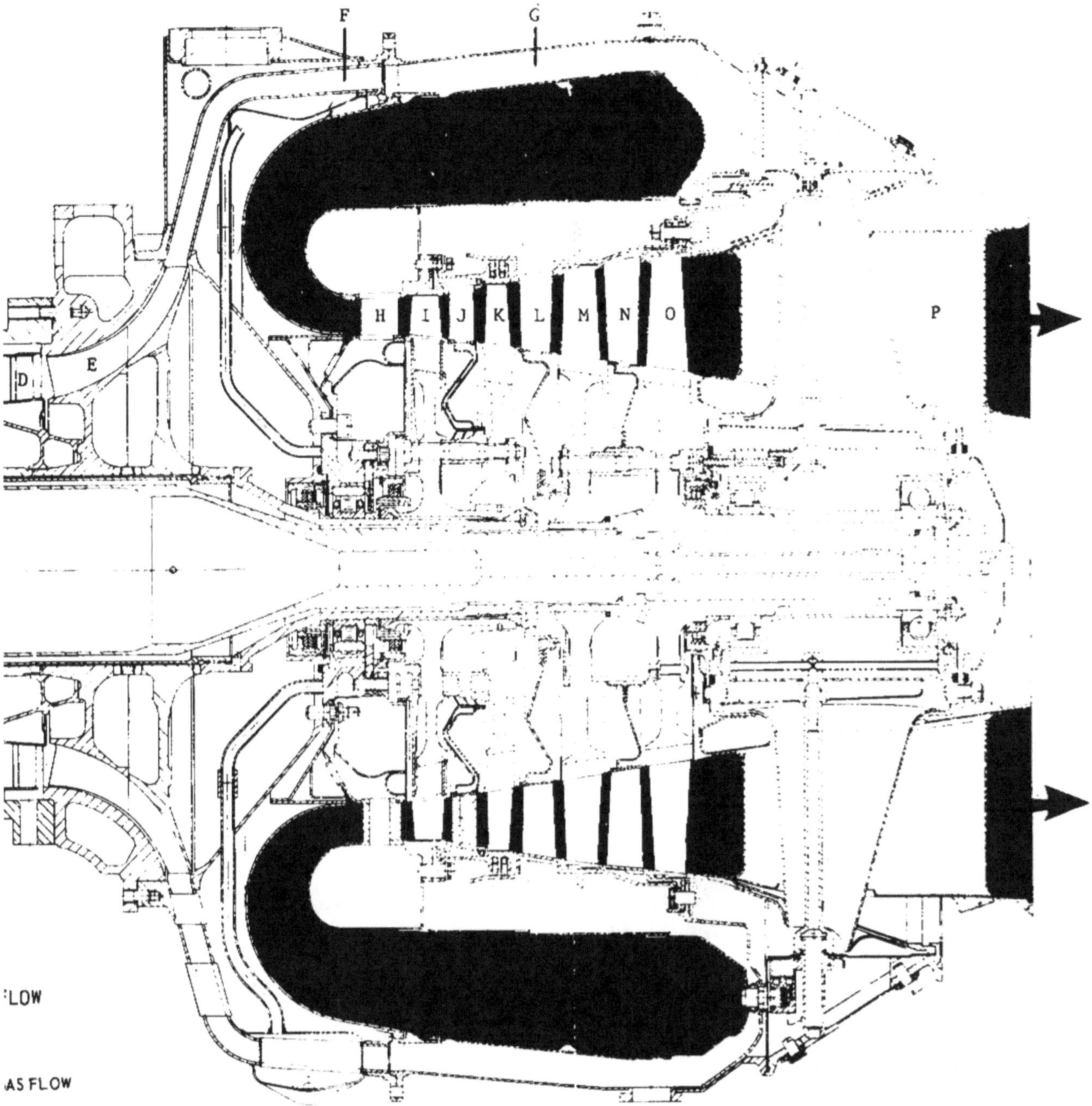

(continued from previous page)

Wexford
2008

b. Operation. The engine is started by energizing the starter, the starting fuel solenoid valve, and the ignition system. Starting fuel flows into the combustion chamber through four starting fuel nozzles and is ignited by the four igniter plugs adjacent to the starting fuel nozzles at the 2, 4, 8, and 10 o'clock positions. At 8 to 13 percent N_1 speed, the fuel regulator valve opens, and main fuel flows into the combustion chamber through 22 fuel atomizers and is ignited by the burning starting fuel. As compressor rotor speed (N_1) increases, the additional fuel mixes with compressed air and burns.

When compressor speed increases to 40 percent N_1 speed, the starter, starting fuel solenoid valve, and ignition system should be de-energized. Combustion gases pass through the gas producer nozzle assemblies; impinge upon (strike) the blades of the gas producer rotor assemblies; How through the power turbine nozzle assemblies; and impinge upon the blades of the power turbine rotor assemblies. Approximately 60 percent of the gas energy passing from the combustion chamber is extracted by the N_1 turbine rotors to drive the compressor, while the remaining energy is extracted by the N_2 power turbines to drive the power shaft. The power turbine rotor assemblies are splined to the power shaft and secured by the power-shaft bolt. The power shaft is splined into the sun-gear shaft, which drives the output reduction gearing and, in turn, the power output gear shaft.

4.5. MODEL COMPARISON

The Lycoming T53 engine design was submitted to the military, and in 1952 a contract was awarded to the Army for development of the present T53 gas turbine engine. The basic T53 powerplant is a free turbine consisting of two mechanically independent stages -- a compressor and a power turbine. From the initial prototype version, many models of the T53 have been developed. The following models are currently being used by the Army.

T53-L-13 -- a second-generation shaft turbine engine having all the improvements developed for the T53- L-11/11A/11B versions. It introduces combustion chamber atomizers (replacing the vaporizer tubes), two gas producer and power turbine rotors instead of one of each, and four atomizing type starting fuel nozzles and four igniter plugs.

T53-L-13A -- Engine model T53- L-13 containing additional modifications is designated as model T53-L-13A. This model incorporates the following improvements: a 34-blade second stage compressor disc assembly; No. 2 bearing forward and aft seals, bearing housing and retaining plate; oil-lubricated fuel-control drive splines, an air diffuser and accessory gearbox with improved oil scavenge capability; and a 6-probe exhaust gas temperature harness assembly.

T53-L-15 -- a turboprop equivalent of the T53-L-13, the T53-L-15 has been flat-rated and is equipped with a torque limiter to prevent engine torque from exceeding the limitations imposed by the airframe manufacturer. Other differences include a 6-probe, 12-point exhaust-gas thermocouple harness, a fuel heater, fuel filter, and bypass fuel filter. Also, the manual operating feature of the fuel control has been deleted.

T53-L-701 -- the newest turboprop addition to the Lycoming T53 series engines is designated the T53-L-701. The most dynamic feature of the T53-L-701 is a newly developed Lycoming split-power reduction gear assembly, using an electric torquemetering system. This split-power reduction gear assembly permits development of the full mechanical and thermodynamic capabilities of the T53-L-701 by allowing operation to 1,451 shp. Very accurate torque measurements on the T53-L-701 are provided by an electric torquemeter system.

4.6. SPECIFICATION SUMMARY

Specifications for the T53-L-13 and 701 engines used in Army aircraft are summarized in the following chart.

	Turboshaft	Turboprop
Military Designation	T53-L-13	T53-L-701
Lycoming Designation	LTCIK-4	LTCIF-5
Compressor Stages	5 Axial, 1 Centrifugal	5 Axial, 1 Centrifugal
Combustor	Annular Reverse Flow	Annular Reverse Flow
Turbine (GP)	2 Stage	2 Stage
Turbine (PT)	2 Stage	2 Stage
Power Outlet	6600 rpm	Propeller Shaft 1,700 rpm
Fuel Control	Auto and Manual	Auto and Manual
Ignition System	High-Energy, 4 Torch Igniters	High-Energy, 4 Torch Igniters
Lubrication System	External Oil Tank, Cooler	External Oil Tank, Cooler
Diameter & Length	23.0 in., 47.9 in.	23.0 in., 59.4 in.
Weight	549 lb.	688 lb.
Fuel	MIL-T-5624/JP4	MIL-T-5624/JP4
Oil	MIL-L-23699	MIL-L-23699
Rating (Military) (Normal)	1,400 shp 1,250 shp	1,451 shp 1,297 shp
Application	UH-1H/M, AH1G	OV-1 (2 Engines)

4.7. DIRECTIONAL REFERENCES AND ENGINE STATIONS

The diagrams in figures 4.3 and 4.4 show directional references and engine stations. Notice that power output is taken from the front end and that exhaust gas is expelled from the rear end. Right and left sides for the engine are determined by viewing the engine from the rear. The engine's bottom is determined by the accessory drive gearbox's location. The top of the engine is directly opposite, or 180 degrees from the accessory drive gearbox. Rotational direction is determined by viewing from the rear of the engine. The compressor rotor and gas producer turbines rotate in a counterclockwise direction. The power turbines and output gearshaft rotate in a clockwise direction.

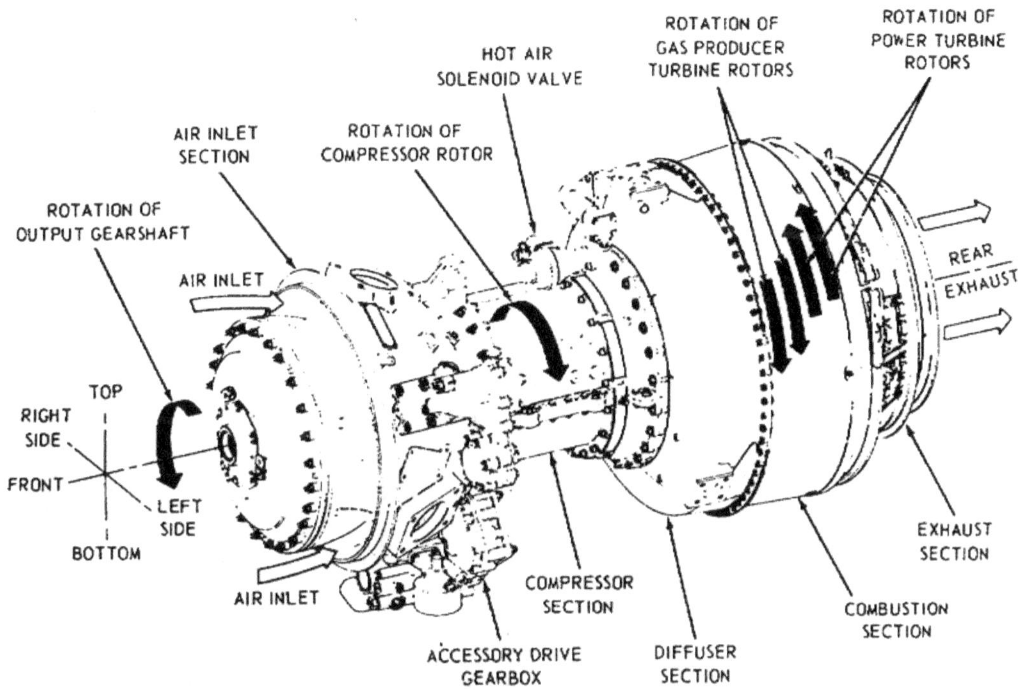

Figure 4.3. Engine Orientation Diagram (T53-L-13).

The T53 engine has 14 ports for test measuring, and it is divided into stations to designate temperature (T) and pressure (P) measuring locations. Engine stations for the T53-L-13 are shown in Figure 4.4. They are identified on the drawing as 1.0, 2.0, on up to 9.0, but in practice they are spoken of as 1, 2, 3, and so forth. Station 1, on the inlet housing, is for ambient air. Stations 2 and 3 are for compressor and diffuser discharge air. Station 4 is located at the combustor section. Stations 5 and 7 designate turbine inlets N_1 and N_2. Station 9 is the location for the exhaust diffuser. No stations are shown for 6 and 8, because these numbers are not used.

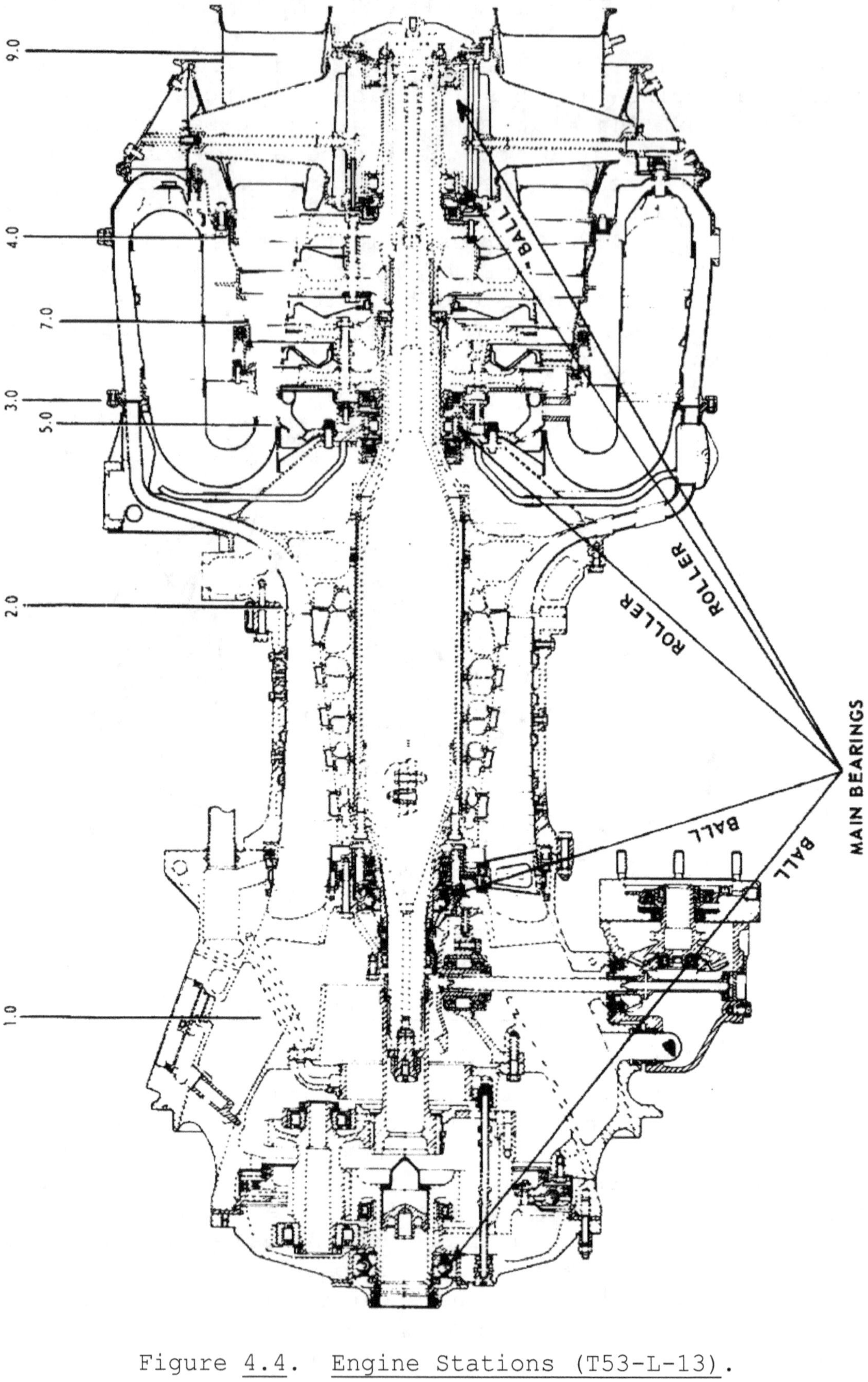

Figure 4.4. Engine Stations (T53-L-13).

4.8. SUMMARY

Following the air path through the engine shows how air is drawn into and moved through the engine's Eve sections. Air brought into the compressor section is accelerated into the diffuser-housing air passageway. As the air moves through the passageway, its velocity changes to pressure and, under pressure, the air enters the combustion chamber to mix with injected fuel. The flow of the hot gases across the turbine rotors produces mechanical energy to drive the compressor and propel the aircraft.

The various models of the T53 include the L-13, L-13A, L-15, and L-701. Some of the specifications differ in each model.

Section II. Major Engine Systems and Assembles

4.9. GENERAL

Starting at the front of the engine and working rearward we discuss the major engine assemblies. Systems such as fuel, oil, and electrical are covered in their entirety, after the major engine assemblies. Also keep in mind that the T53-L-13 is a turboshaft engine, and the T53-L-701 is a turboprop engine. The turboprop engine has a propeller reduction gear assembly in the inlet housing where the turboshaft engine has a smaller output reduction gear assembly.

4.10. INLET HOUSING ASSEMBLY

The forward structural support of the engine is provided by the inlet housing assembly shown in figure 4.5. The outer housing, supported by six hollow struts, forms the outer wall of the annular air inlet and houses the anti-icing manifold. The inner housing forms the inner wall of the inlet area.

Enclosed in the inlet housing is the output reduction carrier and gear assembly, the oil transfer support assembly, and the accessory drive carrier. A torquemeter valve and cylinder, power shaft support bearing, and No. 1 main bearing are also mounted internally. At the rear of the housing the inlet guide vanes are installed in the airflow path to direct air at the proper angle onto the first stage of the compressor rotor.

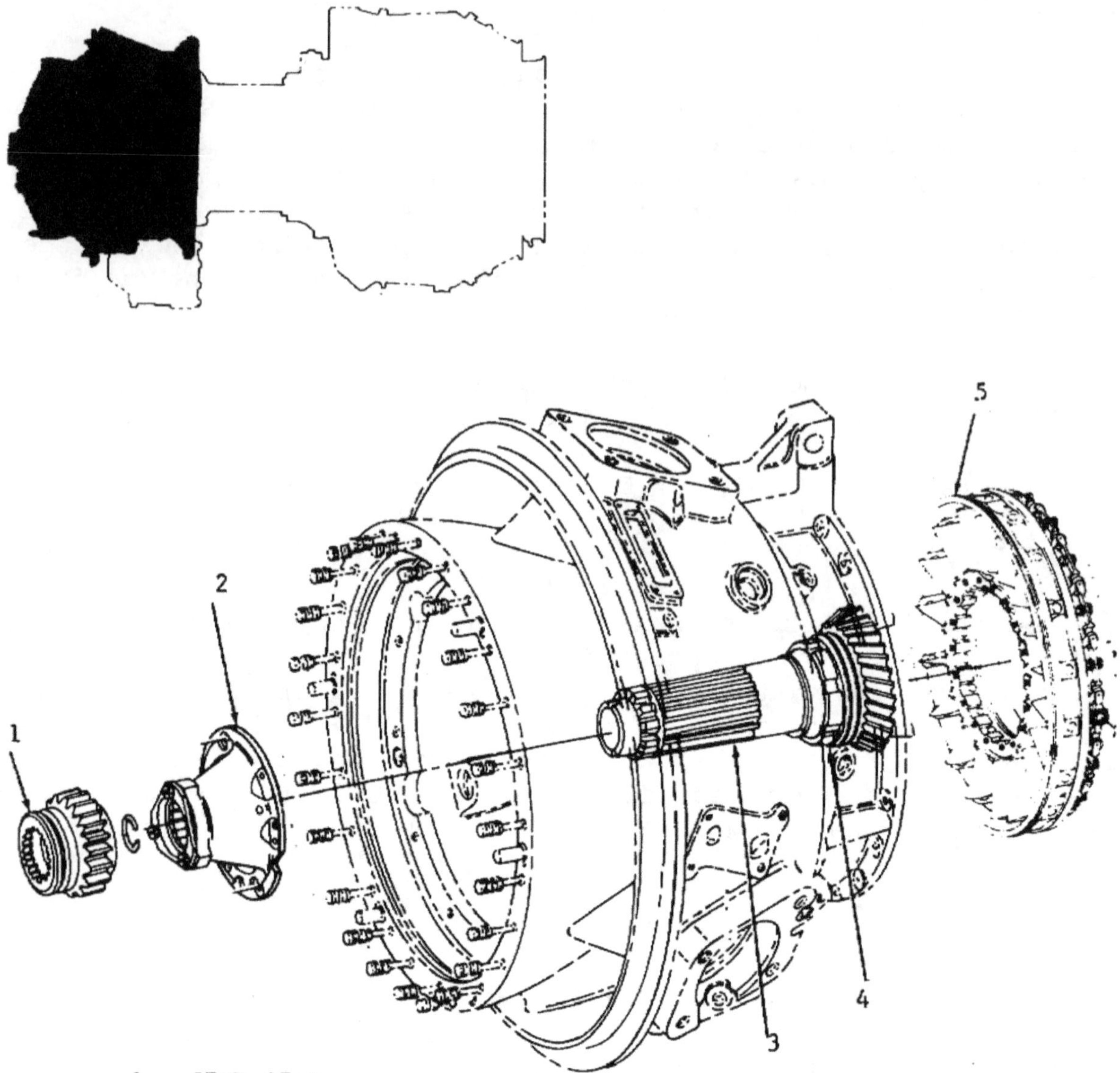

1. SPUR GEAR
2. POWER SHAFT SUPPORT
3. POWER SHAFT
4. INTERSHAFT SEAL
5. INLET GUIDE VANES

Figure 4.5. Inlet Housing.

Externally the inlet housing provides mounting points for the overspeed governor and tachometer drive assembly, and the N_1 and N_2 accessory drive gearbox assemblies. The housing also has engine mounting pads, a hoisting eye, and engine and airframe accessory mounting pads. The entire one-piece, cast-magnesium housing is coated with a heat-applied epoxy paint (HAE) to prevent erosion and corrosion.

The gas producer compressor assembly is supported at its forward end by the No. 1 main bearing, figure 4.6, which is a ball bearing to absorb thrust and radial loads, mounted within a leaf spring retainer that dampens minor torsional vibrations. The aft side of the bearing is sealed by a positive contact carbon seal aided by springs and pressurized air. A radial labyrinth seal is located forward of the carbon seal, operating on pressurized air through bleed holes; it assists in positive sealing of the bearing area. The forward end of the power shaft is supported by an unnumbered roller bearing within the inlet housing. All main bearings may be seen by referring back to figure 4.4.

The following subparagraphs give some information about the T53-L-13 output reduction carrier and gear assembly and the T53-L-701 propeller reduction carrier and gear assembly.

a. Output reduction carrier and gear assembly on the turboshaft engine in the T53-L-13 is located in the inner inlet housing as shown in figure 4.7. It consists of the support housing (1), carrier assembly (2), three planetary gear assemblies (3), oil transfer tubes (4), an output gear shaft (5), and a torquemeter assembly (not shown). The sun gearshaft is splined and bolted to the forward end of the power shaft and drives the three planetary gears, which in turn drive the output gearshaft. Reduction ratio of the turboshaft engine is 3.2 to 1.

b. Propeller reduction carrier and gear assembly used in the T53-L-701 engine utilizes a revolutionized reduction gear system called split power gearing. This type of power gear system has the ability to absorb greater torque loads, which permits the delivery of increased horsepower. Figure 4.8 is a cross sectional view of the split power reduction gear system. The power turbine speed reduction is accomplished within the split power gearing by primary and secondary drive systems, with power being transmitted to the propeller shaft through each of these systems.

4.11. ACCESSORY DRIVE ASSEMBLY

This assembly provides drive for both the N_1-driven accessory gearbox and the N_2-driven overspeed governor and tachometer. The numbers in parentheses in the following paragraphs are used in figure 4.9 to show the accessory gear drives.

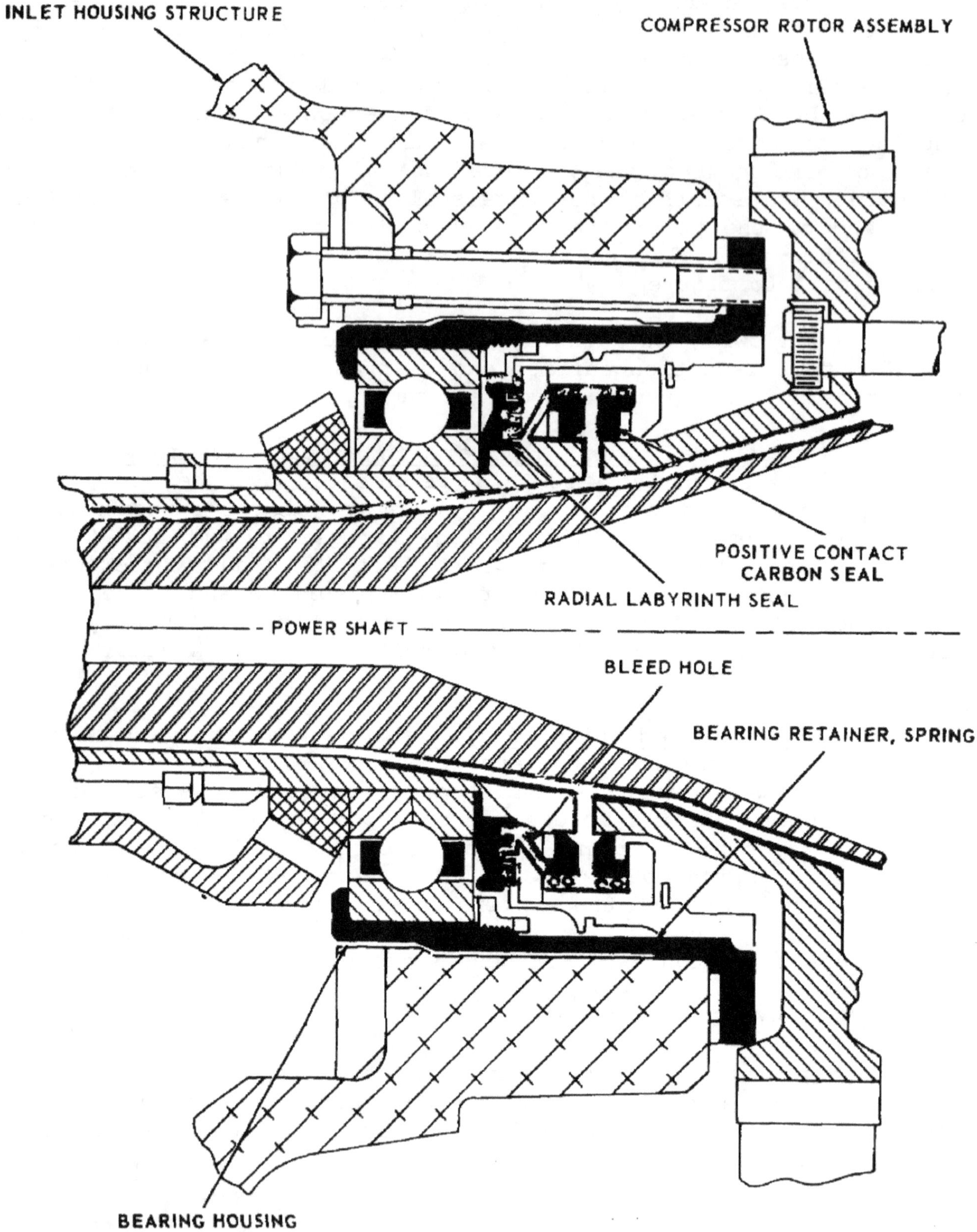

Figure 4.6. No. 1 Bearing and Seal Area.

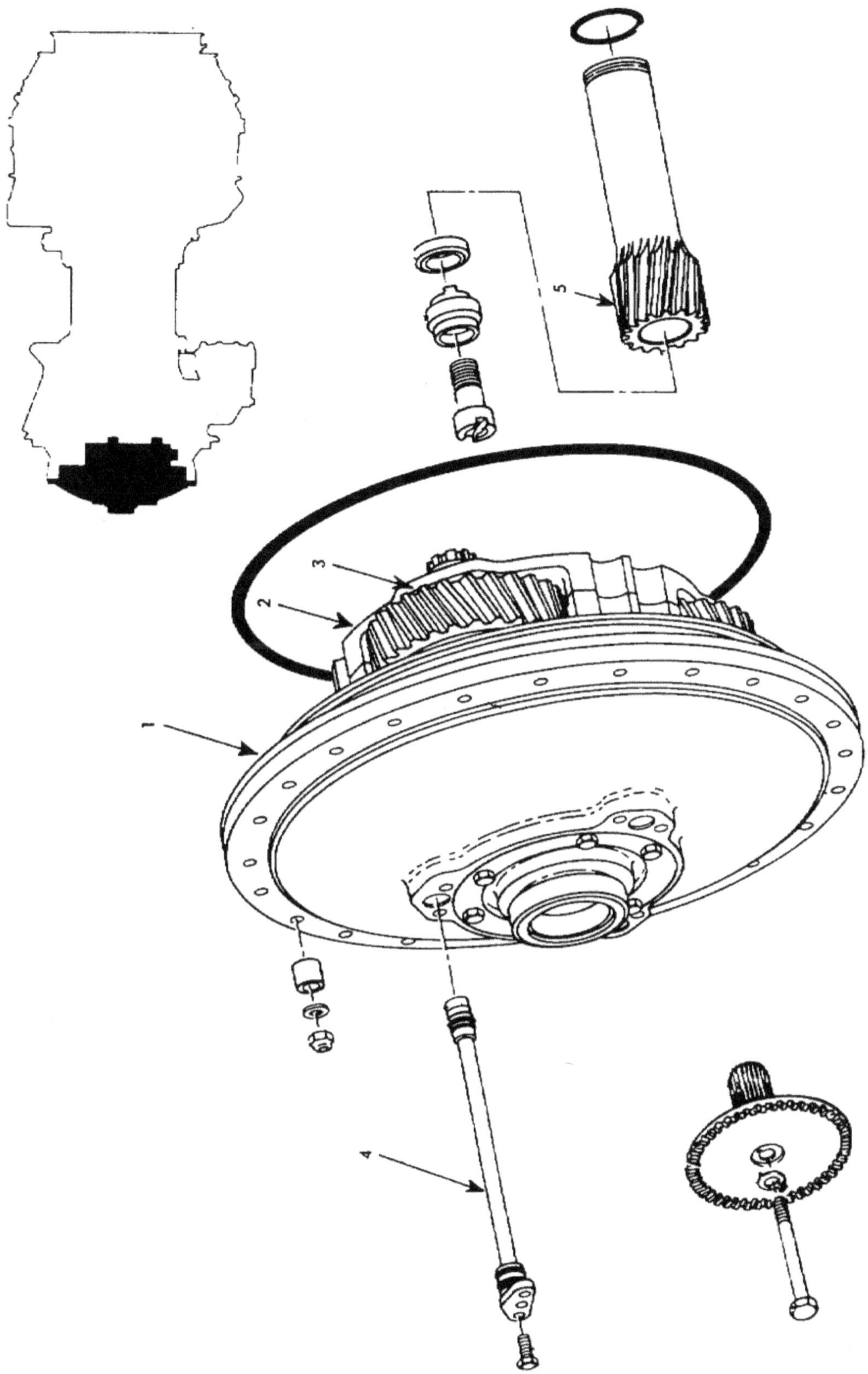

Figure 4.7. Output Reduction Carrier and Gear Assembly.

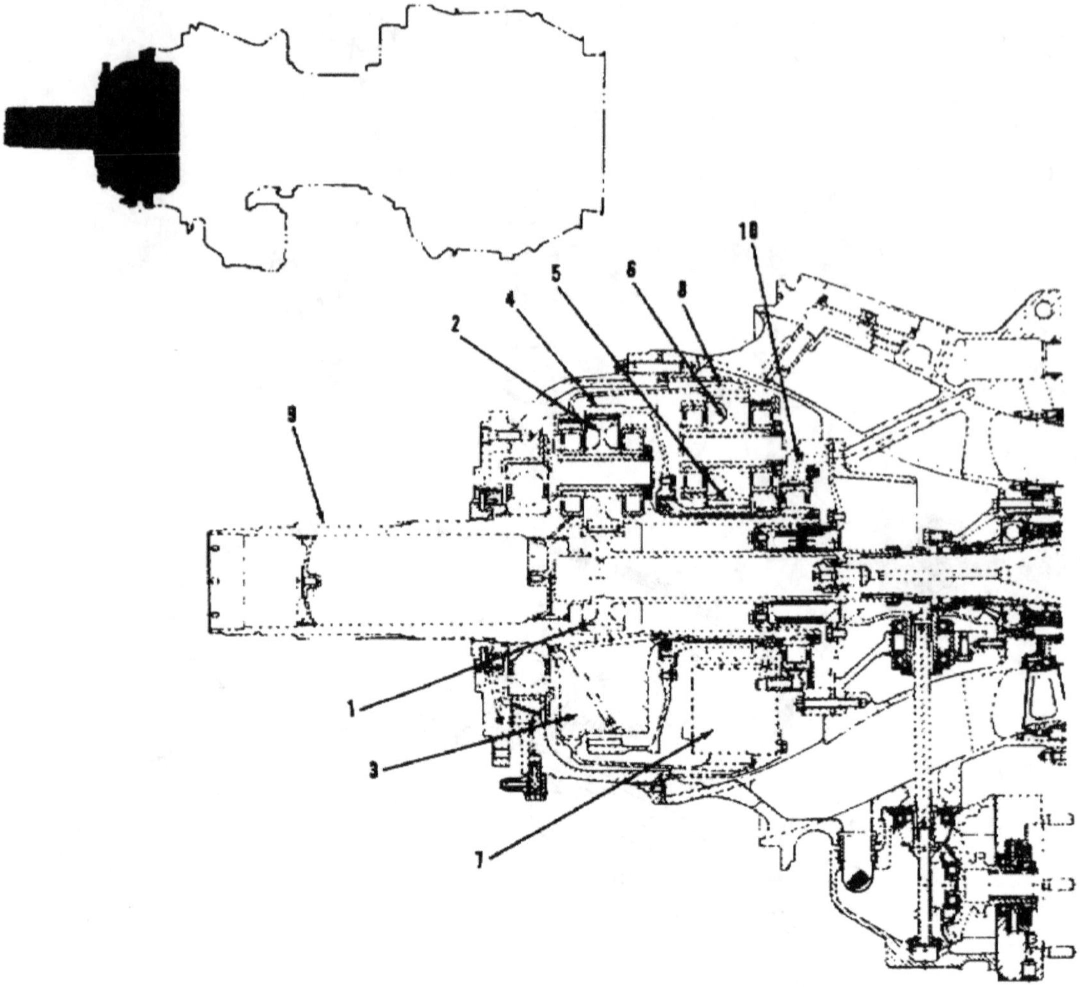

1. Primary sun gear	6. Secondary planet gear
2. Primary planet gear	7. Secondary gear carrier
3. Primary gear carrier	8. Secondary ring gear
4. Primary ring gear	9. Prop shaft
5. Secondary sun gear	10. Oil transfer assembly

Figure 4.8. Split Power Reduction Gearing (T53-L-701).

N_1 drive is provided from a pinion gear (9) mounted on the forward end of the compressor rotor shaft, driving two bevel gears (10 and 19) located within the accessory gear carrier. The bevel gear located at the six o'clock position within the carrier, being the accessory gearbox drive gear (10), is splined internally to accept the accessory gearbox shaft (18). This drive shaft connects the gear carrier to the accessory gearbox through the 90° pinion gear (16) which in turn is splined directly to the starter-generator drive gear (15). The starter-generator drive gear provides drive to all subordinate gears located within the accessory gearbox housing.

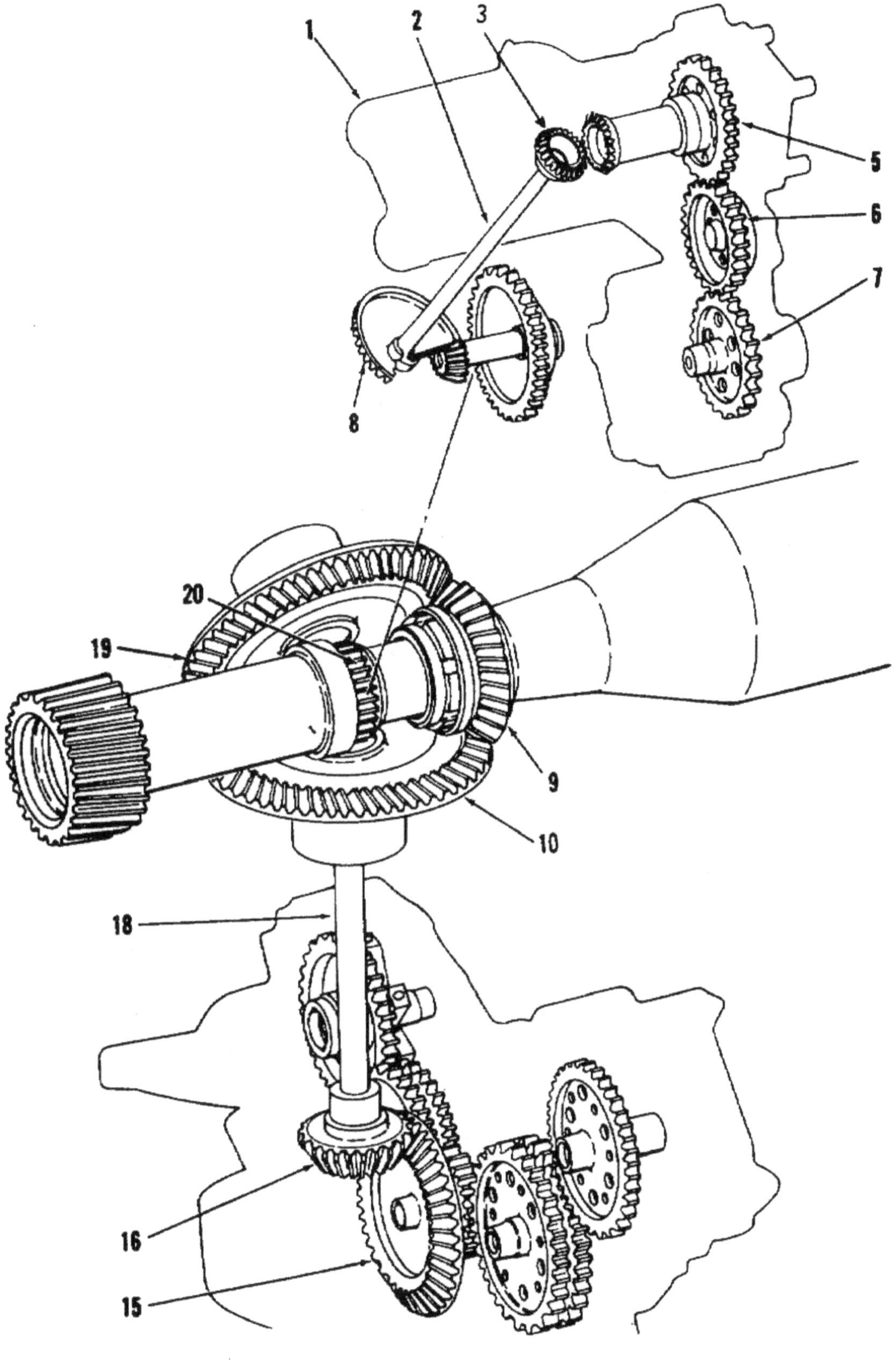

Figure 4.9. Accessory Drives.

The power takeoff drive is provided through the second bevel gear (19) located within the accessory gear carrier, and it is used to drive airframe accessories.

The N_2-driven overspeed governor and tachometer drive gearbox (1) receives its drive from a spur gear (20) pressed to the power shaft aft of the sun gear. This gear engages the N_2 drive and driven gear package (8) located within the accessory gear carrier. This package, a series of three gears, provides an internally splined drive for the drive shaft (2) which passes up through the ten o'clock inlet housing strut and into the gearbox (1).

The drive shaft then engages the internal splines of the upper drive gear (3) which provides drive to the tachometer gear (5). This gear meshes directly with an idler gear (6) which in turn transmits the drive to the combination torquemeter boost pump and overspeed governor drive gear (7).

a. N_1 accessory drive gearbox assembly, shown in figure 4.10, is mounted on the underside of the engine inlet housing and is driven through bevel gears from the front end of the compressor rotor. Drive pads are provided on rear of the gearbox for the fuel control, the starter-generator, and the gas producer (N_1) tachometer generator. The gearbox front side has mounting for the rotary oil pump, and also has an unused drive pad with connection for the torquemeter pressure transmitter vent line. Oil scavenge lines are connected at right rear on the gearbox which is an oil collector sump, kept practically empty by the pump. A chip detector plug is located in the lower right side, and the oil filter is on the left side.

b. N_2 overspeed governor and tachometer drive assembly, shown in figure 4.11, is a gearbox mounted on the engine inlet housing at the upper left side and is driven from the power shaft. The drive assembly provides mounting and drive pads for the power turbine (N_2) tachometer generator and the torquemeter boost pump, (except on the T53-L-701) and also drives the fuel control overspeed governor. A relief valve, on the drive housing, allows adjustment of torquemeter oil pressure. An internal filter and metering cartridge lubricates the gear train.

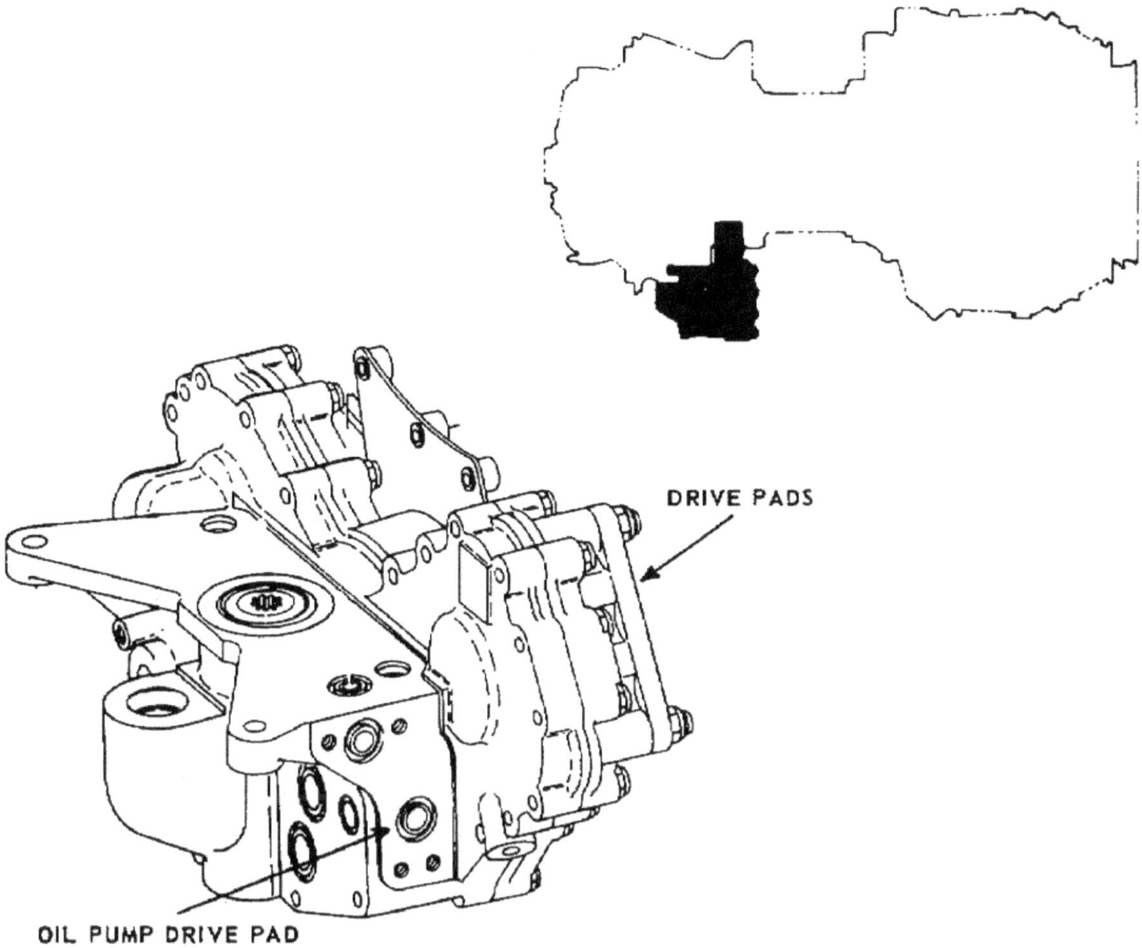

Figure 4.10. N_1 Accessory Drive Gearbox.

4.12. COMPRESSOR ASSEMBLY

The compressor and impeller housings, figure 4.12, consist of two matched halves constructed of cast magnesium and coated with HAE like the inlet housing assembly. The compressor housings provide alignment and support between the inlet housing forward and the diffuser housing aft. The housings enclose the five-stage axial compressor and the single-stage centrifugal compressor impeller. The stator vanes are located in lands (areas) between the compressor rotor disks when the housings are installed. The stators convert air velocity, from the rotating compressor, into pressure. The stators also direct airflow, at the proper angle, on to the following set of rotating compressor blades. The fifth stator vane assembly includes a row of exit guide vanes which direct airflow on to the centrifugal compressor impeller.

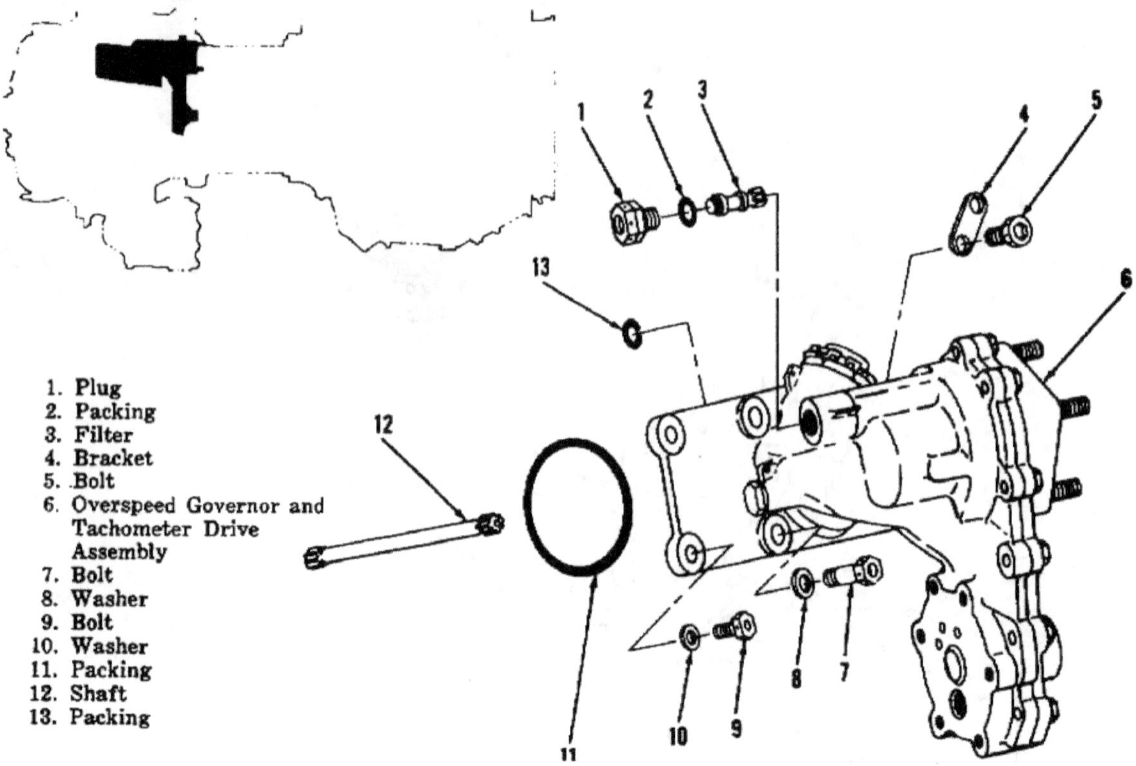

Figure 4.11. N_2 Overspeed Governor and Tachometer Drive Assembly.

Stainless steel inserts are mounted between the stator vane rows of stages two through five to reduce compressor housing erosion from sand and foreign objects. At the rear of the axial compressor housings, a series of machined passages are provided to allow bleeding of compressor air. This bleeding of air is controlled by an interstage airbleed system. The centrifugal impeller housings have a hollow core which allows compressor bleed air to flow through them to the customer bleed air and anti-icing systems.

Externally the housing provides mounting points for limited engine and airframe accessories. Because of the structural support provided by the compressor housing, only one half may be removed at a time.

The dual compressor rotor assembly, figure 4.13, consists of five axial compressor rotor disks and one centrifugal impeller. The axial compressor blades are mounted in dovetail slots, machined into the rotor disks. Roll pins and lock plates, which act as shims, secure the blades to the disk. The centrifugal impeller is constructed of titanium for a high strength-to-weight ratio. The compressor rotor assembly is attached to the rear shaft and connects to the gas

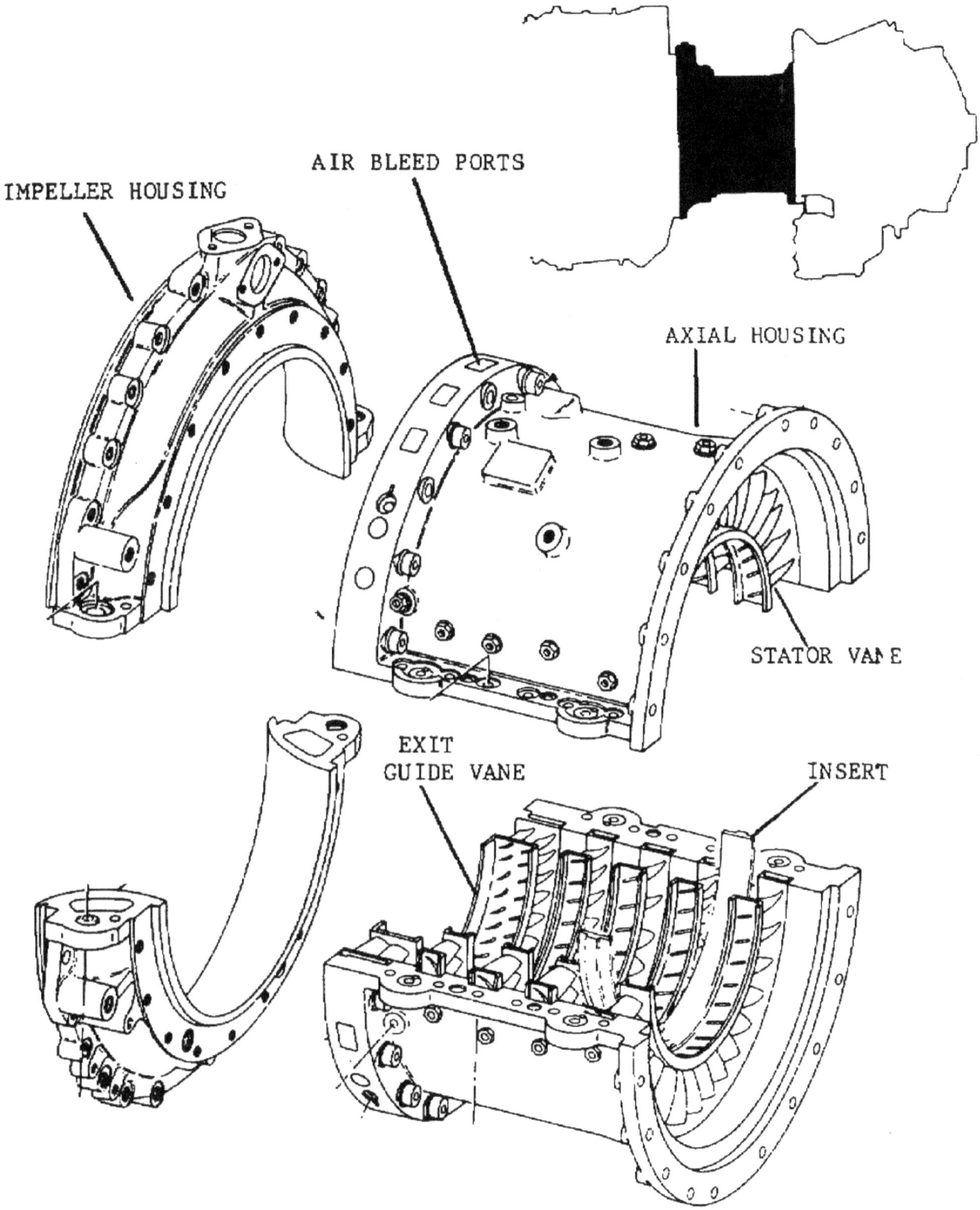

Figure 4.12. Compressor and Impeller Housing.

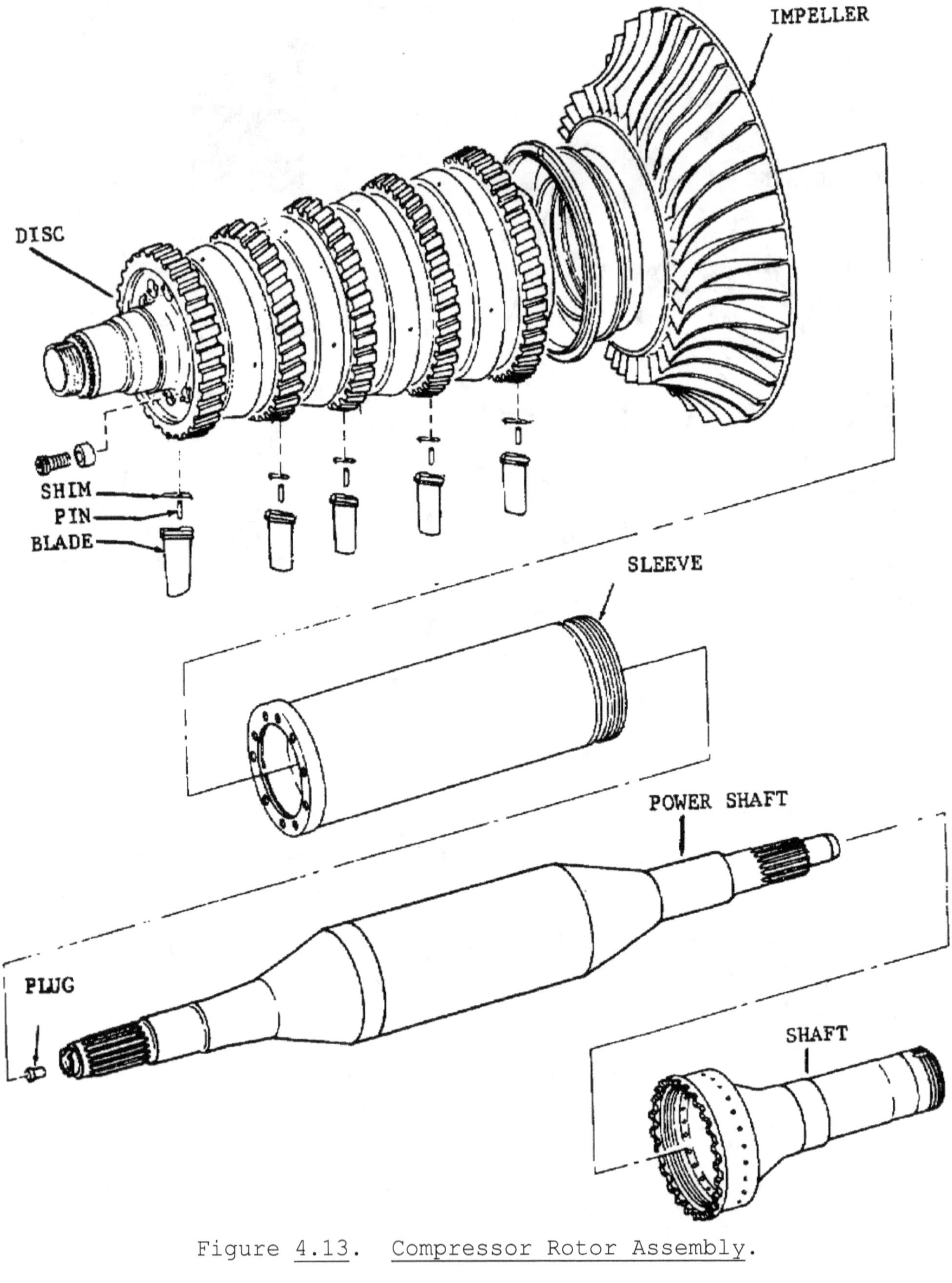

Figure 4.13. Compressor Rotor Assembly.

producer turbine. The hollow steel powershaft passes through and rotates independently of the compressor rotor. Splines on the forward end of the powershaft mate with the sun gearshaft that drives the output reduction gears. Splines at the aft end of the powershaft mate with the power turbine. The powershaft is supported at the forward end, within the inlet housing, by a roller type bearing. The aft end of the shaft is supported along with the power turbines (N_2) by the No. 3 and 4 main bearings.

4.13. DIFFUSER HOUSING ASSEMBLY

The diffuser assembly housing shown in figure 4.14 is made of steel and is located aft of the compressor section.

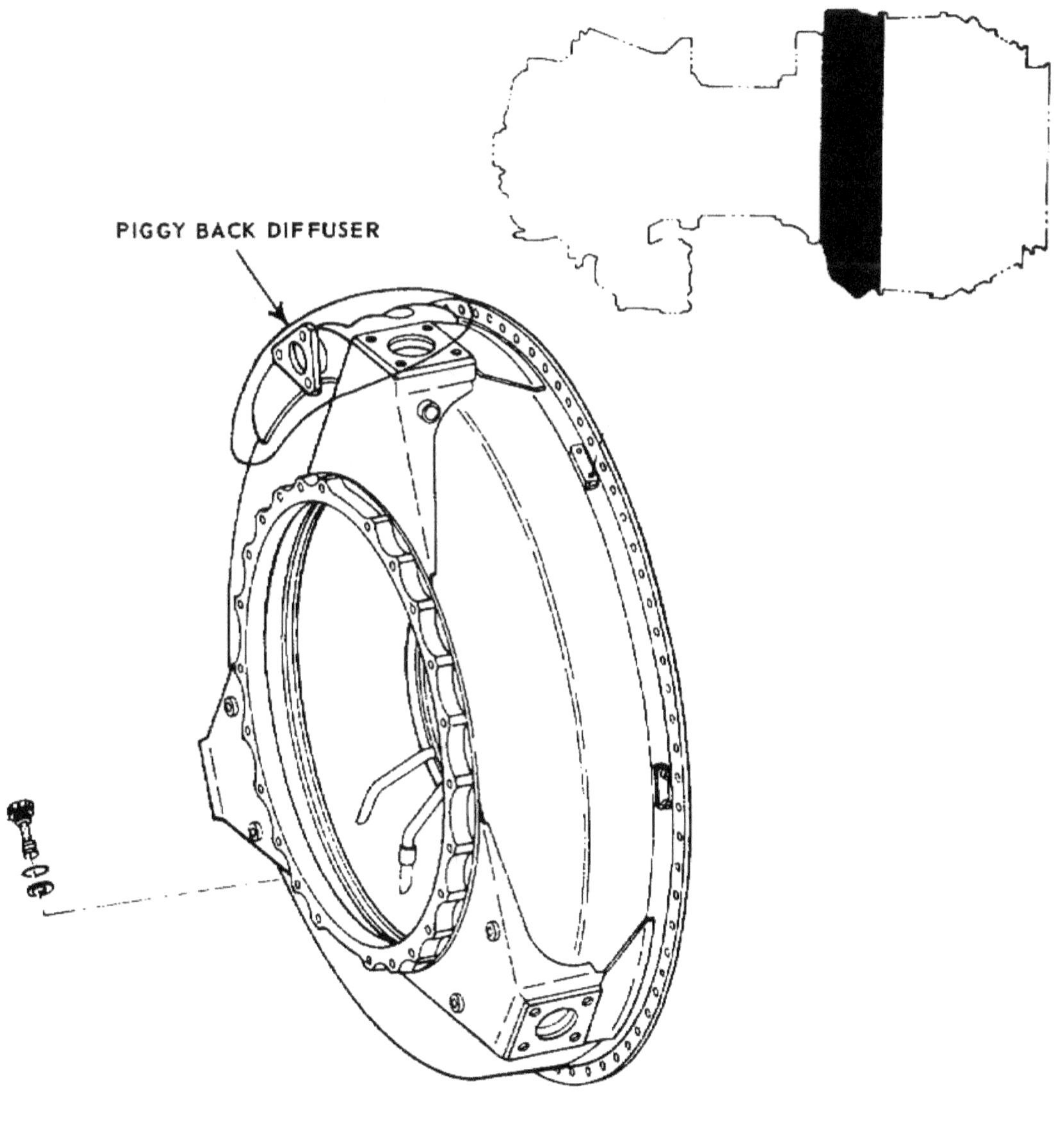

Figure 4.14. Diffuser Housing.

The diffuser receives high velocity air from the tip of the centrifugal impeller. The function of the diffuser is the decrease velocity and increase air pressure in this area. Air pressure at the diffuser discharge is at its highest value, with air temperature in the vicinity of 500° F. The diffuser also provides a means for bleeding a portion of the high temperature air for required engine and airframe use, such as engine anti-icing and cockpit heat. Later versions have an external air manifold, known as a piggyback diffuser, where air is extracted for use as required.

Externally, the diffuser provides engine mounting points at the four, eight, and twelve o'clock positions. Hoisting provisions are incorporated in the top or twelve o'clock mount. A port for extracting pressurized air for use as the pneumatic force for operation of an interstage bleed system actuator is provided at the three o'clock position. Mounting points for required engine accessories are provided on the external housing.

4.14. COMBUSTOR TURBINE ASSEMBLY

Located aft of the diffuser housing, the combustor turbine assembly consists of the combustion chamber housing and liner, gas producer turbines, power turbines, and exhaust diffuser. The combustor assembly is an external annular reverse-flow type. Although this design increases the diameter of the engine to a degree, it significantly reduces its overall length. It is classified an external annular reverse-flow type, in that the circular combustion chamber is located outside of and encloses the turbine area.

As shown back in figure 4.2, compressed air flowing aft from the diffuser enters the combustor (25 percent primary air) and mixes with fuel and supports combustion, within the combustion liner. The hot expanding gas flows forward within the liner; it is diluted and cooled by the remaining compressed air (75 percent secondary air).

Flow direction is changed again to the rear by the stationary deflector mounted within the diffuser inner housing. The gases then flow through the gas producer nozzles which greatly accelerate the gas stream and direct it onto the gas producer turbine (N_1). The N_1 turbine extracts approximately 60 percent of the energy to rotate the compressor assembly. The gases still possessing energy are again accelerated as they pass through the power turbine N_2 nozzles. The gas stream is then directed onto the power turbines where most of the remaining gas energy is extracted to rotate the N_2 power shaft.

The gases are then directed into the exhaust diffuser, and an average temperature of the gas stream is measured in this area (station 9, in figure 4.4). Although this temperature is much lower than that existing in the turbine inlet area (station 5, in figure 4.4), it is relative and indicative of the temperature at station 5. The fuel control automatically programs fuel flow, so the maximum turbine inlet temperature is not exceeded during normal operation.

a. Combustion chamber housing. The T53 has an annular combustion housing Which is constructed of steel. A flange at the forward end mates with the aft flange of the compressor diffuser housing. It is at this point that the engine is split to perform a hot-end inspection. A combustion chamber drain valve is located at the 6 o'clock position. This valve is spring loaded in the open position to drain any unburned fuel from the combustor during engine shutdown after a false or aborted start. During engine operation, compressed air flowing through the combustion chamber automatically closes the valve when chamber pressure exceeds outside pressure by approximately 2 psi. If the drain valve fails to close during engine operation a power reduction from air loss will occur. The stainless-steel, annular, combustion chamber liner is shown in figure 4.15. The liner contains a series of holes and louvers which vary in size, regulate the flow of pressurized air into the inner area to support combustion, and form a cooling air blanket on the liner surface. The T53-L-13 and later version have twenty-two swirl cups on the aft end that provide fuel nozzle access into the burner zone. Slots in the swirl cups direct airflow in a pattern to provide proper fuel atomization and flame control.

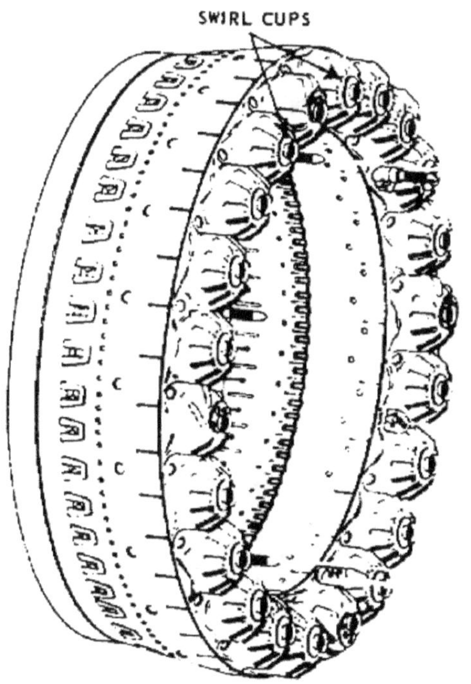

Figure 4.15. Combustion Chamber Liner.

b. Turbine assembly. As the gases flow rearward from the defector, item 1 in figure 4.16, they contact the first stage gas producer turbine nozzle. The gases are accelerated by the nozzles to impinge upon the open tip blades of the N_1 turbine, causing them to rotate at high speed in a counterclockwise direction. As the gases pass from the trailing end of the blades, an additional force is imparted to the turbine by the reaction to this flow.

1. COMBUSTION CHAMBER DEFLECTOR
2. FIRST STAGE GAS PRODUCER NOZZLE ASSEMBLY
3. SUPPORT PLATE
4. WASHER
5. BOLT
6. BOLT
7. AFT OIL RING
8. SHIM
9. FORWARD CONE
10. SEALING RING
11. FIRST STAGE GAS PRODUCER TURBINE ROTOR ASSEMBLY
12. SPACER
13. BOLT
14. SEAL RING
15. EXPANDERS
16. BOLT
17. NUT
18. TAB WASHER
19. SECOND STAGE GAS PRODUCER ROTOR
20. SEAL RING
21. EXPANDER
22. SECOND STAGE GAS PRODUCER NOZZLE
23. RING
24. SPACER
25. RETAINING PLATE
26. PIN
27. SECOND STAGE GAS PRODUCER CYLINDER
28. NUT
29. LOCKRING
30. REAR CONE

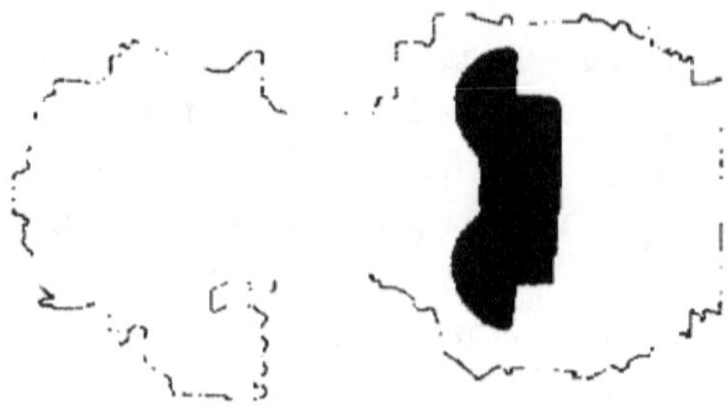

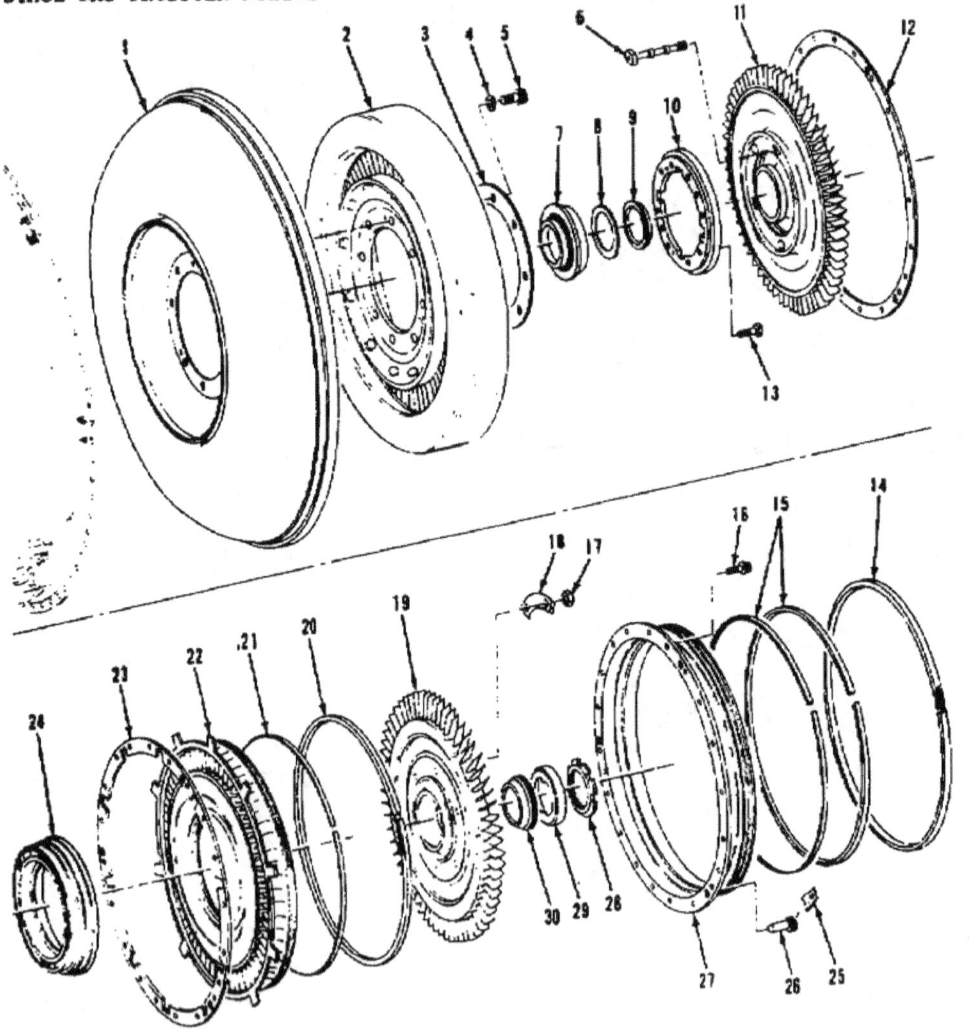

Figure 4.16. Gas Producer Assembly, N_1.

The first stage gas producer turbine (GP1) and the second stage gas producer turbine (GP2) are mechanically joined together and rotate as one assembly (N_1). The dual turbine design in later versions of the T53 permits more lightly loaded turbine blades than previous single turbine models and has a power output increase of 20%.

As the gases flow from the GP2 they pass on to the first stage power turbine nozzle (PT1). The gases are again accelerated and flow across the two N_2 power turbines. Both turbine rotors have tip shrouded blades to prevent air losses and excessive vibrations. The power turbines are supported at the aft end by No. 3 (roller) and No. 4 (ball) main bearings. An N_2 power turbine assembly is shown in figure 4.17.

c. Exhaust diffuser. The welded steel diffuser, shown in figure 4.18, forms a divergent flow path for the exhaust gases. The diffuser consists of an inner and outer housing separated by four hollow struts. It is mounted to the aft inner flange of the combustor housing. Support for the aft section of the diffuser is provided by a support cone, 20 back in figure 4.1, that is secured by a "V" band coupling to the aft outer flange of the combustor housing. Located between the exhaust diffuser and support cone is a stainless steel fire shield, 19 in figure 4.1. During operation ambient air flows between the outer and mid cones of the diffuser. This air passes through a series of holes on the forward area of the outer cone and into the chamber formed by the diffuser and the fire shield. Ambient air then lows through the hollow struts to cool the bearing housing mounted within the diffuser and aft face of the power turbine (PT2). Mounts for an exhaust gas temperature harness are located on the diffuser midcone. The aft flange on the diffuser midcone is the mounting point for an airframe furnished tailpipe. The tailpipe routes the exhaust gas stream to the atmosphere.

4.15. DESCRIPTION OF FUEL SYSTEM

The T53 series engines are designed to operate primarily on MIL-J-5624 grade JP-4 fuel. The fuel system consists of the components shown in figure 4.19.

Fuel flow is maintained between components by flexible or rigid lines. An airframe-mounted boost pump supplies fuel to the fuel control inlet port. During the starting sequence fuel flows through an external line from the hydromechanical fuel control to the starting fuel solenoid inlet, shown in figure 4.20. The cockpit-controlled

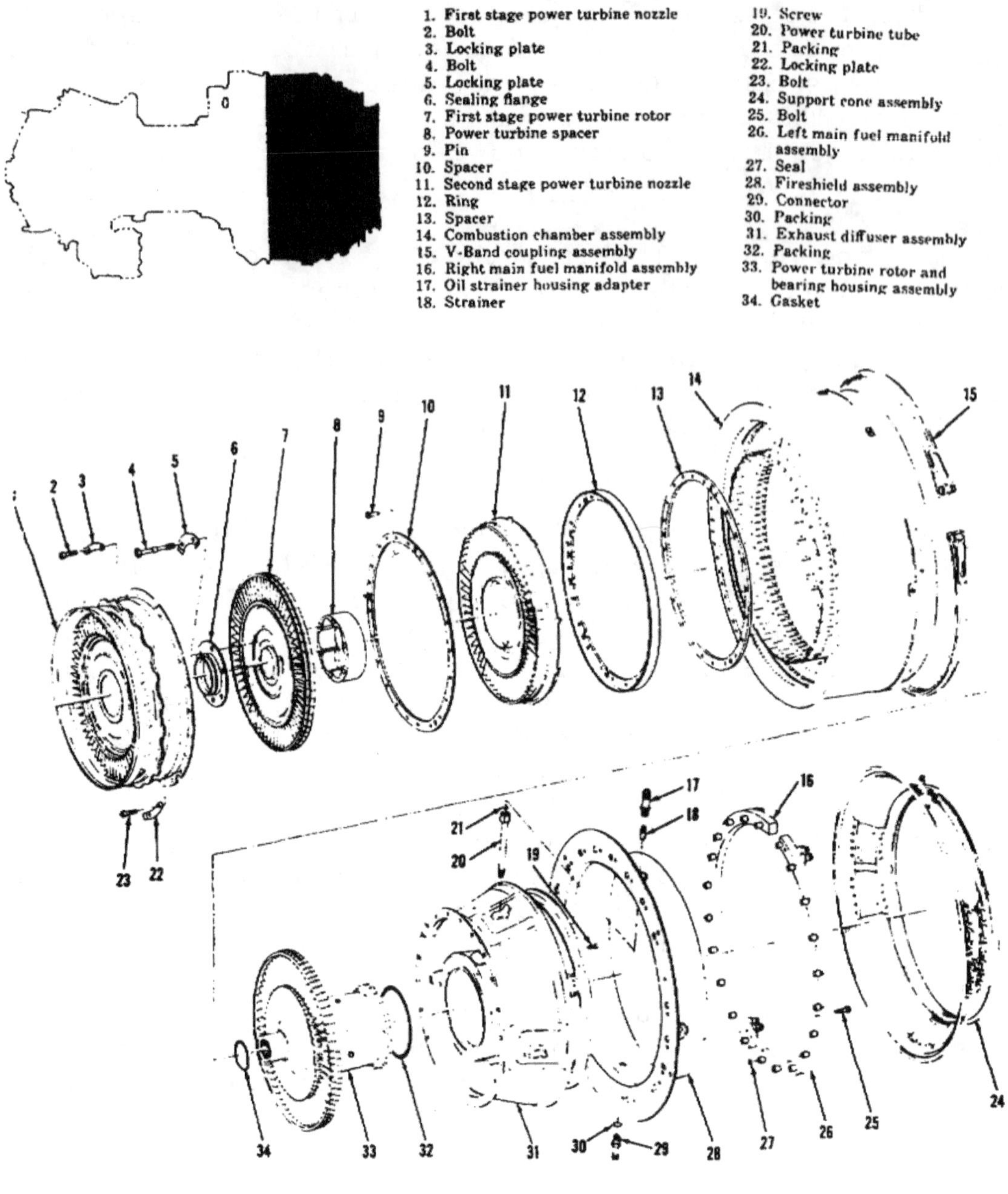

Figure 4.17. N_2 Power Turbine.

solenoid is a two-position valve electrically opened and spring loaded closed. With the valve energized to the open position, fuel flows through an external line to the starting fuel manifold. The start fuel manifold is a two-piece assembly with four rigidly-mounted starting fuel nozzle attaching points. Fuel flows into the starting nozzles located at the 2, 4, 8, and 10 o'clock positions at the rear of the combustor housing. The nozzles inject atomized fuel into the combustion chamber during the starting sequence. The following subparagraphs discuss the two fuel flow systems.

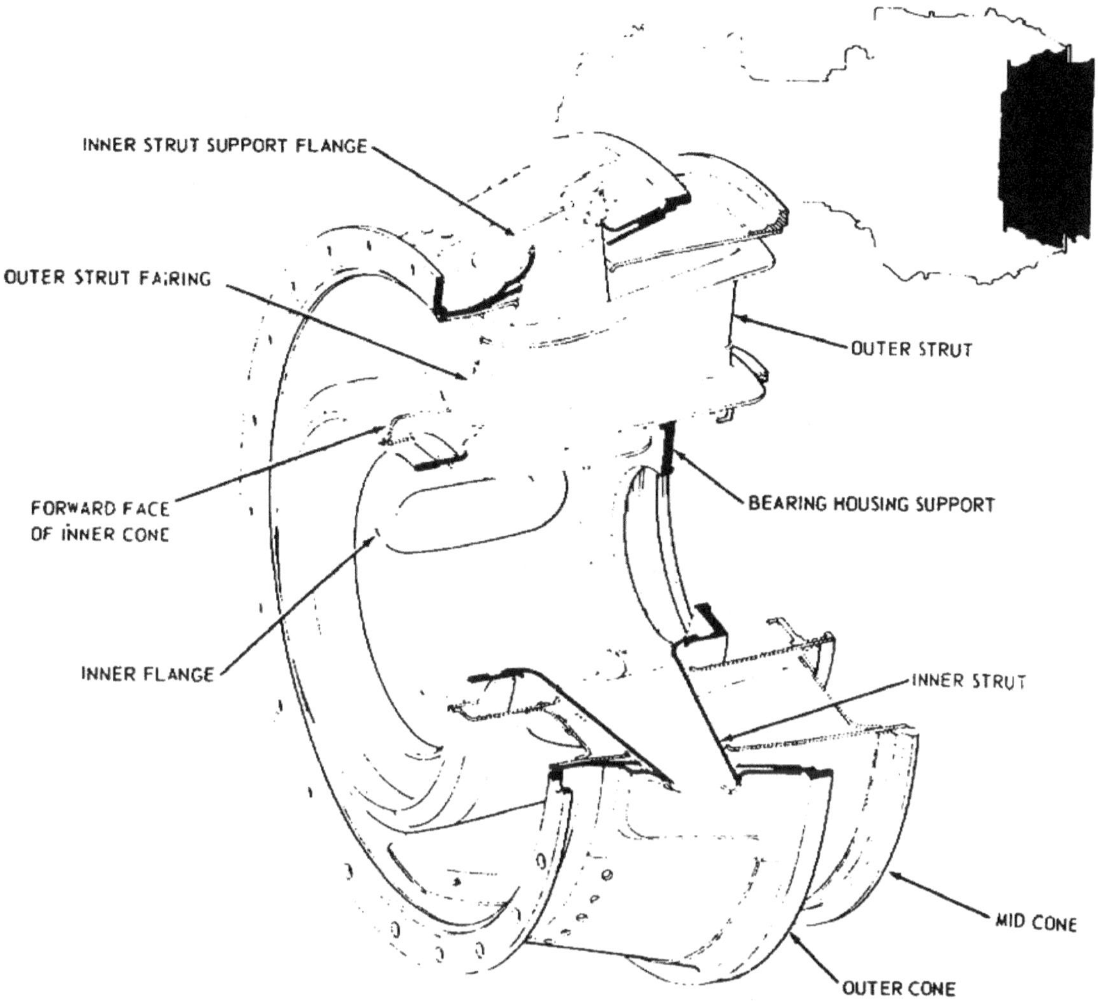

Figure 4.18. Exhaust Diffuser.

a. Normal fuel flow. The fuel flow continues in the fuel control from the main metering valve to the main system outlet port.

An external line carries the fuel from the port to the flow divider and dump-valve assembly. With the introduction of the atomizing combustor configuration with dual orifice nozzles, a flow divider dump-valve assembly was installed to achieve correct fuel atomization throughout the engine operating range.

The flow divider meters fuel to the engine nozzles according to a predetermined schedule of secondary flow versus primary flow. A dump valve is incorporated as an integral unit to drain the fuel trapped in the manifold and fuel lines when the engine is shut down. The flow divider and dump valve assembly is designed to function with fuel pressure up to 1,200 psi with an ambient temperature of 250° F (121° C) and a fuel supply temperature from -65° F

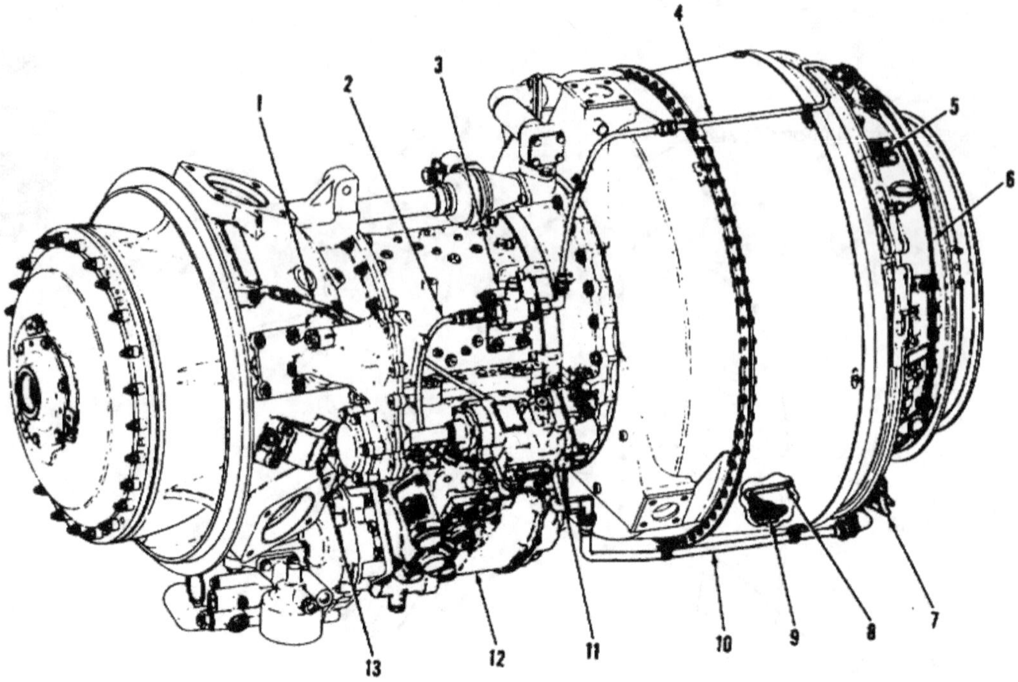

1. Air Pressure Sensing Hose
2. Starting Fuel Hose
3. Starting Fuel Solenoid Valve
4. Starting Fuel Hose
5. Main Fuel Manifold
6. Starting Fuel Manifold
7. Flow Divider and Dump Valve Assembly
8. Flow Divider Drain Hose
9. Combustion Chamber Drain Valve
10. Main Fuel Hose
11. Overspeed Governor
12. Fuel Regulator
13. Temperature Sensing Element

Figure 4.19. Fuel System Components.

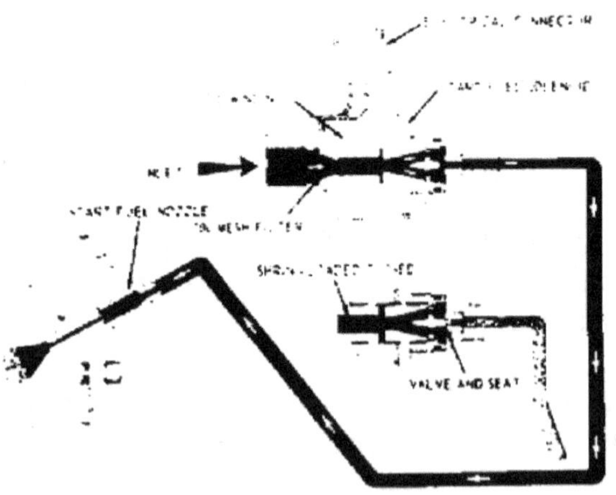

Figure 4.20. Starting Fuel Flow Sequence.

(-54° C) to 200° F (93° C). The flow divider assembly, shown in figure 4.21 has a lower housing containing the dump valve and an upper housing for the flow divider. These housings are machined from corrosion resistant steel casting, and sharp edges are maintained on all metering slots and ports.

When fuel pressure at the inlet reaches a predetermined value, the dump valve plunger moves toward

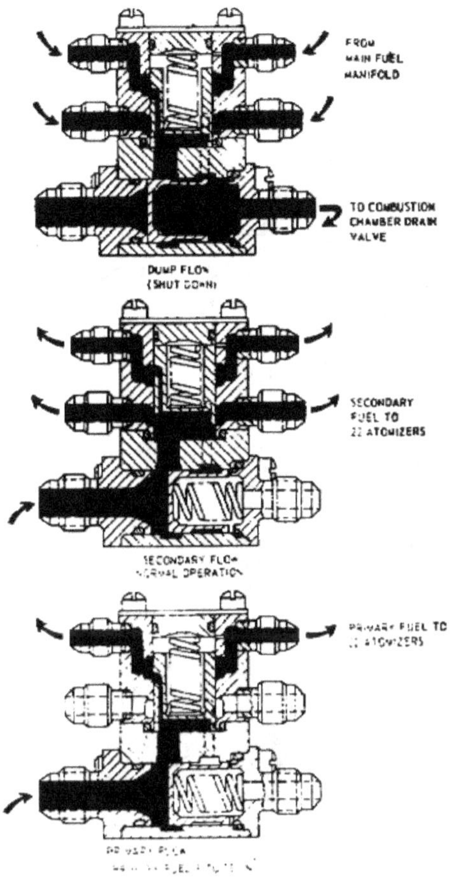

Figure 4.21. Fuel Flow Divider.

From the flow divider, fuel flows through primary and secondary lines to the main fuel manifold assembly. The manifold is a two-section, dual-channeled assembly with eleven outlets in each section. Each manifold half is interchangeable, requiring only a minor hardware adjustment to make the change. The 22 fuel atomizers are attached directly to the manifolds, which discharge the atomized fuel into the combustion chamber.

The fuel atomizer, as shown in figure 4. 23 is a dual orifice injector designed to accommodate the separate primary and secondary fuel flow functions. The separate orifices spray

the closed position, allowing fuel to enter the flow divider valve. When the inlet pressure has reached the minimum engine operating pressure, the dump valve plunger is in the fully closed position. The drain valve seal prevents fuel from the primary and secondary manifolds from draining.

Fuel passes the flow divider series orifice in the flow divider plunger en route to the primary manifold, creating a pressure drop. This pressure drop across the orifice is sufficient to move the plunger off its stop. As the plunger is displaced, the secondary flow metering ports in the plunger are progressively opened, allowing fuel to pass to the secondary manifolds. From the flow divider, fuel flows through primary and secondary lines to the main fuel manifold assembly shown in figure 4.22.

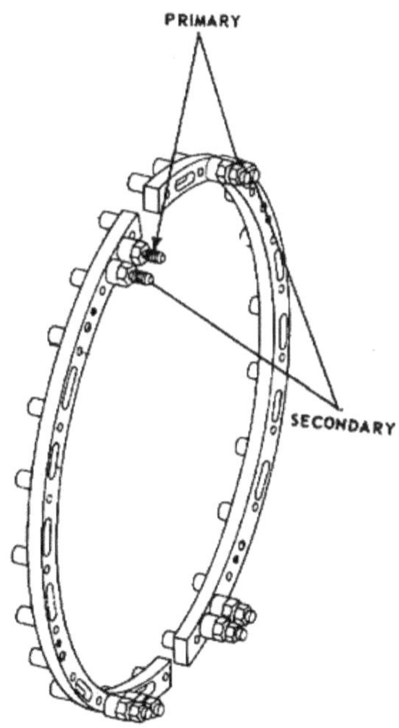

Figure 4.22. Fuel Manifold.

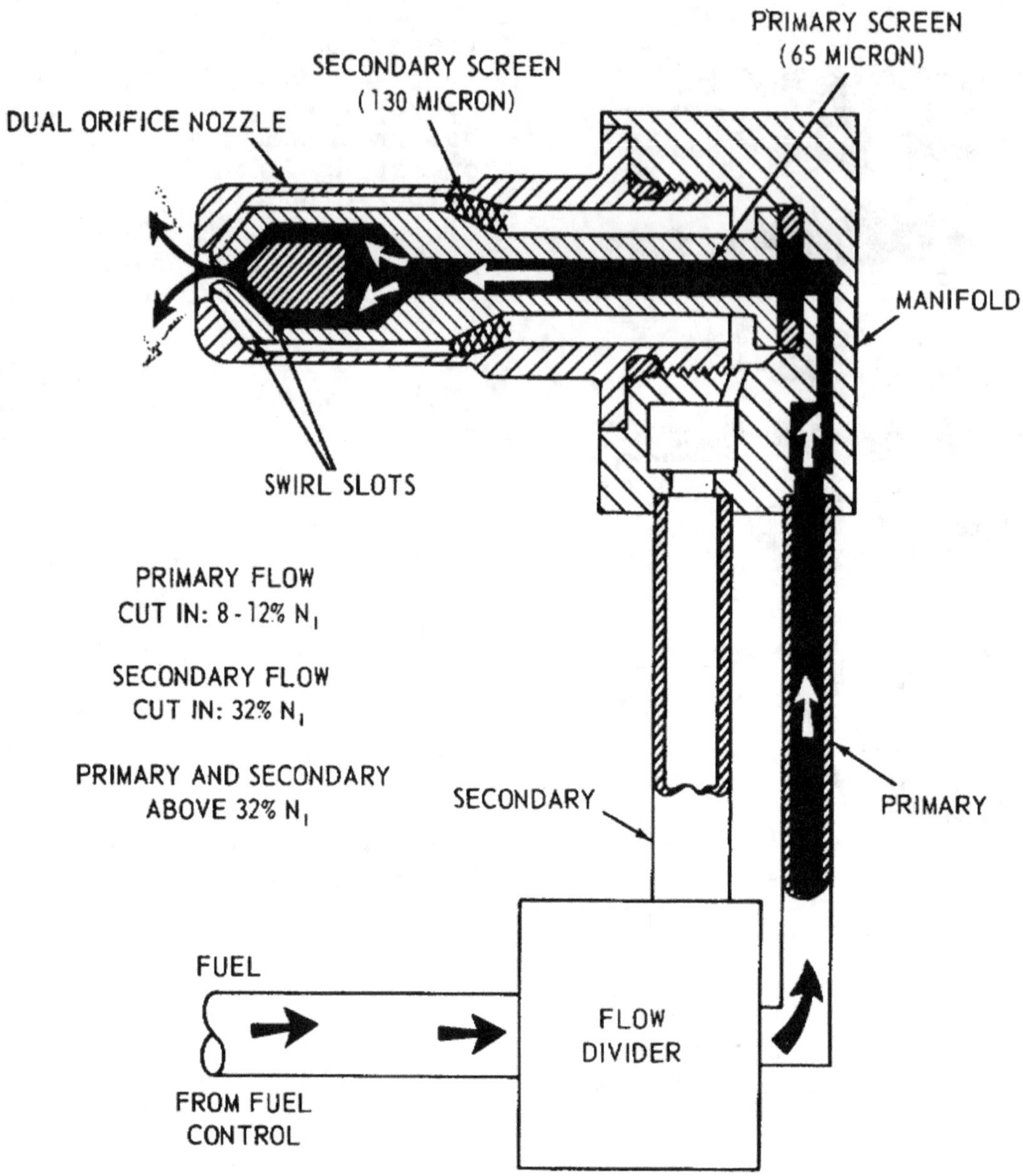

Figure 4.23. Fuel Atomizer.

fuel into the combustion chamber through the action of the flow divider. Fuel entering the primary section of the atomizer passes through the primary screen and continues to flow through the center of the nozzle to the swirl chamber located internally in the head of the nozzle. Here it passes through three swirl slots and is discharged into the combustion chamber at a 90° spray angle. The fine spray density established by the primary slots is required to start and run the engine. Higher N_1 speeds require additional fuel and a heavier density spray pattern; therefore, secondary fuel is introduced through the nozzle at speeds above 32% N_1. The secondary flow enters the outer shell of the nozzle and passes through the secondary screen and into the secondary swirl slots. The secondary swirl slots, being slightly larger in size than the primary swirl slots, allow a greater volume of fuel to be discharged into the combustor. The combination of primary and secondary flow which is delivered to the combustor at an-optimum spray angle of 90° is sufficient to operate the engine at all power settings above 32% N_1.

To further atomize the fuel entering the combustor liner, swirl cones located at the aft end of the combustor liner assembly allow combustion air to enter the liner and swirl in the opposite direction to that of the fuel being injected by the atomizer. This additional swirling air establishes a definite flame pattern at the end of each atomizer. Additionally, air is routed through the air shroud to cool the atomizers and assist in establishing this flame pattern. When fuel flow is cut off at engine shutdown, the inlet pressure falls below the dump-valve opening pressure, and the dump-valve plunger is moved by its spring to the fully open position. The fuel in the primary manifold then flows to drain through the main flow passage between the flow divider and drain valve. The secondary manifold is drained via a small drain port in the upper housing leading into the dump-valve cavity.

b. Manual fuel flow. The manual or emergency fuel flow sequence is the same as normal fuel flow sequence except for the fuel-control changeover valve. When the changeover valve is actuated to the manual position, fuel is redirected to the manual-system metering valve, which is mechanically linked to the main power control in the cockpit. This flow path bypasses the main metering valve. Other than that, the manual fuel-flow sequence is the same as the normal. The fuel system on the T53-L-13 is similar to the one on the T53-L-701. However, the T53-L-701 turboprop engine has a fuel heater to prevent fuel from icing by using engine lubricating oil to heat the fuel. The fuel heater supplies fuel in temperature ranges between 35 to 70° F when fuel inlet temperatures are in the -65 to -70° F range.

4.16. INTERNAL COOLING AND PRESSURIZATION SYSTEM

The internal cooling system provides cooling air to the internal engine components and pressurizes the number one, two, and three main bearing seals. Cooling and pressurization air is obtained from five parts of the engine. The following numbers in parentheses, such as (1), (2), and so on in the discussion, correspond to similar numbers in figure 4.24 and refer you to that particular portion of the engine. Air flows down through the fourth stage spacer (1) into the

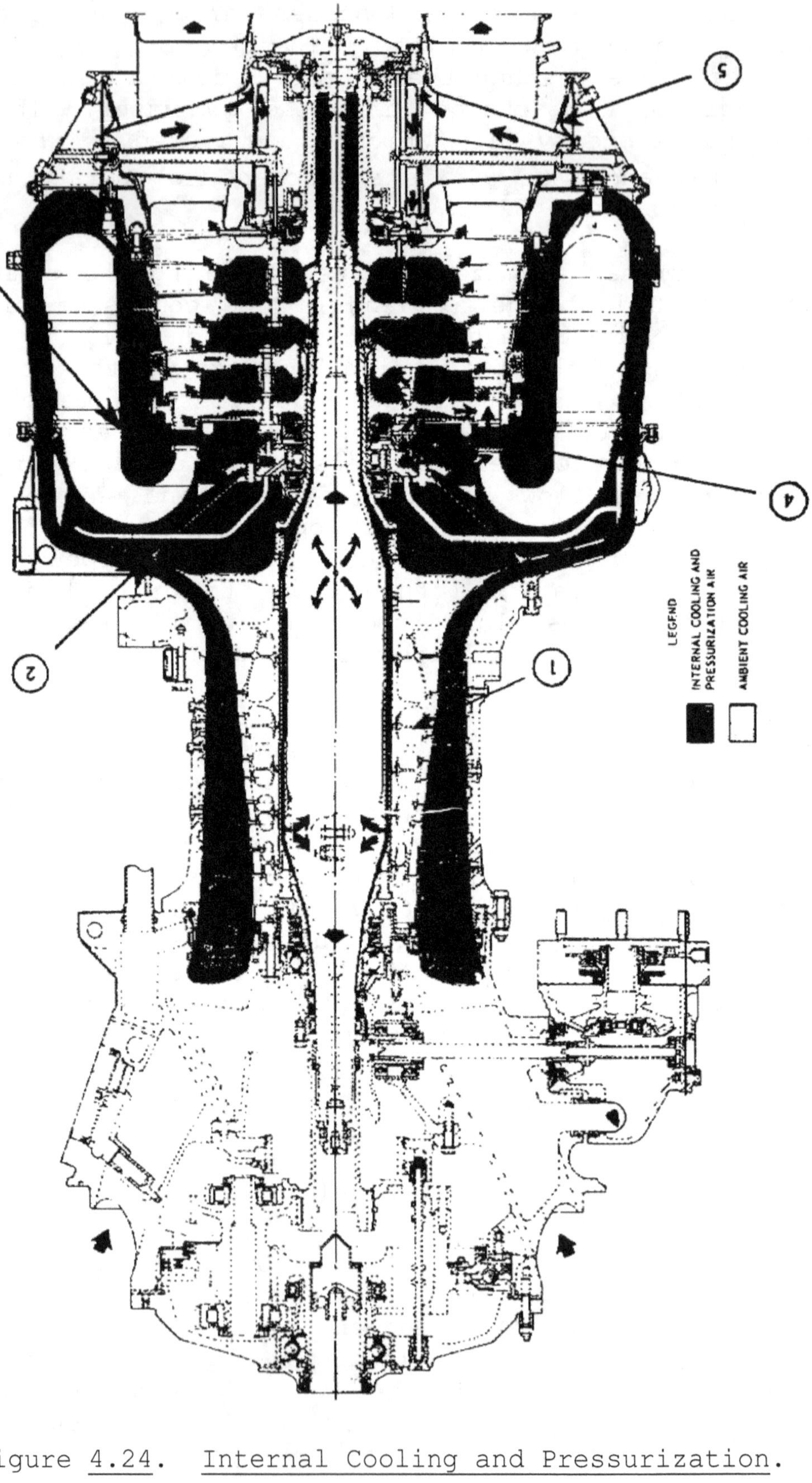

Figure 4.24. Internal Cooling and Pressurization.

area between the compressor rotor sleeve and the rotor disk inside diameters, then forward to the first stage rotor disk where it is bled back into the compressor airstream through holes in the first stage spacer. This airflow cools the three aluminum disks in the compressor rotor assembly. Compressed air bled from the tip of the centrifugal compressor impeller (2) cools the forward face of the diffuser housing and pressurizes the No. 2 bearing forward seal, and continues rearward through transfer tubes in the bearing housing to pressurize the No. 2 aft oil seal. It also passes through a series of holes in the rear compressor shaft into the space between the rotor assembly and the power shaft. At this point, it separates into three flow paths. Part of the compressed air, used for seal pressurization, flows forward and through a series of holes in the compressor front shaft. This air fills the area between the carbon elements of the No. 1 bearing seal. The intershaft seal, located forward of the No. 1 bearing, prevents flow of pressurized air into the inner inlet housing area. A portion of this compressed air flows aft over the power shaft and emerges at the aft end of the rear compressor shaft to cool the rear face of the second gas producer (GP) rotor, the forward face of the first power turbine (PT) rotor, and the first stage PT nozzle. The air then passes into the exhaust stream. The remainder of the compressed air flows through a series of holes in the power shaft. This air flows aft, inside the power shaft, through holes drilled in the hollow power-shaft through bolt, and into the interior of the second PT rotor assembly. Air then passes through a series of holes in the turbine hub and the turbine spaces to cool the rear surface of the first PT rotor assembly, the forward surface of the second PT rotor assembly, and both faces of the second PT nozzle.

Compressed air, bled through slots in the mating surfaces of the combustion chamber deflector and the air diffuser (3) cools the forward face of the deflector and the No. 2 bearing housing. Then the air is split into paths to cool the blade roots of the first stage GP rotor assembly. The air then flows through holes in the inside diameter of the GP turbine spacer to cool the hub area of the rear face of the first GP turbine and the forward face of the second GP turbine. The air then flows rearward where it joins the cooling air being discharged from the aft end of the compressor shaft where it is expelled into the exhaust stream.

Compressed air is then directed through the first stage GP nozzle and cylinder assembly (4) to cool the rear face of the first stage GP rotor and then into the exhaust stream.

Ambient air is used to cool the No. 3 and No. 4 bearing housing. The air enters the exhaust diffuser struts (5) and moves forward between bearing housing walls to cool the rear face of the second stage PT rotor assembly. As the ambient air passes the forward face of the No. 3 bearing seal, it helps pressurize the seal.

4.17. VARIABLE INLET GUIDE VANE SYSTEM

To ensure a compressor surge margin, the angle of incidence of the inlet air to the first compressor rotor must be within the stall-free operating range of the compressor blades, and, because this stall-free operating range varies with compressor speed (N_1), it is necessary to vary the angle of attack with changes in N_1 speed. This is done by varying the angle of the inlet guide vanes. The variable inlet guide vanes (VIGV) are located in front of the first compressor rotor as shown in figure 4.5.

At low N_1 speeds, a high angle of attack is required, while at higher N_1 speeds, the angle of attack decreases. Refer to the blocks in figure 4.25 for the angle of attack at high and low N_1 speeds.

The VIGV's are positioned by the inlet guide vane actuator pilot valve, located in the fuel control, which monitors N_1 speed and compressor inlet temperature (T1). While setting the desired position of the VIGV's, the actuator relays their position back to the fuel control through an external feedback control rod to nullify the fuel pressure signal so that at any steady-state N_1 speed between 80 and 95 percent, the inlet guide vanes will assume a constant position. The VIGV actuator is mounted on the right side of the compressor housing assembly, shown in figure 4.26. The actuator is controlled by main fuel pressure from the fuel control. Two fuel lines carry the fuel from the fuel control to the VIGV actuator. This fuel pressure acts upon the piston inside the actuator to move the VIGV's. The VIGV's are positioned by the inlet guide vane actuator control rod through a synchronizing ring.

4.18 INTERSTAGE BLEED SYSTEM

The interstage bleed system, shown in figure 4.27, consists of a bleed band, an actuator assembly, and air hoses and connectors. The function of the system is to improve compressor acceleration characteristics. The system automatically unloads the compressor of a small amount of compressed air (about one tenth) during the period in the engine acceleration cycle when faster compressor acceleration is more desirable than the slight loss in engine power due to the air bleed.

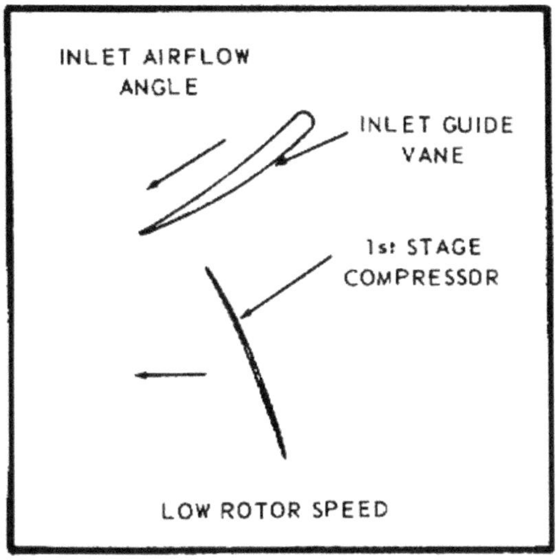

BLOCK A

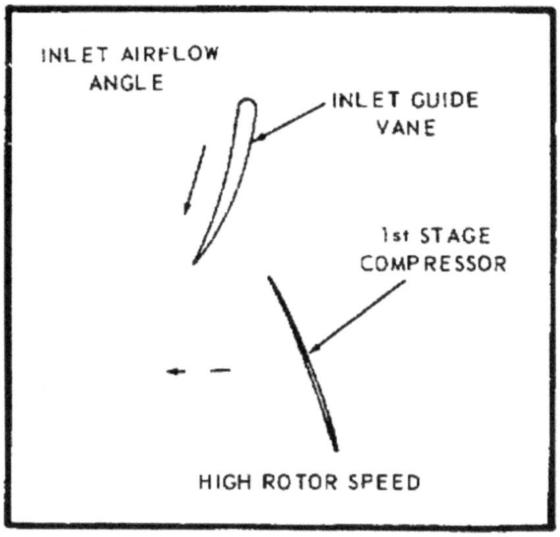

BLOCK B

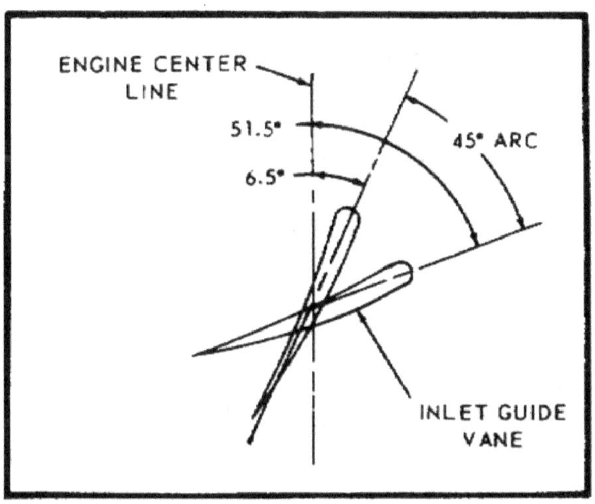

BLOCK C

Figure 4.25. Variable Inlet Guide Vane Angle of Attack.

The air bleed actuator, shown in figure 4.28, operates by compressor discharge air (P3)which is extracted from a port on the right side of the air diffuser housing.

The air entering the actuator assembly passes through a filter to the underside of the relay valve diaphragm. A small portion of this air, which is under the diaphragm, is bled through an orifice in the base of the relay valve assembly to an external line which directs it to a slide valve located on the fuel regulator housing.

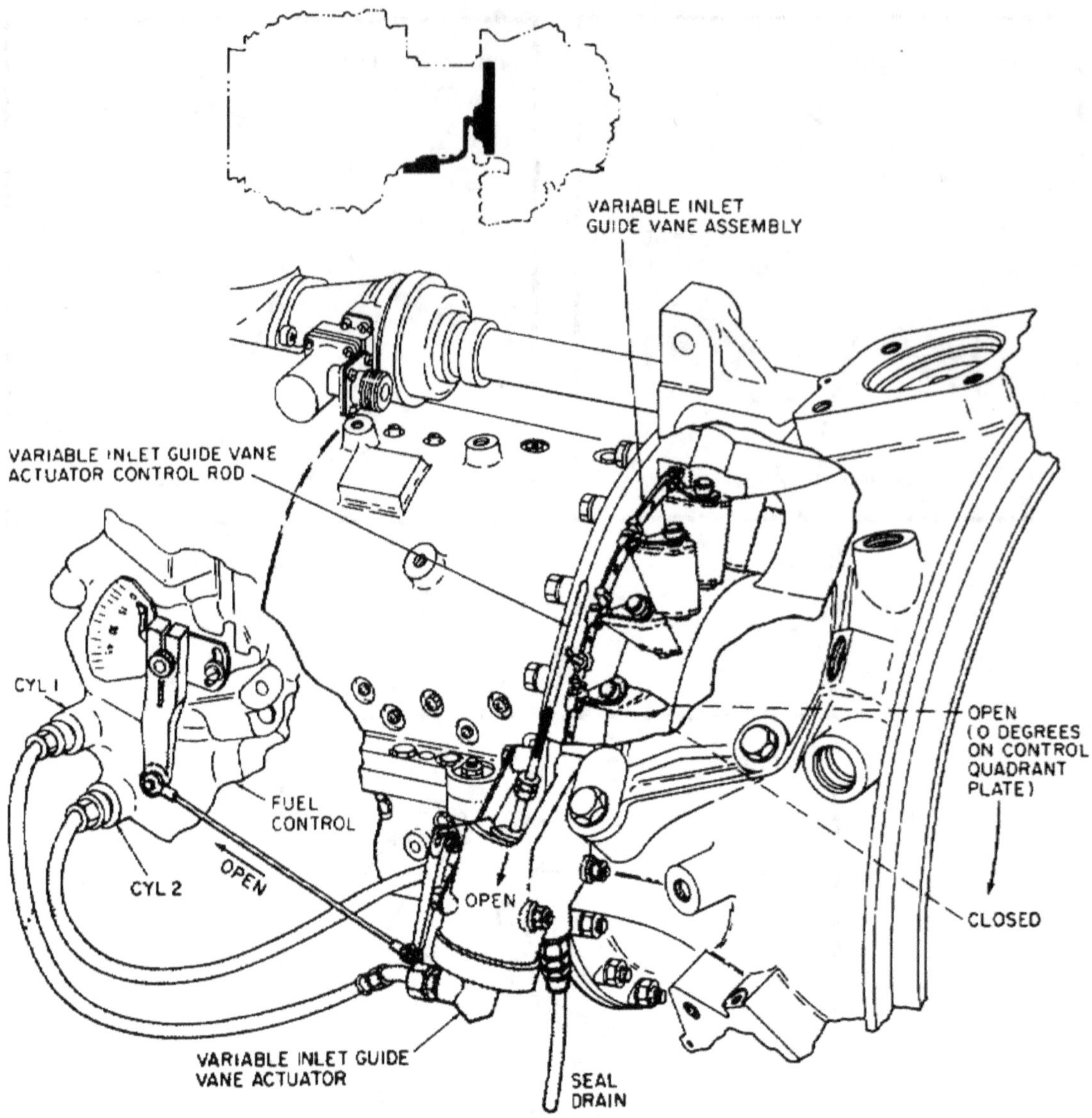

Figure 4.26. Variable Inlet Guide Vane System.

With the slide valve in the open position, this air (P_m) is vented overboard, reducing pressure at the top surface of the diaphragm. Simultaneously, air is being bled overboard through the open actuator valve; this reduces pressure at the bottom surface of the diaphragm. This equalization of pressure on both surfaces of the diaphragm causes it to remain in a neutral position holding the relay valve in its open position. With the actuator valve open, the majority of the P_3 air that enters the actuator assembly is vented to the atmosphere. When the P_3 pressure is vented, the actuator spring, located on top of the actuator piston, expands and pushes the piston downward. This causes the bleed band to open and remain open as long as the slide valve on the fuel control is in the open position.

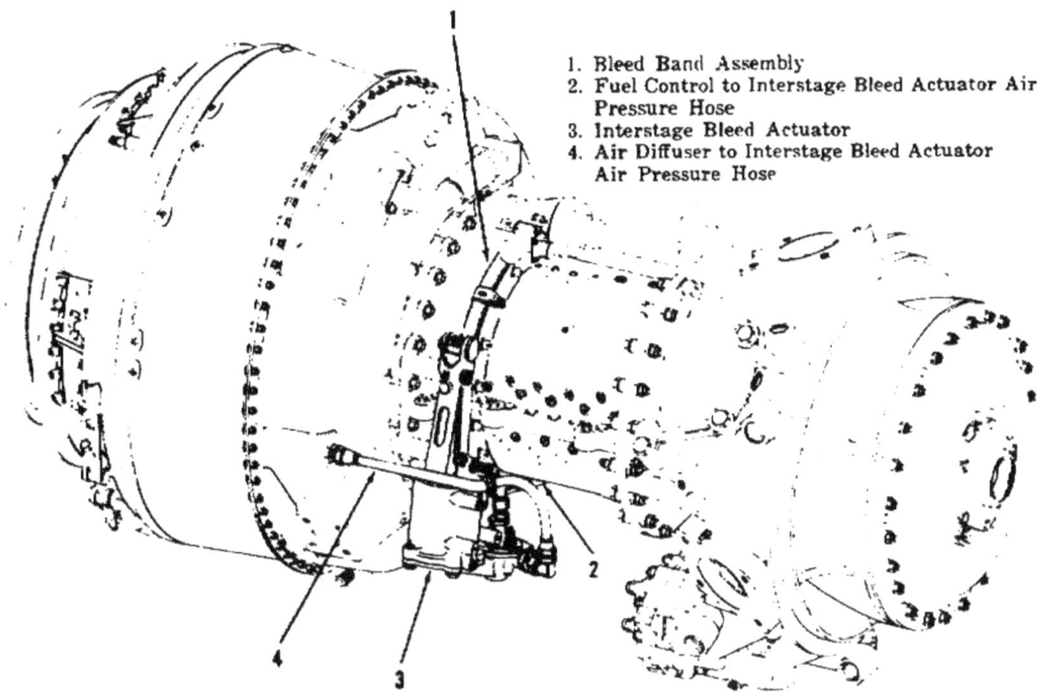

Figure 4.27. Interstage Bleed System, External Components.

When the slide valve is closed, it follows then that the bleed band will be closed. This is accomplished by a buildup of pressure on the top side of the relay valve diaphragm which forces the relay valve down, closing off the overboard vent. With the overboard vent closed, the P3 pressure is now routed into the actuator piston assembly to move upward. This causes the bleed band to close around the compressor bleed ports.

The entire sequence of operation is controlled by the fuel control which senses gas producer (N_1) speed, fuel flow and pilot demand, therefore ensuring proper opening and closing of the interstage air bleed.

4.19. ANTI-ICING SYSTEM

The engine anti-icing system, shown in figure 4.29, supplies hot air, under pressure, to prevent icing of the inlet housing areas and inlet guide vanes when the engine is operating under icing conditions. Pressurized hot air from the air diffuser flows through the holes in the trailing edge of the diffuser vanes and collects in the bleed air diffuser manifold, where it is passed to an external bleed-air manifold located at the 1 o'clock position on the diffuser housing. An elbow and tube are connected to the external bleed-air manifold and to an adapter located on top of the impeller housing. The tube and elbow pass air through the impeller housing to the hot air solenoid valve.

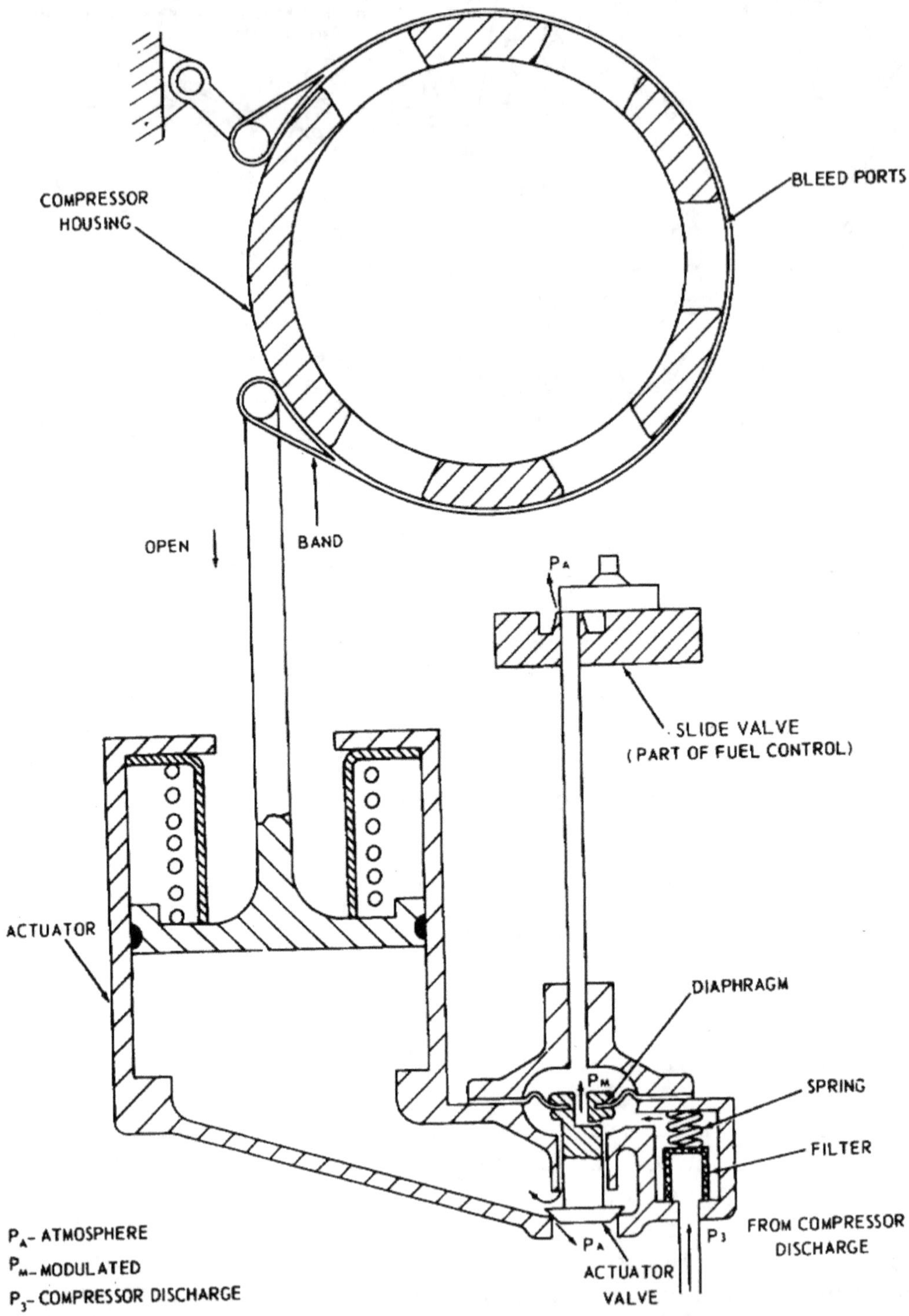

Figure 4.28. Interstage Bleed System, Cutaway View.

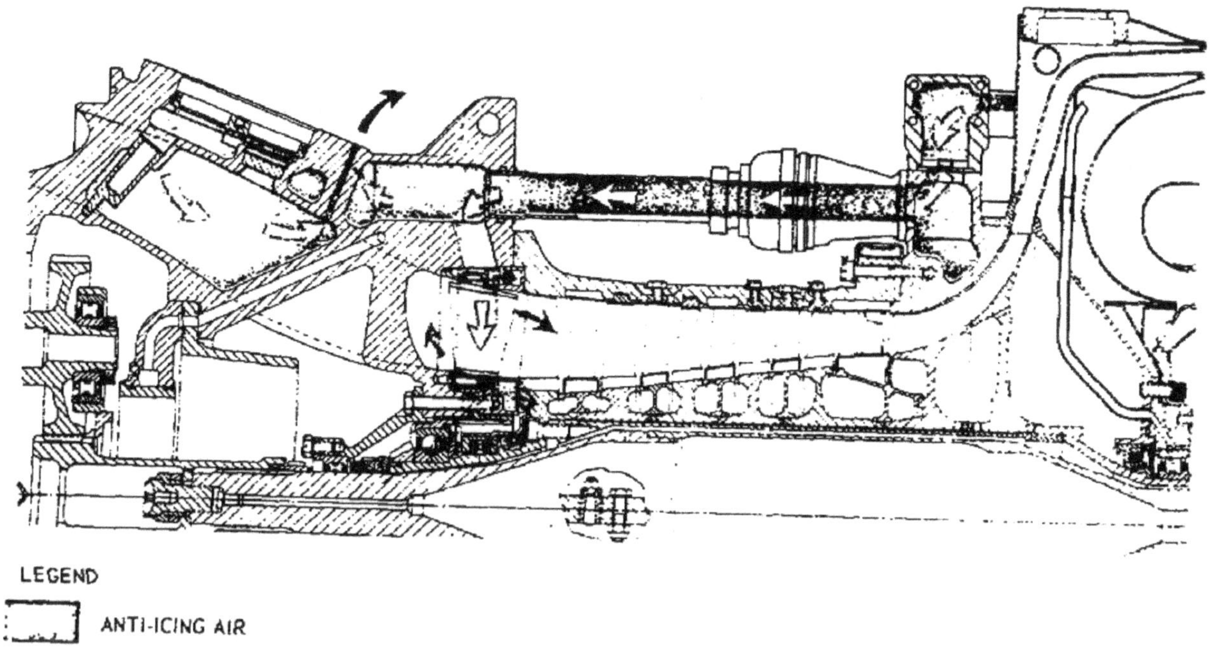

Figure 4.29. Anti-icing System Diagram.

The hot air solenoid valve is mounted on top of the compressor and impeller housing. The solenoid-operated valve controls the flow of anti- icing hot air from the diffuser to the inlet housing. It is an electrically controlled, pneumatically operated, fail-safe valve and will open in the event of an electrical failure.

During engine operation, the solenoid is generally energized and the valve remains closed. When anti-icing air is needed, the solenoid is deenergized by activating a switch in the cockpit. This vents one side of the valve to atmospheric pressure, and the differential pressure between diffuser pressure and atmospheric pressure overcomes spring tension and allows anti-icing air to flow to the inlet area. The valve will anti-ice at gas producer speeds (N_1) above flight autorotation (68% - 72% N_1).

After leaving the hot air solenoid valve, anti-icing air flows forward through a tube into the port on top of the inlet housing. This anti-icing air is then circulated through five of the six hollow inlet housing support struts to prevent ice formation in the inlet housing area. Anti-icing air also flows into the rear of the inlet housing where it passes through the hollow inlet guide vanes to prevent icing. After passing through the inlet guide vanes, the air exits in front of the inlet guide vanes and flows into the compressor area. Hot scavenge oil, draining through the strut at the 6 o'clock position of the inlet housing prevents ice formation in the bottom of the inlet housing area.

4.20. DESCRIPTION AND OPERATION OF LUBRICATION SYSTEM

The engine lubrication system consists of the main oil pressure supply system and the oil scavenge system. The principal components of the lubrication system are the oil filter assembly, power-driven rotary oil pump, power-driven rotary boost pump, and associated oil lines and internal passages. Figure 4.30 shows the internal lubrication system for the T53-L-13. The operation of the oil system is covered in the following subparagraphs.

a. Main oil pressure supply system. Engine lubricating oil is supplied from an aircraft-mounted oil tank. Oil enters the power-driven rotary oil pump, which is mounted on the N_1 accessory drive gearbox, along with the main oil filter, shown in figure 4.31A. Filter oil is directed into two main flow paths. Oil is directed through internal passages in the inlet housing to supply lubricating oil to the front section of the engine, including the reduction gearing, torquemeter, accessory drive gearing, the No. 1 main bearing, and the power shaft forward bearing. The second oil path is through the external oil pressure lines to the rear section of the engine to lubricate the No. 2, 3, and 4 main bearings.

In the inlet housing section, oil is directed through the accessory drive carrier flanges into the main oil transfer assembly located in the rear support flange of the carrier, as illustrated in figure 4.31(A). Oil from this passage is directed to the oil transfer assembly for forced-feed spray lubrication of the reduction gears through three oil transfer tubes. Oil flows through internal passages in the output reduction carrier liner, under pressure, to three jets in the liner. One jet sprays oil forward, lubricating the main output shaft bearing runner, the second lubricates the reduction gear forward bearing, and the third sprays aft, lubricating the output gear shaft bearing.

Oil from the transfer tube sprays against the output shaft plug deflector. This deflector is manufactured with a predetermined angle to splash the oil rearward, to lubricate the sun gear and power shaft splines. Three oil jets located 120 degrees apart in the main oil-transfer assembly, shown in figure 4.31(A), direct oil to the rear planetary support bearings. The main oil-transfer support assembly also houses an oil jet positioned so that high pressure oil is directed to impinge or the power shaft bearing runner, thus lubricating the bearing. Machined oil grooves in the accessory drive carrier assembly, illustrated in figure 4.31A(B), transport oil through an internal strainer to an oil nozzle located in the power shaft support bearing retainer. The oil nozzle has three machined jets. The first

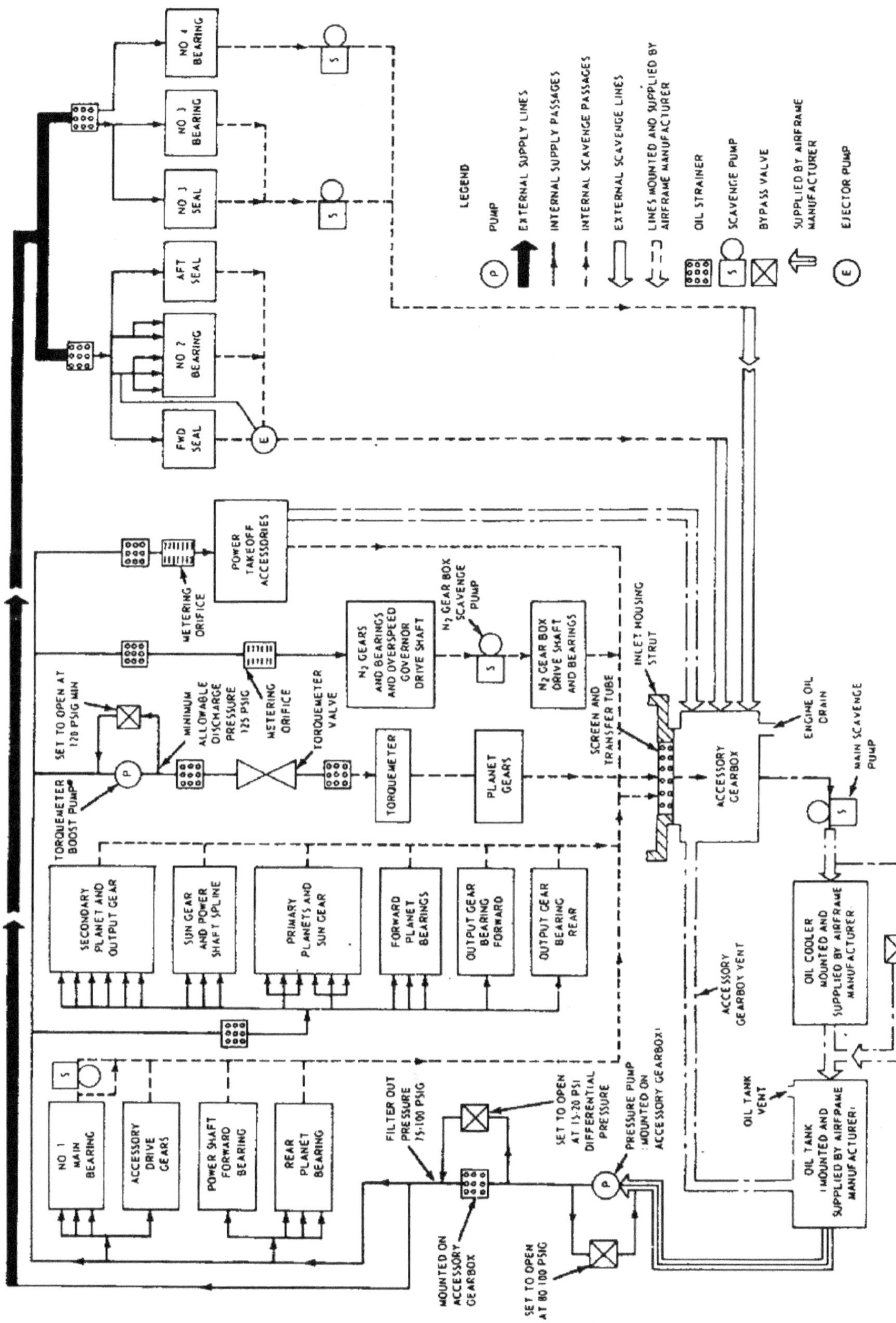

Figure 4.30. Internal Lubrication (T53-L-13).

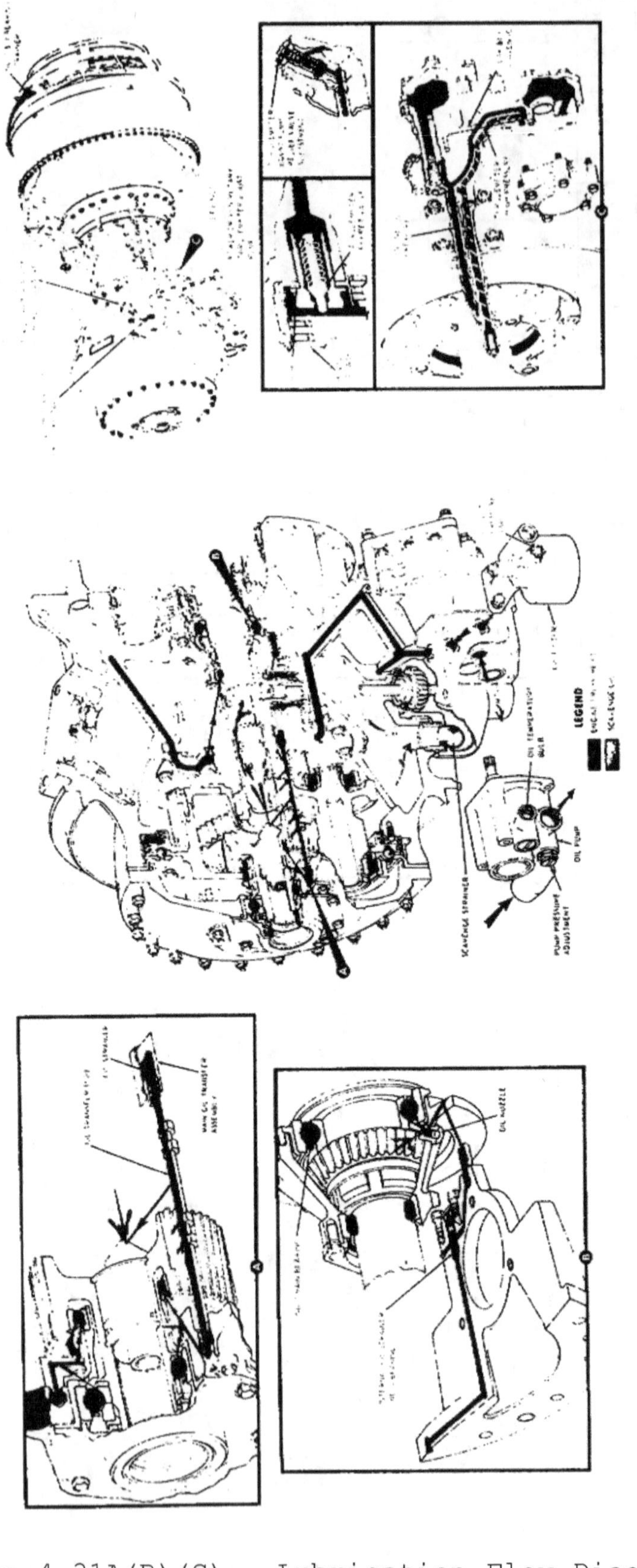

Figure 4.31A(B)(C). Lubrication Flow Diagram.

jet is positioned to lubricate the accessory drive pinion gear, the second jet is designed to lubricate the number one main bearing runner, and the third jet is positioned to lubricate the main bearing. Surplus oil from the number one main bearing and pinion gear also lubricates the accessory drive shaft support bearings. The number one main bearing and the accessory drive pinion gear are lubricated by oil from a transfer tube located in the accessory drive carrier assembly. Oil under a constant pressure from the transfer assembly lubricates the power shaft support bearing.

Oil from a third transfer passage is directed from the main transfer assembly up through an inlet housing strut to the power-driven rotary booster pump. This pump is mounted on the overspeed governor and tachometer drive assembly. The assembly includes a pressure regulating valve that governs the output pressure of the rotary torquemeter boost pump by circulating the excess pressurized oil back to the inlet housing. The pressurized oil from the rotary booster pump is directed back through an inlet housing strut to the torquemeter cylinder, shown in figure 4.31A(C).

An offset passage in the overspeed governor mounting flange directs engine oil to the strainer and metering cartridge in the overspeed governor gearbox. An additional transfer passage from the main transfer support assembly directs oil through internal passages in the inlet housing to the power takeoff mounting flange.

This oil passes through a strainer and metering orifice, which lubricates accessories driven by the engine, mounted on the power takeoff. Oil flow to the rear section of the engine is supplied from an oil pressure port at the 5 o'clock position in the inlet housing through an external hose to a pressure manifold. The manifold is mounted on the forward face of the diffuser housing. Oil is directed from the bottom of the manifold through a strainer mounted on the diffuser housing, shown at L in figure 4.31B, to the No. 2 main bearing. Oil is directed from the top of this manifold through an external hose and strainer, shown in H in figure 4.31B, through the upper strut in the exhaust diffuser and directed through the power turbine oil tube at F in figure 4.31B. The oil tube consists of two jets; one directs oil to the forward face of the No. 4 bearing runner, and the second jet lubricates the aft face of the No. 3 bearing runner. Oil is also directed through a horizontal tube forward to the No. 3 bearing seal.

Wexford
2008

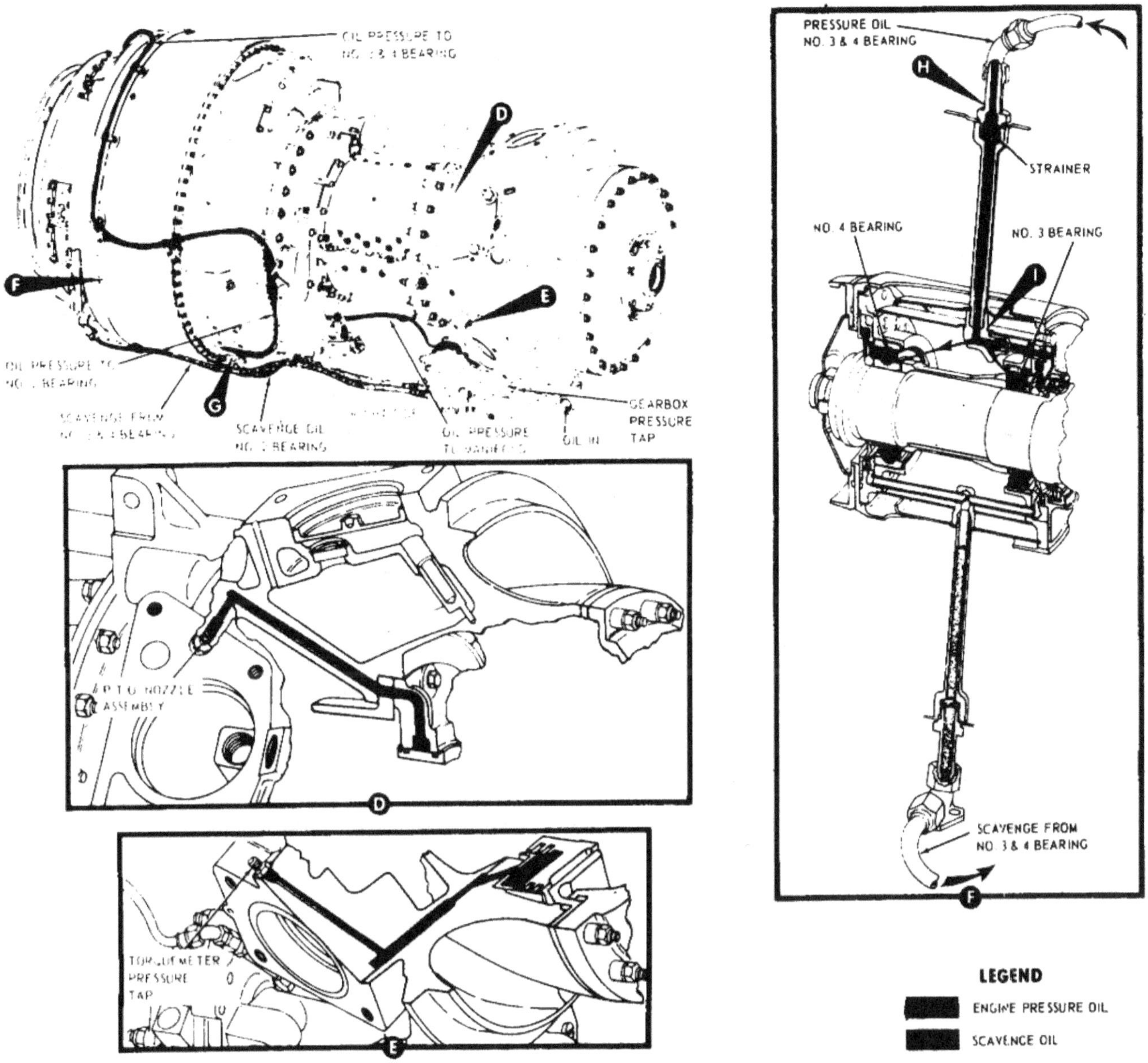

(continued on next page)

Figure 4.31B. Lubrication Flow Diagram.

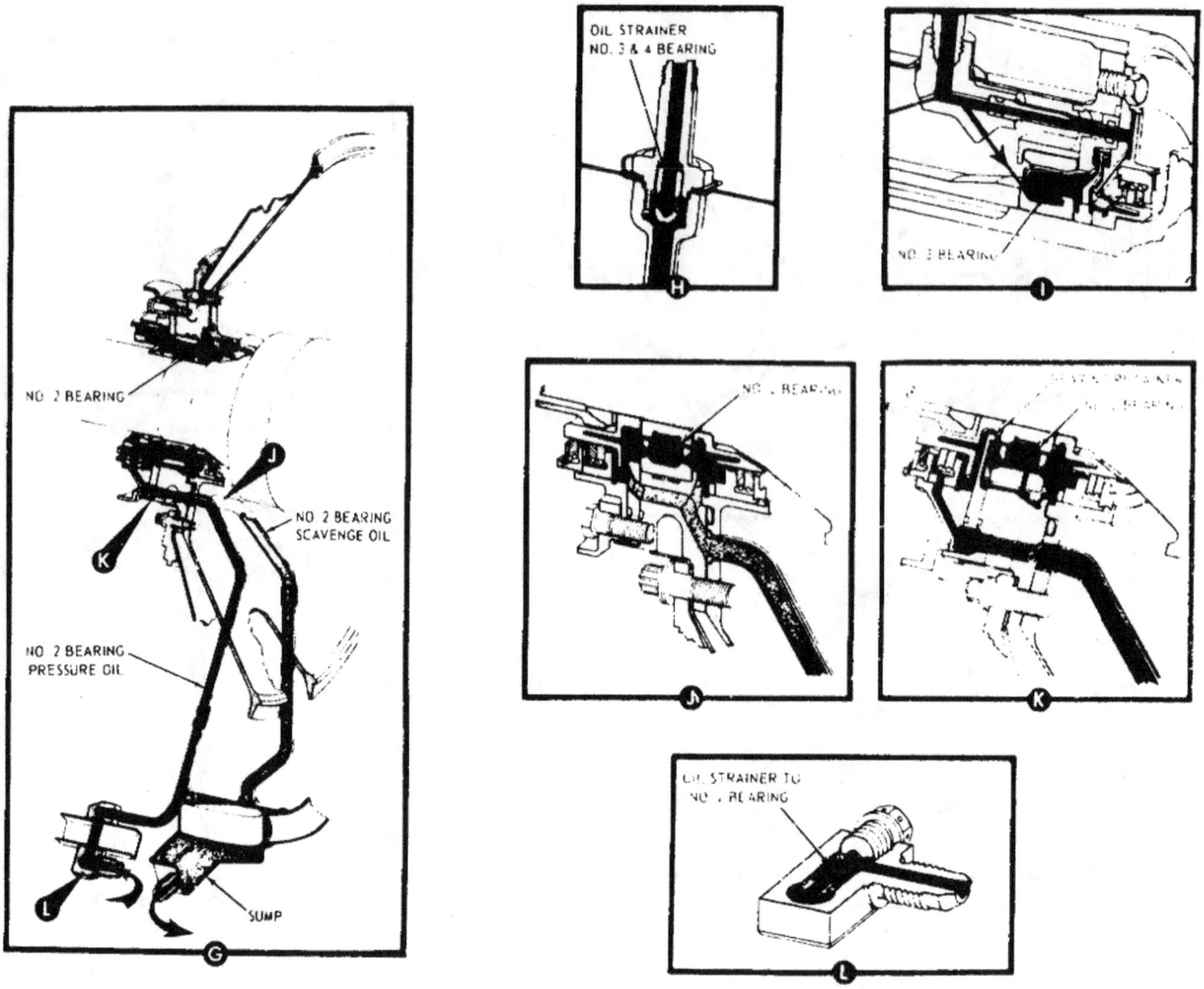

(continued from previous page)

Wexford
2008

b. Oil scavenge system. All internal scavenge oil from the inlet housing section drains through a hollow support strut to the bottom of the inlet housing through a scavenge strainer and transfer tube, and into the accessory drive gearbox. Scavenge oil from the output reduction carrier and gear assembly flows by gravity into the hollow inlet housing struts.

Scavenge oil from the No. 1 main bearing is pumped to the inlet housing struts by an impeller or paddle pump located on the rear of the bearing. Scavenge oil from the No. 2 main bearing flows through a scavenge oil tube, illustrated at G in figure 4.31B, in the diffuser housing and is directed to the accessory drive gearbox by an external scavenge oil hose assembly. Scavenge oil from the No. 3 and 4 bearings, as shown in figure 4.31B(F), flows through an oil tube that extends through the bottom of the exhaust diffuser and is directed to the accessory drive gearbox by an external oil scavenge hose assembly. The scavenge portion of the power-driven rotary oil pump returns scavenge oil from the accessory drive gearbox through the aircraft oil cooler to the aircraft oil storage tank.

4.21. TORQUEMETER SYSTEM

The torquemeter shown in figure 4.32 is used on the T53-L-13; it is a hydromechanical torque-measuring device located in the reduction-gear section of the inlet housing. It uses boosted engine oil to measure engine torque effort; the measurement is read in the cockpit as torque oil pressure in psi. Although this system uses engine oil, it is not a part of the lubrication system. The following numbers in parentheses correspond to the numbers in figure 4.32.

The mechanical portion of the torquemeter consists of two circular plates. One is attached to the inlet housing and is identified as the stationary plate (1). The second, or movable plate (2) is attached to the reduction gear assembly (6). The movable plate contains front and rear torquemeter sealing rings (12), which enable it to function as a piston in the rigidly mounted cylinder (3). The cylinder assembly houses the variable-opening torquemeter (poppet) valve (4). The movable plate maintains the fixed-orifice metered bleed (13), which functions in relation with the poppet valve. The movable plate is separated from the stationary plate by steel balls (5) positioned in matched conical sockets machined in the surfaces of both plates. When the engine is not operating, the torquemeter movable plate is a position forward and clear of the torquemeter valve plunger, allowing the spring-loaded valve to remain in the closed position. With the engine operating and a load applied to the output shaft (14), the torque

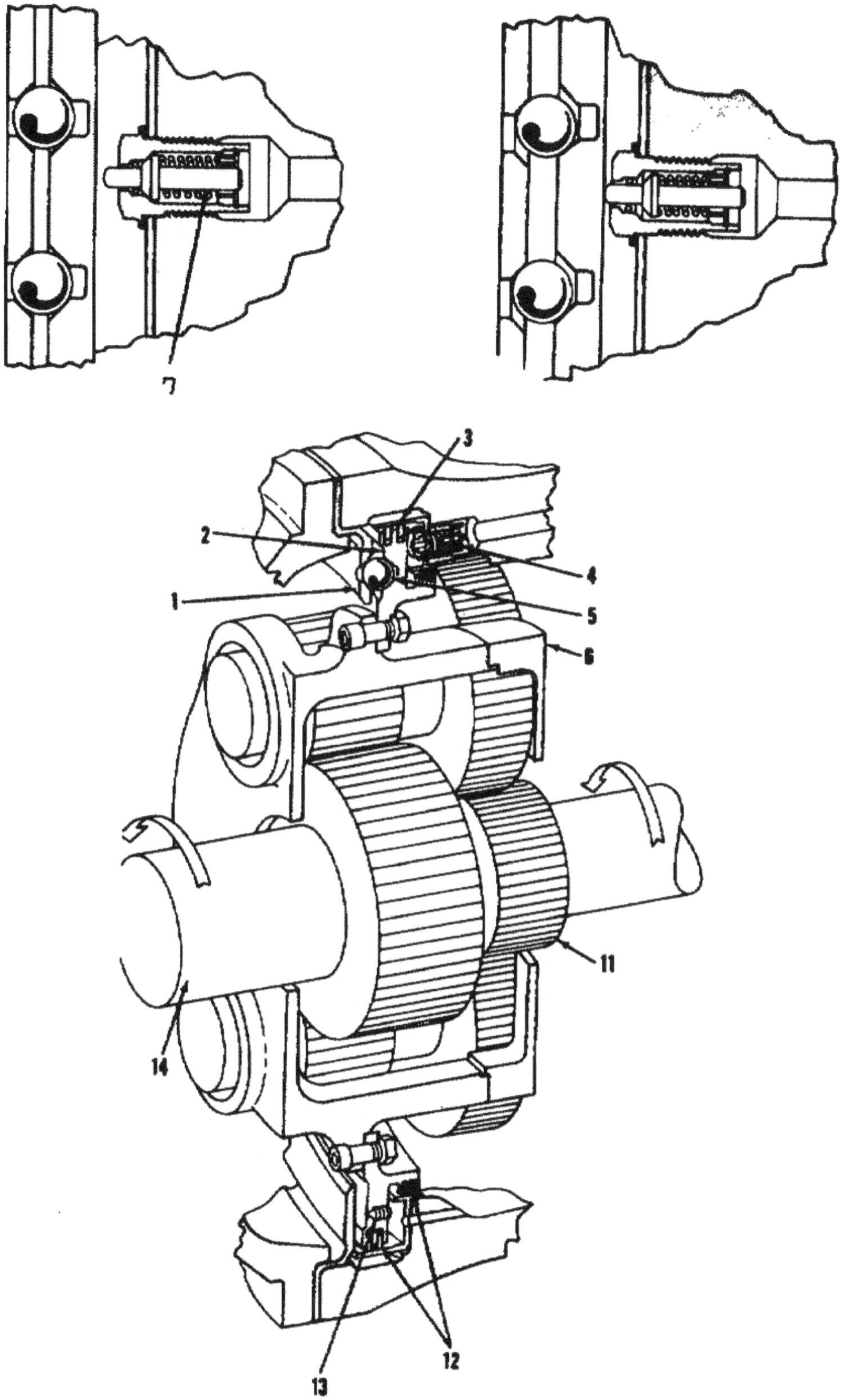

Figure 4.32. Torquemeter Diagram.

developed in the engine to drive the shaft is transmitted from the sun gear (11) through the reduction gear assembly. The attached movable plate tends to rotate with the assembly. However, this mechanically limited radial movement positions the steel balls against the conical sockets of both plates, resulting in the movable plate being axially directed rearward in the assembly.

The plate, moving rearward, contacts the torquemeter valve plunger, opening the valve and allowing oil to flow into the cylinder. This contact is maintained during all engine operation, and the size of the valve opening varies as the plate moves rearward or forward. As torque continues to increase and the torquemeter valve opens further, oil pressure increases in the cylinder but will not exceed the boost pump pressure because of the metered bleed (13).

The oil pressure developed in the cylinder exerts pressure against the piston (movable plate), restraining the rearward movement. With the engine operating in a steady-state condition the cylinder oil pressure and movement of the plate hold in an equalized position.

A factor affecting torque indications is the air pressure that develops in the inlet housing at high power settings. This air pressure produces a force oh the forward face of the piston (movable plate) inducing a higher torque indication than is actually being delivered. From the port on the forward face of the accessory drive gearbox, the air pressure is vented to the airframe-mounted torquemeter transmitter.

The pressurized oil from the torquemeter cylinder is also directed to the transmitter from a port at the 3 o'clock position of the inlet housing. The transmitter cancels the air pressure effect, resulting in a true torquemeter indication at the instrument in the cockpit.

A power-driven rotary (booster) pump, containing pressure and scavenge elements is mounted on and driven by the overspeed governor and tachometer drive assembly. Each element is an individual pumping unit and draws oil from a separate source. The pressure element receives engine lubricating oil and delivers it, at a boosted pressure, to the torquemeter valve. Excess oil flows back to the inlet side of the pump. A relief valve in the overspeed governor and tachometer drive assembly sets the power-driven rotary (booster) pump outlet pressure. The scavenge element receives oil from the overspeed governor and tachometer drive gear housing and delivers it to the oil return passages in the inlet housing assembly.

The Lycoming T53- L-7 01 uses an electric torquemeter system to monitor power output. The Avco Lycoming electric torquemeter is a refined torque measuring system which measures torque imposed on the engine power output shaft. The torque signal is a result of tension and compression stresses changing the magnetic reluctance of the shaft.

The torque system is comprised of five components and their interconnecting wiring. Two of these, the power supply and the indicator, are airframe mounted components and may be replaced at any time without recalibration. The remaining three, the power output shaft, the head assembly (transformer), and junction box, must be replaced only as a pre-calibrated set.

The transformer consists of one primary and two secondary windings. The primary generates a constant magnetic field which penetrates the transformer core attached to the primary sun gear. The voltage induced in the two secondaries varies with the tension and compression stresses imposed on the shaft. This difference in secondary voltage is transmitted through the junction box and is read on the cockpit indicator as psi of torque. Figure 4.33 is a cross section of the inlet housing to show relationship of power output shaft, torquemeter head assembly, and torquemeter junction box.

4.22. ELECTRICAL SYSTEM

The engine electrical system consists of the main wiring harness and connectors for electrical components as shown in figure 4.34. The airframe and engine wiring diagram is shown in figure 4.35. The following subparagraphs briefly discuss the operation of the engine electrical system.

a. Ignition system. The high-energy, medium-voltage, capacitor-discharge ignition system consists of an ignition-exciter unit, output leads, spark-splitter coil, and four surface-gap igniter plugs. The system is activated simultaneously with the start fuel solenoid valve and starter by a switch in the cockpit. The ignition system is used only for engine starting and not for sustaining combustion.

Power from the 28v dc electrical system is stepped up in the exciter unit to 2,500 volts and discharged through igniter plugs in the combustion chamber at a spark rate of two to eight per second.

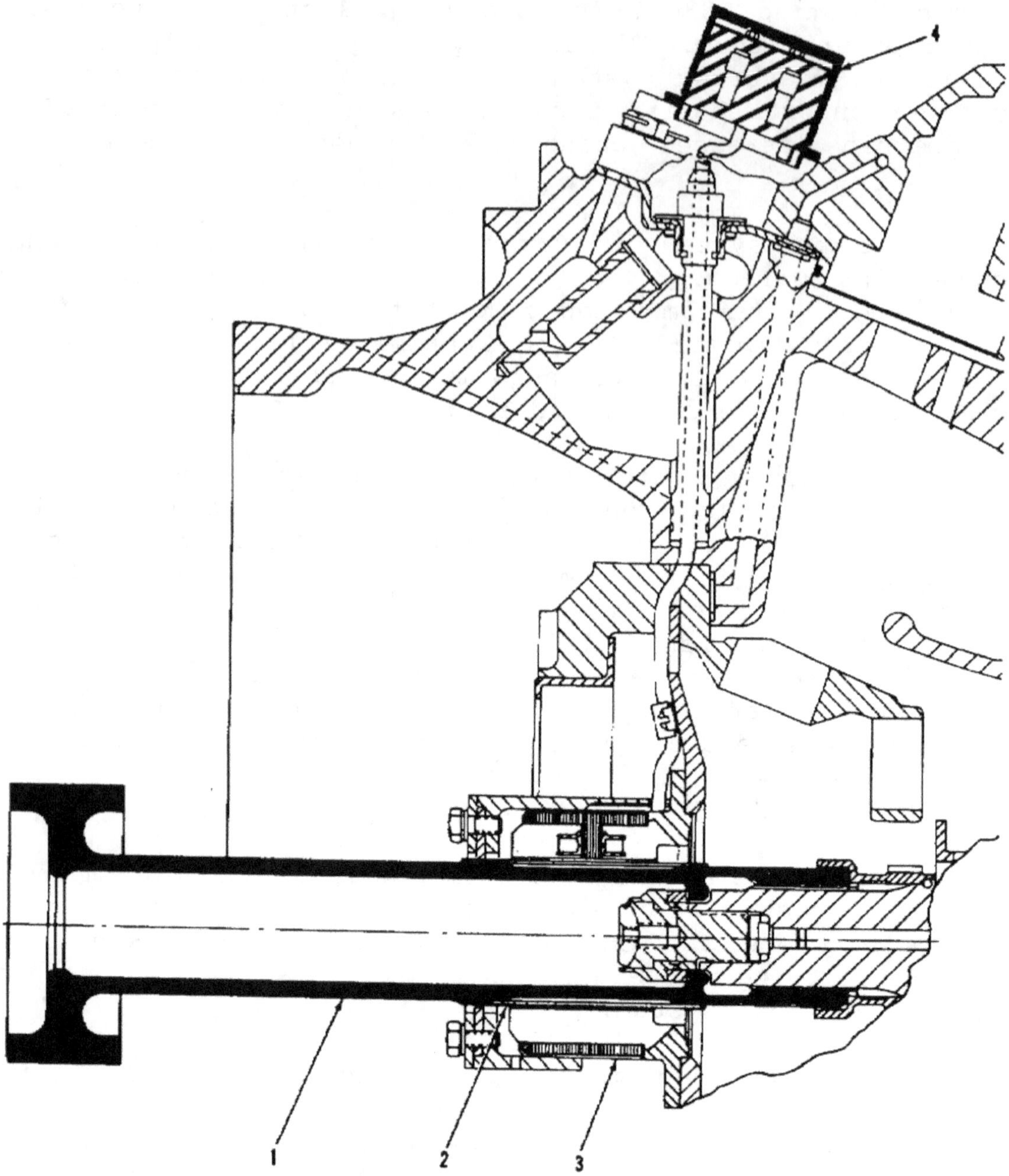

1. Sun Gear
2. Torquemeter Sleeve
3. Head Assembly
4. Junction Box

Figure 4.33. Torquemeter System (T53-L-701).

b. Exhaust gas temperature harness. An exhaust thermocouple harness, consists of an electrical connector, a shielded manifold, and six thermocouples, as shown in figure 4.36.

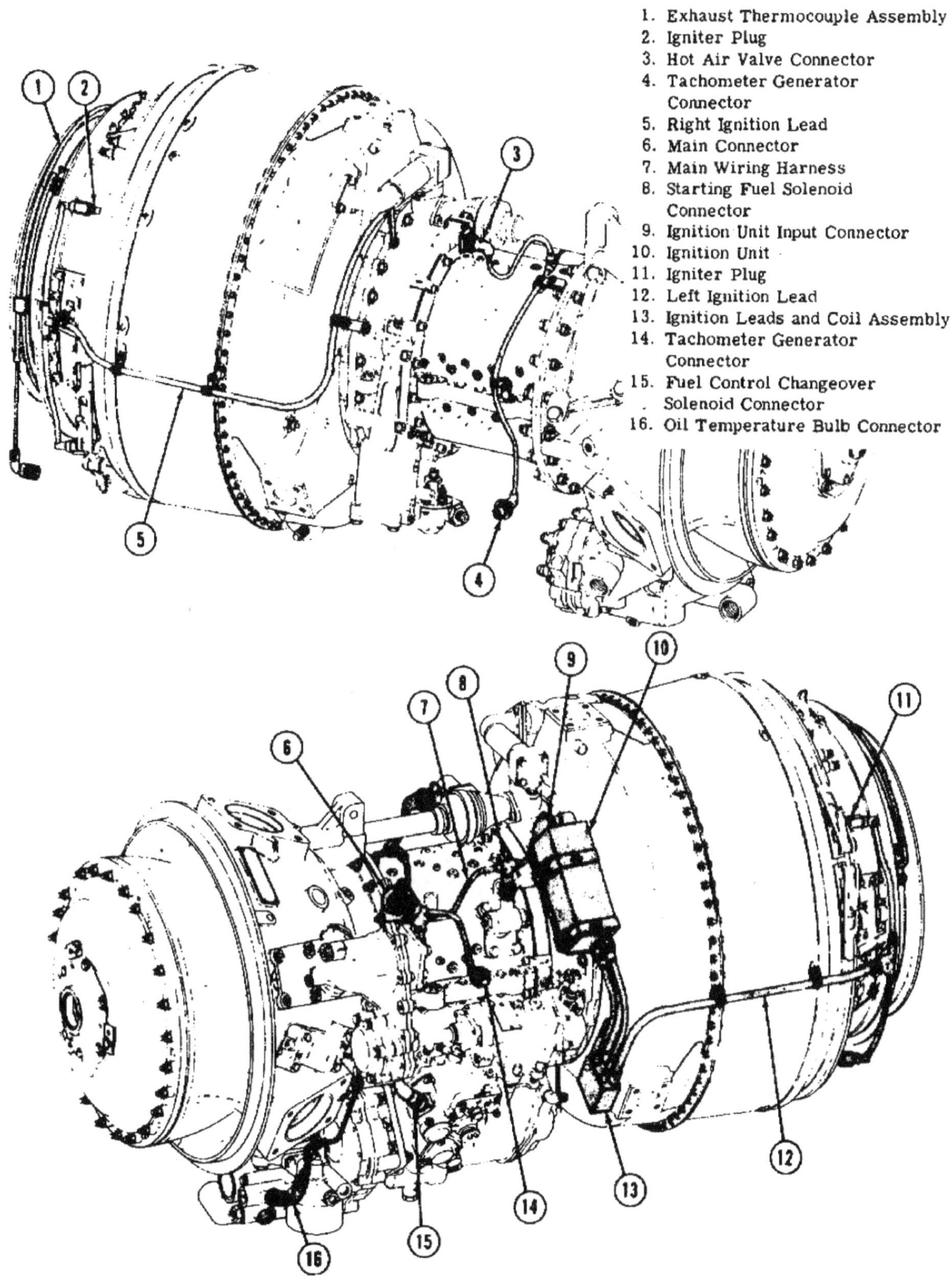

Figure 4.34. Electrical System.

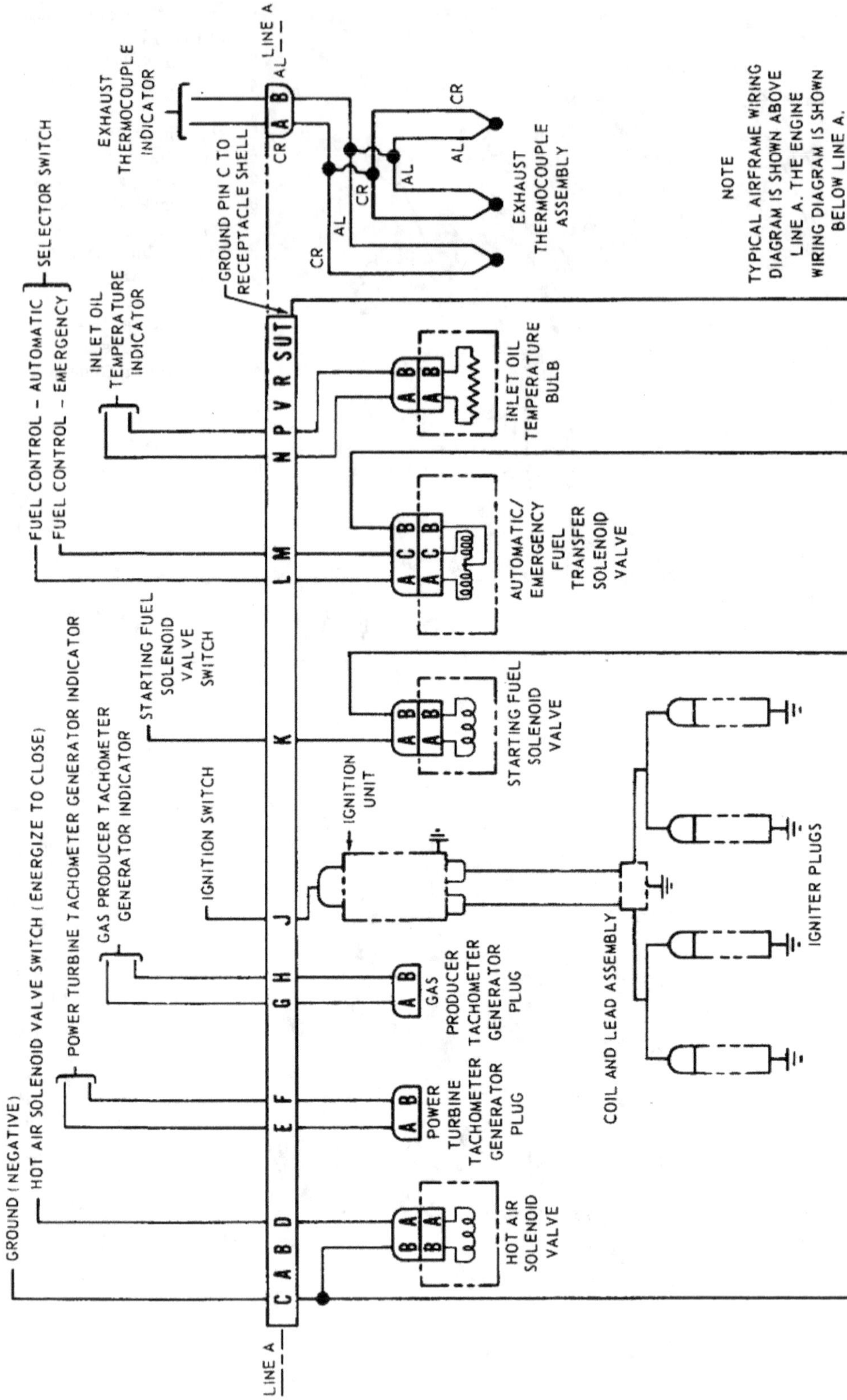

Figure 4.35. Electrical Wiring Diagram.

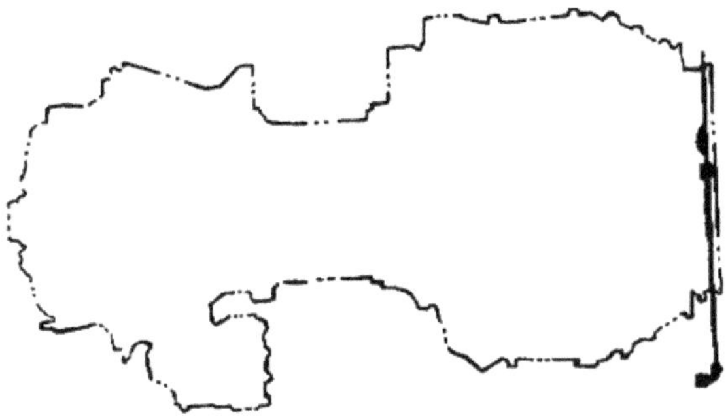

Figure 4.36. Exhaust Thermocouple Harness.

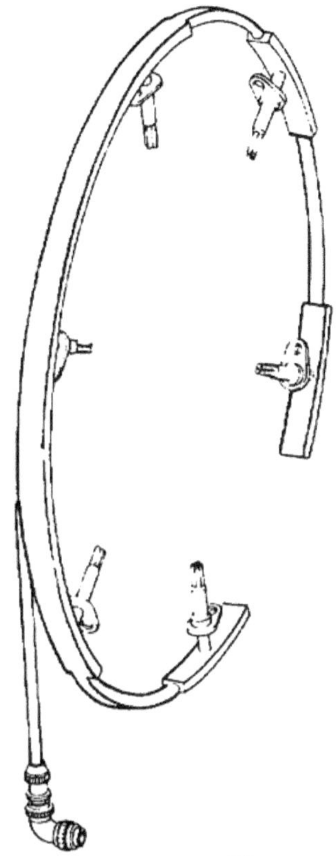

The thermocouples are inserted through the exhaust diffuser into the exhaust gas flow. When heated by the exhaust gas, an electromotive force(emf) is generated. Any two dissimilar metals placed in contact with each other generate a small voltage if heated at that junction. The amount of voltage produced varies with the metals used and the temperature they are heated to. Electrically, the thermocouples are connected in parallel with each other. This results in an average temperature reading. If one thermocouple becomes inoperative an average reading of the remaining ones will result. The thermocouples and their leads are made of chromel and alumel. Chromel is an alloy of nickel and chromium. Alumel is an alloy of nickel, manganese, aluminum, and silicon. The accuracy of the thermocouple harness is +5° at 1292° F (700° C).

4.23. SUMMARY

The major engine assemblies are the inlet, accessory gearbox, compressor, diffuser, and combustor turbine assembly. The inlet housings on both the T53-L-15 and T53-L-701 have a propeller reduction gear assembly.

Fuel is metered by a hydromechanical fuel control to twenty-two atomizers. The fuel control has two modes of operation: normal, and manual fuel flow. The internal cooling and pressurization system provides cooling air to internal engine parts and pressurizes the main bearing seals. The engine has variable inlet guide vanes to ensure a compressor surge-free margin. The interstage bleed system automatically unloads the compressor of a small amount of air during engine acceleration. The anti-icing system supplies hot air, under pressure, to prevent icing of the inlet housing areas and inlet guide vanes. Oil is supplied for lubrication by the oil pressure supply system and is scavenged by the scavenge oil system. The torquemeter is a hydromechanical, torque-measuring device that indicates engine torque in psi. Ignition for starting is produced by a high-energy, medium-voltage, capacitor-discharge system. Exhaust gas temperature is measured by six thermocouples inserted in the exhaust gas flow.

Chapter 5

LYCOMING T55

5.1. INTRODUCTION

This chapter covers the Lycoming T55 gas turbine engine. Section I gives an operational description of the T55, covering the engine's five sections. Section II covers in detail each of the engine's sections and major systems.

Basically, all models of the T55 are of the same design. The major difference between the T5 5-L-7C and L-11 engines is that the L-11 has a two-stage gas producer turbine and the L-7C uses a single-stage gas producer turbine. However, the description and information given in this chapter are applicable to all models of the T55 except where noted. The Lycoming T55 engine is used to power the CH-47 Chinook helicopter.

Section I. Operational Description of the T55 Gas Turbine Engine.

5.2. GENERAL

This section describes in detail the inlet, compressor, diffuser, combustion, and power turbine sections, and the differences between models and specifications are compared.

Except for the discussion comparing models and specifications, the section's coverage is limited to the T55-L-7C and -11A engines.

5.3. DESCRIPTION

The T55 gas turbine engine is a direct-drive, annular reverse flow, free-power turboshaft engine developed for use in rotary-wing aircraft. An exploded view of the T55-L-11A engine is shown in figure 5.1.

The engine consists of an air inlet, accessory drive, compressor, diffuser, combustion, and power turbine sections. All the sections, except the accessory drive, form an annular (circular) flow path for air and hot gases. in addition, all the sections are structurally interdependent. The T55 is a direct drive engine,

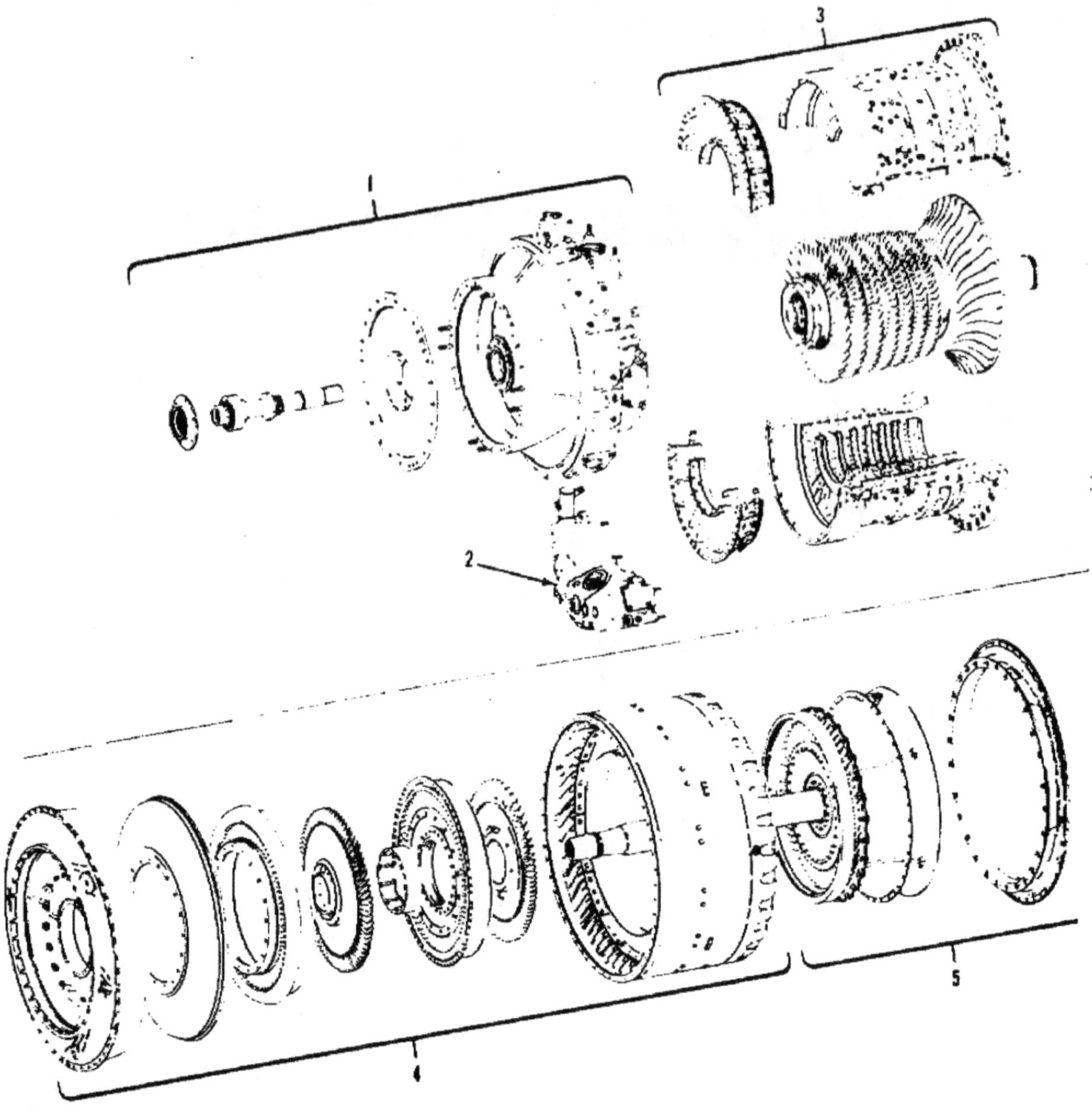

1. Inlet Housing Section
2. Accessory Gear Box and Drive Section
3. Compressor Section
4. Diffuser and Combustion Section
5. Power Turbine Section

Figure 5.1. Exploded View of T55-L-11A Engine.

meaning it has no gear reduction, and the power output shaft speed is the same as the power turbine speed. The airflow paths through the T55-L-7 and L-11A engines are shown in figures 5.2 and 5.3 respectively.

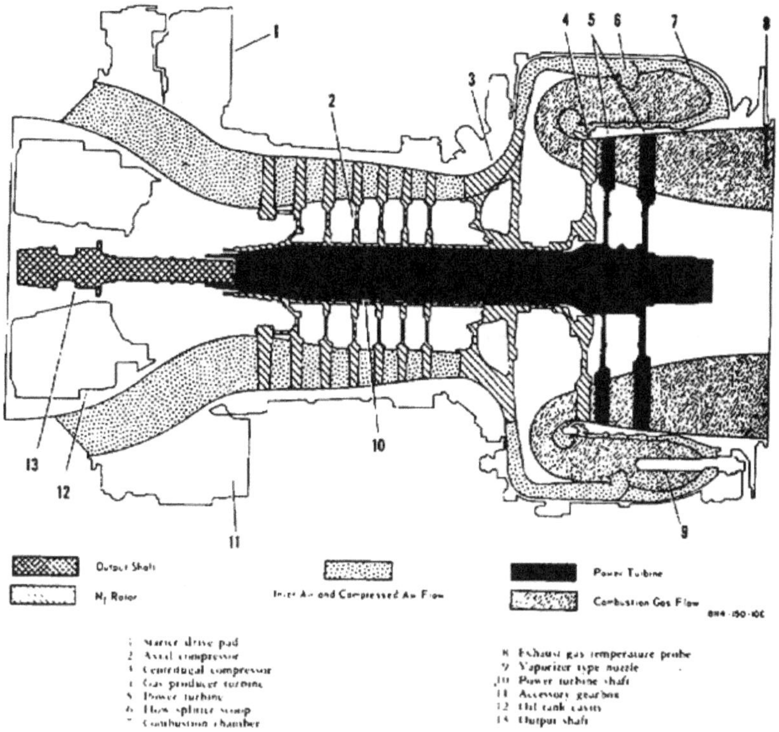

Figure 5.2. Engine T55-L-7 Series.

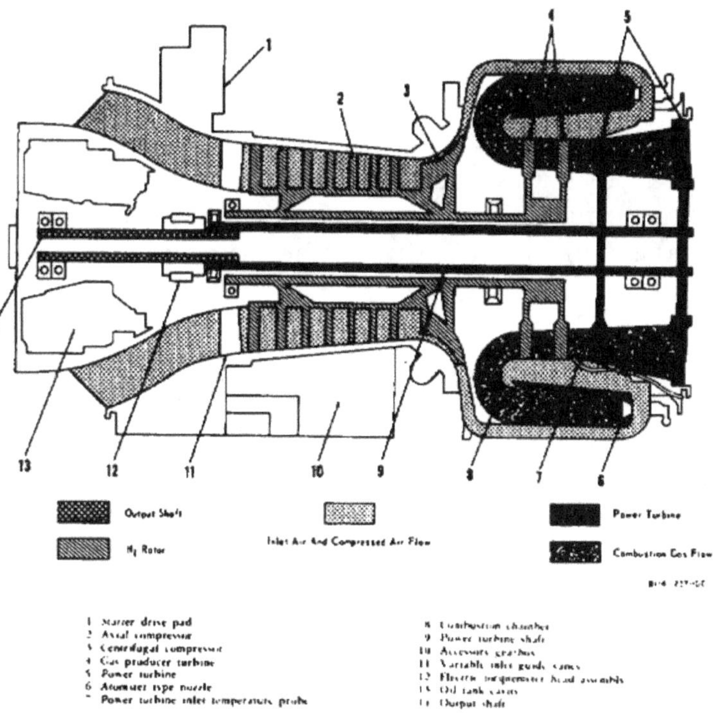

Figure 5.3. Engine T55-L-11 Series.

At the front of the engine, the inlet housing forms an annular airflow path to the inlet guide vanes (IGV) which are variable on the T55-L-11A engine.

An airflow path is between the inner diameter of the housing and the base contours of the compressor stages. The compressor rotor has seven axial and one centrifugal stage compressor. The diffuser channel contains radial vanes which straighten the airflow from the centrifugal compressor. These vanes increase the static air pressure to its highest pressure at the diffuser exit. Air passes into the combustor section where its flow direction is changed twice. Flowing rearward from the diffuser, the air surrounds and enters the combustor liner through holes and louvers, and 28 swirl cups reverse the airflow direction for the first time. Each swirl cup contains a dual-orifice, fuel-atomizing nozzle. As combustion occurs, the hot expanding gases move forward to the curl assembly, again reversing the flow. The T55-L-7 uses 14 "T" cane vaporizer tubes to inject the fuel for combustion.

The gas flow is directed by the first and second stage nozzles to the two-stage gas-producer (GP) turbine. The T55-L-7C engines use a single-stage GP turbine. These turbines are attached to the rear of the compressor rotor shaft and extract the energy required to turn the compressor.

From the second stage turbine of the L-1A engine, the gas flow continues through the power turbine (PT) section. Most of the remaining energy of the hot gases is extracted by the two-stage PT and is transmitted by the power shaft to the power output shaft. After the expanding gases have passed through the PT section they are exhausted to the atmosphere through the exhaust duct.

5.4. MODEL COMPARISON

The T55 series gas turbine engines are used to power the CH-47 Chinook helicopter. The CH-47B is powered by either the Lycoming T55-L-7, -7B, or -7C turboshaft engines. The CH-47C is powered by either T55-L-7C or T55-L-11 engines.

Basically, the Lycoming T55 series engines are of the same design and differ only in shaft horsepower and internal details. The T55-L-7 and -7B engines are rated at 2,200 shp at normal power and 2,650 at military power. The T55-L-7C engine is rated at 2,400 shp at normal power, 2,650 at military power, and 2,850 at maximum power. The T55-L-11 engine is rated at 3,000 shp at normal power, 3,400 at military power, and 3,750 at maximum power.

The T55-L-7C engine differs from the other T55-L-7 series engines in the following ways: (1) an atomizing combustor is used as opposed to the vaporizing type; (2) the 14 vaporizer "T" canes are replaced by 28 dual-orifice fuel spray nozzles for injecting fuel into the combustor; (3) four ignitors and start fuel nozzles are installed to ensure optimum ignition and flame propagation during starting. However, only two of the four start fuel nozzles are used; (4) a modified fuel control is used.

In addition, the T55-L-11A differs from the T55-L-7 series in the following: (1)a two-stage gas producer(GP) turbine is used; (2) a more accurate electric torquemeter with an indicator reading in percent of maximum torque; (3) five thermocouple probes, that measure the gas temperature at the power turbine inlet; and (4) variable inlet guide vanes to improve compressor efficiency.

5.5. SPECIFICATION SUMMARY

Specifications for the T55-L-7, -7C and -11 engines used in Army aircraft are summarized in the following chart.

MILITARY DESIGNATION	T55-L-7	T55-L-7C	T55-L-11
Lycoming Designation	LTC 4B-8	LTC 4B-8-C	LTC 4B-11-B
Compressor stages	7 axial, 1 centrifugal	7 axial, 1 centrifugal	7 axial, 1 centrifugal
Combustor	Annular reverse flow 14 vaporizing tubes	Annular reverse flow 28 atomizing nozzles	Annular reverse flow 28 atomizing nozzles
Turbine (GP)	1 stage	1 stage	2 stage
Turbine (PT)	2 stage	2 stage	2 stage
Power outlet	1,500 rpm	1,600 rpm	1,600 rpm
Fuel control	Auto and manual	Auto and manual	Auto and manual
Ignition system	2 torch igniters	4 torch igniters	4 torch igniters
Lubrication system	Oil tank, cooler	Oil tank, cooler	Oil tank, cooler
Diameter and length	24.2 in., 44.0 in.	24.2 in., 44.0 in.	24.2 in., 44.0 in.
Weight	580 lb	590 lb	670 lb
Oil	MIL-L-23699	MIL-L-23699	MIL-L-23699
Fuel	MIL-T-5624/JP4	MIL-T-5624/JP4	MIL-T-5624/JP4
Rating (Military)	2,650 shp	2,850 shp	3,400 shp
(Normal)	2,200 shp	2,400 shp	3,000 shp
(Maximum)		2,850 shp	3,750 shp
Application	2 in CH-47B	2 in CH-47B, C	2 in CH-47C

5.6. DIRECTIONAL REFERENCES AND ENGINE STATIONS

The diagrams in figure 5.4 and 5.5 show directional references and engine stations. Right and left sides of the engine are determined by viewing the engine from the rear. Direction of rotation of the compressor rotor and gas producer turbines is counterclockwise as viewed from the rear of the engine. The power turbine and the output gearshaft rotate in a clockwise direction.

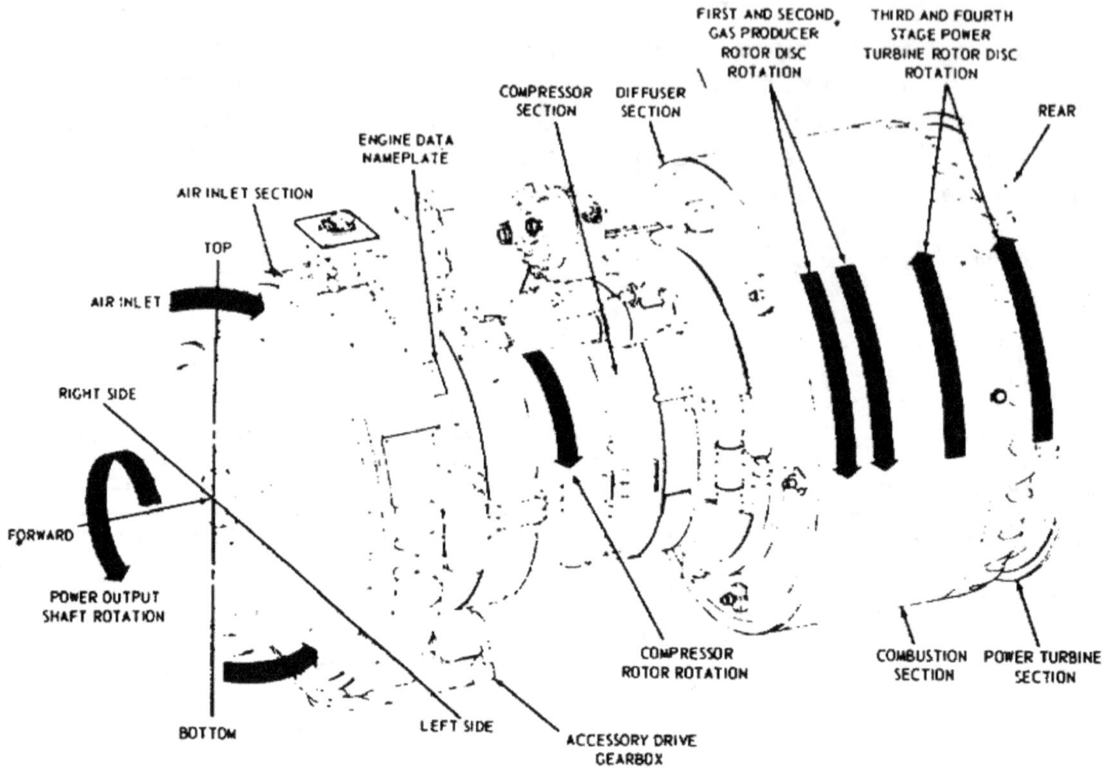

Figure 5.4. T55 Engine Directional References.

Engine stations for the T55-L-11A are shown in figure 5.5. The compressor housing is station No. 1 starting from the inlet guide vanes and extending to the centrifugal compressor. Station No. 2 starts at the beginning of the centrifugal compressor and ends at the air diffuser. Station No. 3 is from the air diffuser exit to the combustor inlet. Station No. 4 runs from the combustor inlet to the gas producer (GP) entrance. Station No. 5 is the GP entrance, station No. 7 is from the GP exit to the PT entrance. Station 7.2 is the PT entrance, and station No. 9 is the PT exit. No stations are shown for 6 and 8, because these numbers are not used.

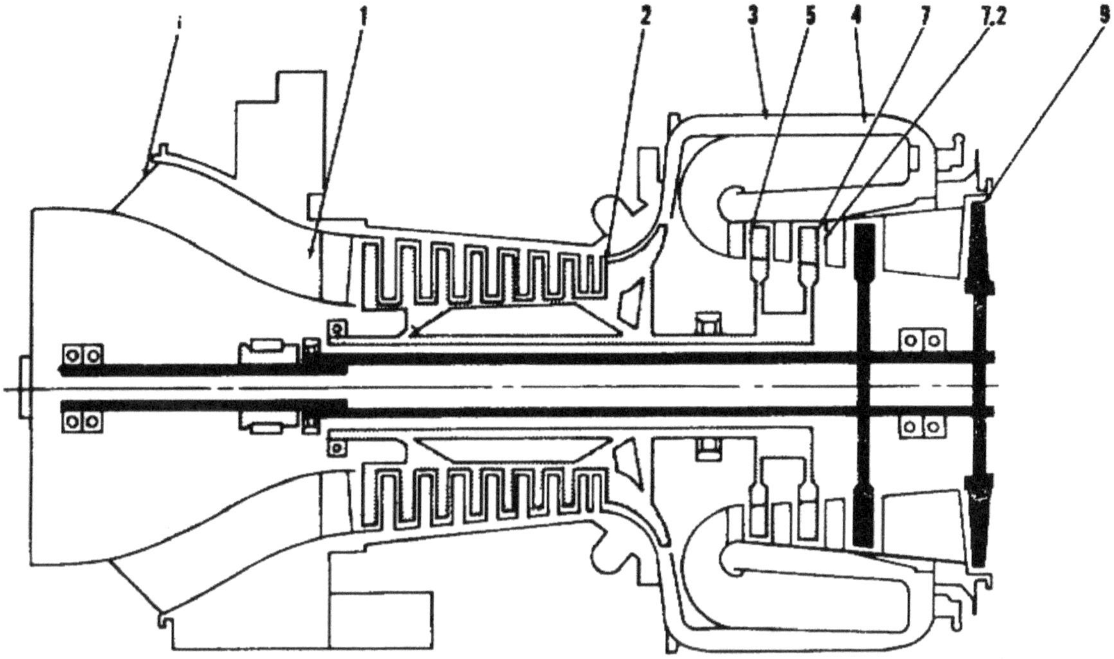

Figure 5.5. T55-L-11 Engine Stations.

5.7. SUMMARY

Basically, all models of the T55 series engine are of the same design and principle. The major difference between the L-7 and L-7C is that the L-7C uses 28 atomizing nozzles instead of the 14 vaporizing tubes in the L-7 engine. The T55-L-11A differs from the T55-L-7C in that it has a two-stage GP turbine where the L-7C has a single-stage, and of course the shp is different for each model engine. The engine consists of an air inlet, accessory drive, compressor, diffuser, and combustion and power turbine sections. The T-55 series engine differs from engines we have previously covered in that it is a direct drive engine. The Lycoming T55 series engine is used to power the CH-47 Chinook helicopter.

Section II. Major Engine Sections and Systems

5.8. GENERAL

This section discusses the five major engine sections, beginning at the front: inlet, compressor, diffuser, combustion, and power turbine; then it continues with descriptions of various engine systems, including fuel, lubrication, and electrical systems. The text and illustrations cover the T55-L-11A engine except where noted.

5.9. INLET SECTION

The air inlet housing is a one-piece magnesium alloy casting, with an inner and outer housing as shown in figure 5.6.

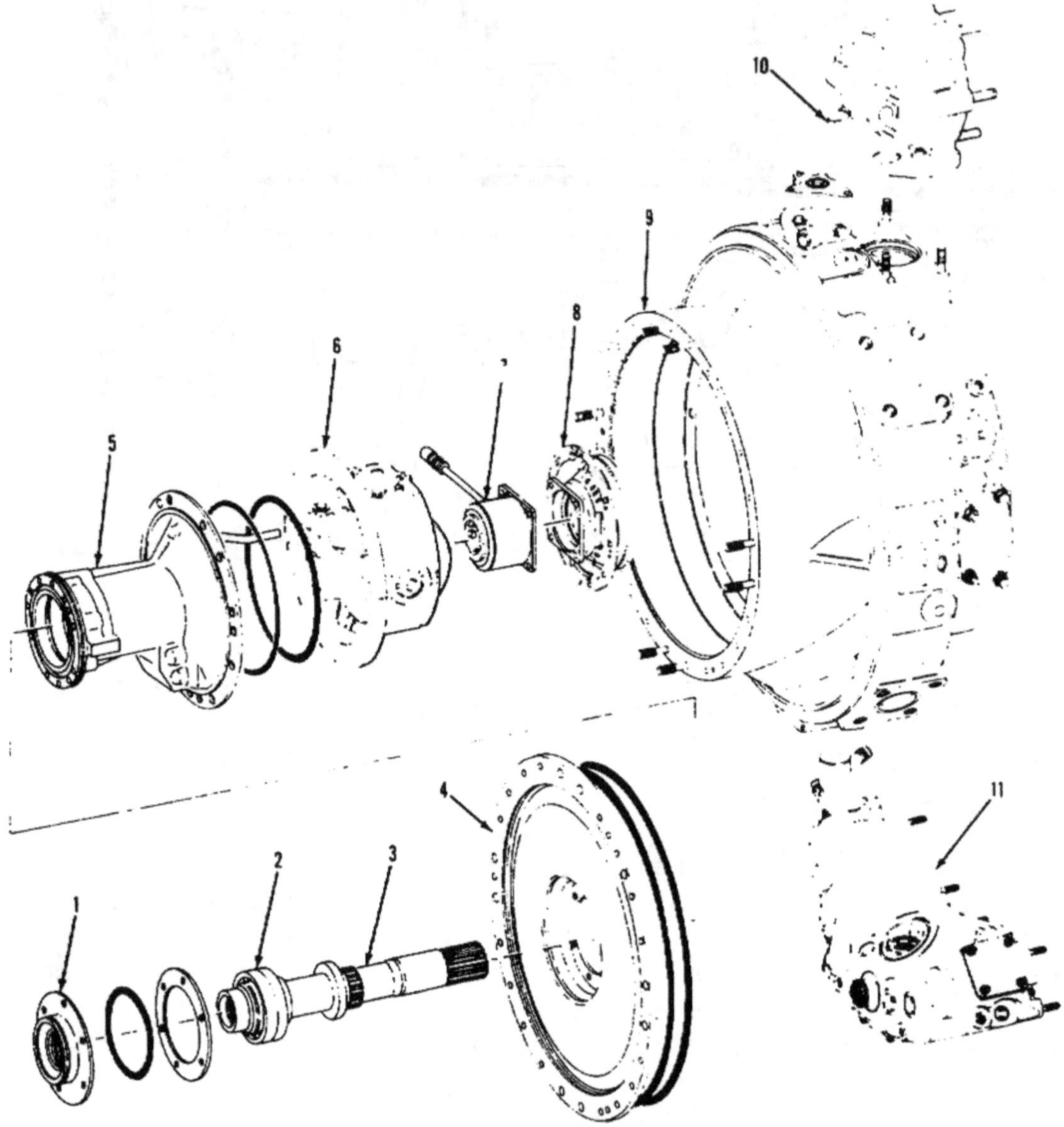

1. Output Shaft Seal
2. Number 6 & 7 Bearing Package
3. Power Output Shaft
4. Oil Tank Cover Plate
5. Output Shaft Support Housing
6. Accessory Drive Gear Carrier
7. Torquemeter Head
8. Support Assembly
9. Inlet Housing
10. Starter Drive Assembly
11. Accessory Drive Gear Box

Figure 5.6. Inlet Housing Assembly.

Four engine mounting pads are on the inlet housing. The front face of the inner housing includes studs to mount the engine oil tank cover plate and a speed reduction gearbox. The inner housing, figure 5.6 item 9, contains the output shaft, the compressor rotor forward bearing support structure, the torquemeter head assembly, and a 3.75-gallon oil tank. The outer housing forms the outer wall for the annular inlet duct. Four struts make the structural connection between the two housings. The struts are hollow and have passages for oil and accessory drive shafts.

5.10. COMPRESSOR SECTION

The compressor rotor assembly, shown in figure 5.7, consists of a seven-stage, all-steel axial compressor and an all-titanium centrifugal compressor. The two compressors form a single rotating assembly providing an 8 to 1 compression ratio.

All axial compressor blades are stainless steel and mounted on steel compressor disks by dovetail roots and spring-loaded locking pins. Seven two-piece stator assemblies are bolted to a two-piece cast magnesium compressor housing, which encloses the axial and centrifugal rotors. The housing is split axially to permit complete access to the compressor rotor for inspection and blade replacement. The internal surfaces of these housings are coated with epoxy phenolic and graphite filler. The external surfaces are treated with epoxy phenolic gray paint.

On the L-111 engine, a variable inlet guide vane assembly is mounted in the front of the compressor housing. The axial compressor housing has an airbleed system consisting of a series of holes and machined passages. The system bleeds air from the sixth stage and thereby improves compressor performance. This system is controlled by an interstage airbleed system covered later in the chapter.

5.11. DIFFUSER SECTION

The air diffuser, shown in item 1 figure 5.8, is constructed of stainless steel. The diffuser receives high-velocity air from the centrifugal impeller. This radial airflow is changed to an axial flow by longitudinal guide. The divergent shape of the diffuser decreases velocity and increases air pressure. Air pressure at the diffuser discharge is at its highest value; air temperature is in the vicinity of 600° F. Temperature and pressure are directly related to rotational speed of the compressor. Internally, the diffuser supports

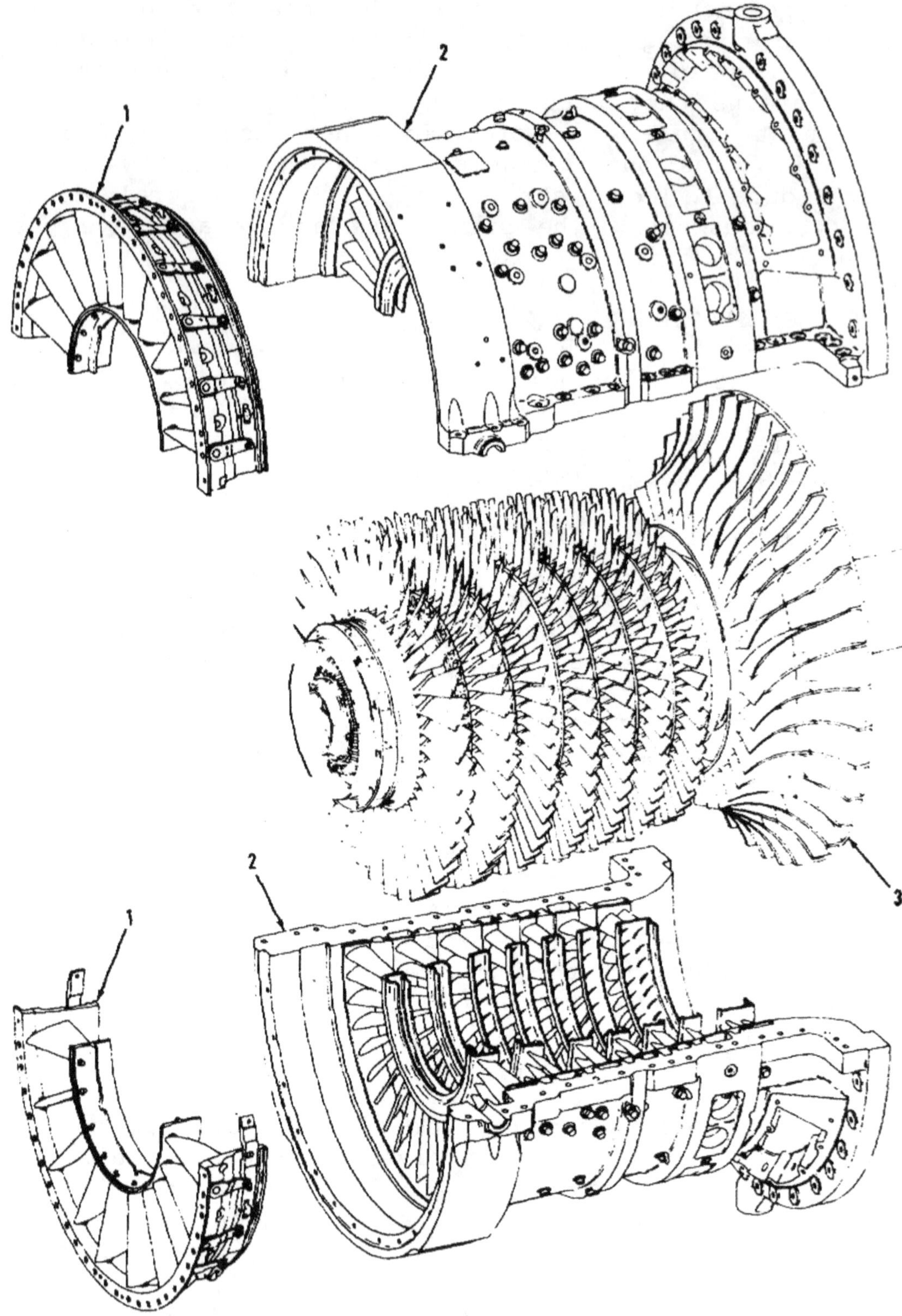

1. Variable Inlet Guide Vanes (V.I.G.V.) 2. Compressor Housing
3. Compressor Rotor

Figure 5.7. Compressor Section.

the rear of the Compressor assembly through the No. 2 main bearing. Also mounted internally, but not a component of the diffuser, are the combustion chamber deflector curl, first and second stage GP nozzles, and turbine rotors.

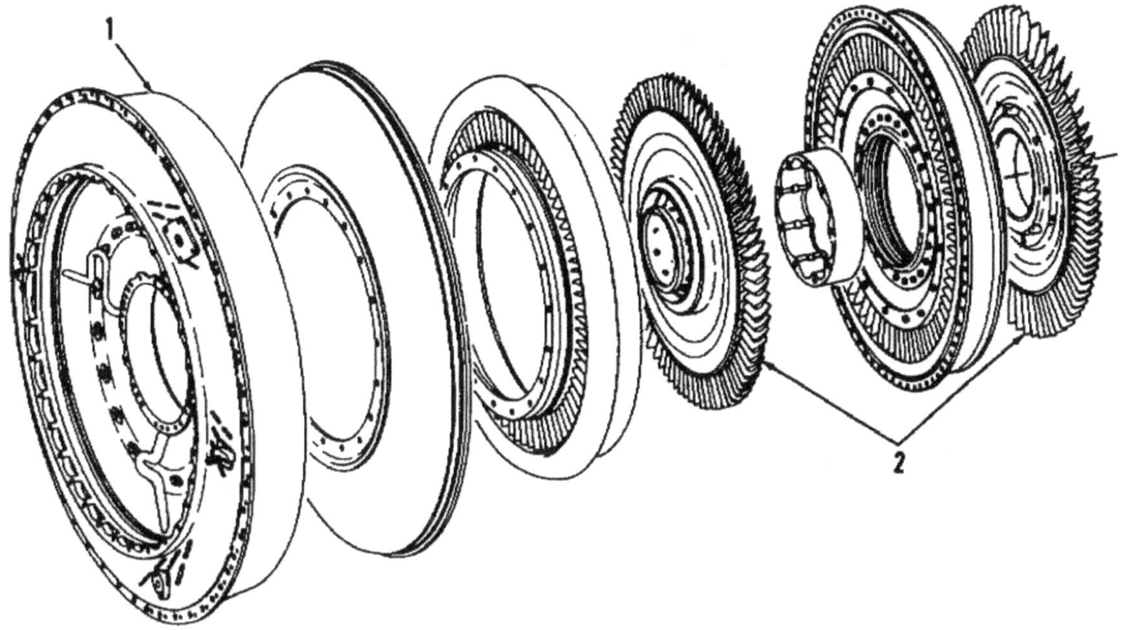

Figure 5.8. Diffuser and Gas Producer Turbine.

5.12. COMBUSTOR SECTION

The combustion chamber is a reverse-flow annular design, which permits maximum use of space and reduces gas producer and power turbine shaft length.

The atomizing combustor has 28 main fuel atomizing nozzles. The nozzles are of the dual-orifice design mounted in two interchangeable, dual-channel, main fuel manifolds, with 14 fuel nozzles in each manifold.

The perforated combustor liner, shown in figure 5.9, is manufactured from a heat-resistant alloy. The perforations are arranged to meter air into the combustor for combustion and cooling. Two combustor drain valves are located at the bottom of the chamber, to drain raw fuel on engine shutdown after a false or aborted start. Fuel is injected directly into the combustor through the atomizing nozzles which are mounted on the fuel manifold at the rear of the combustor.

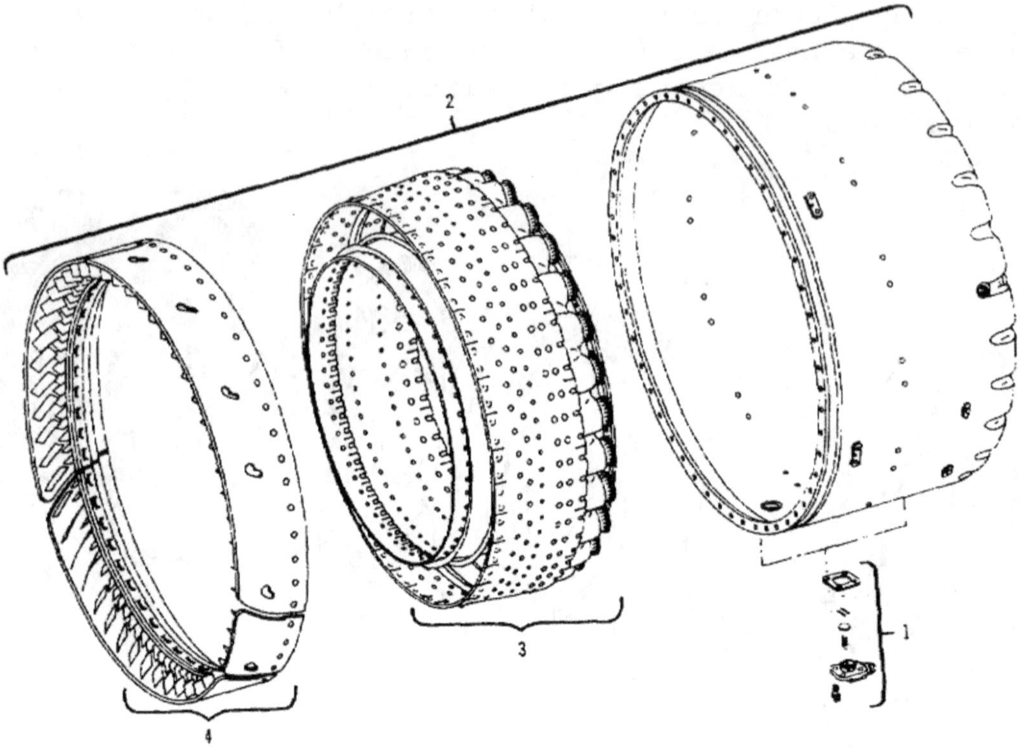

Figure 5.9. Combustion Chamber Housing Assembly.

5.13. TURBINE SECTION

The gas producer turbines on the T5 5-L-11 engine are two-stage, axial-flow turbines coupled together and mounted on the rear of the compressor rotor shaft. The gas producer turbine nozzles and rotors are shown in figure 5.10.

The gas producer turbine blades are air-cooled hollow blades held in place on the disk by pins. Ahead of the first and second stage rotors are the gas producer nozzles, which direct the hot gas onto the turbine rotors. The power turbine assembly is shown in figure 5.11.

The power turbine is a two-stage, axial-flow turbine coupled together and mounted on the power turbine shaft. The power shaft transmits power from the power turbines to the power output shaft. Basically, the function of the power turbine is to extract velocity energy from the hot gases and deliver mechanical power to the output shaft. The GP and PT systems are mechanically independent of each other; however, their speeds are controlled by the fuel control unit.

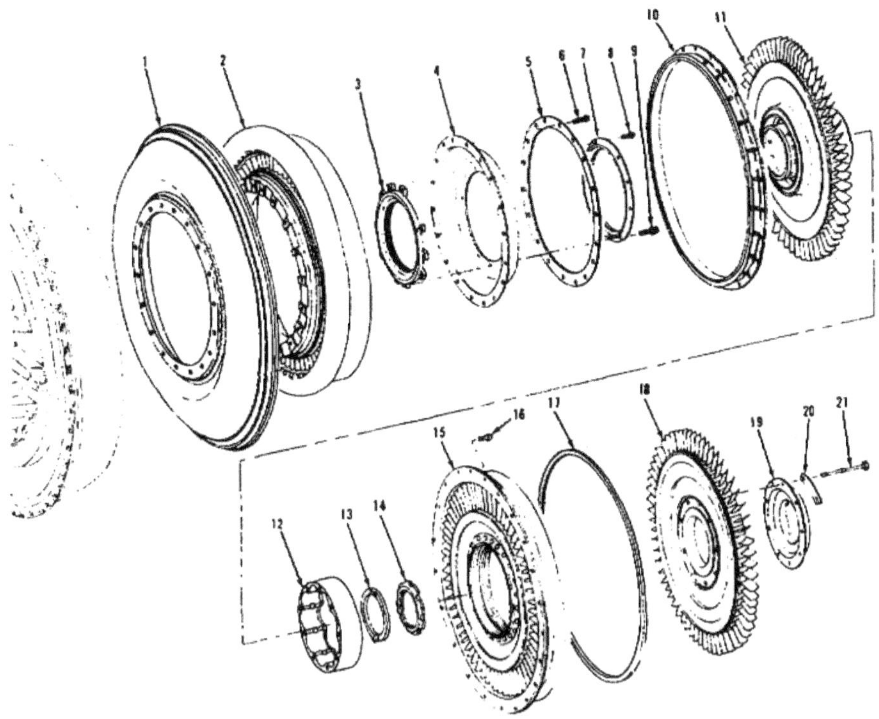

1. Diffuser Curl Assembly
2. First Turbine Nozzle Assembly
3. First Turbine Disc Seal Ring
4. No. 2 Bearing Housing Air Baffle
5. Rear Compressor Bearing Baffle Retainer
6. Bolt
7. First Turbine Cooling Disc Seal
8. Bolt
9. Bolt
10. First Turbine Rotor Cylinder Assembly
11. First Gas Producer Disc Assembly
12. First Gas Producer Turbine Spacer Assembly
13. Washer
14. Nut
15. Second Stage Gas Producer Nozzle Assembly
16. Bolt
17. Second and Third Turbine Nozzle Shim
18. Second Stage Gas Producer Rotor Disc Assembly
19. Second Stage Gas Producer Turbine Seal
20. Nut and Bolt Locking Plate
21. Bolt

Figure 5.10. Gas Producer Assembly.

5.14. DESCRIPTION OF FUEL SYSTEM

The fuel system of the T55- L-11 engine consists of the components illustrated in figure 5. 12. The T55-L-7 engine uses 14 vaporizing tubes in place of the 28 dual atomizing nozzles on the L-7C and L-11 engine. A cross-section view of the flow divider and atomizing nozzles may be seen by referring to figure 4.21 and 4.23 in chapter 4. The following subparagraphs discuss the components in the fuel system.

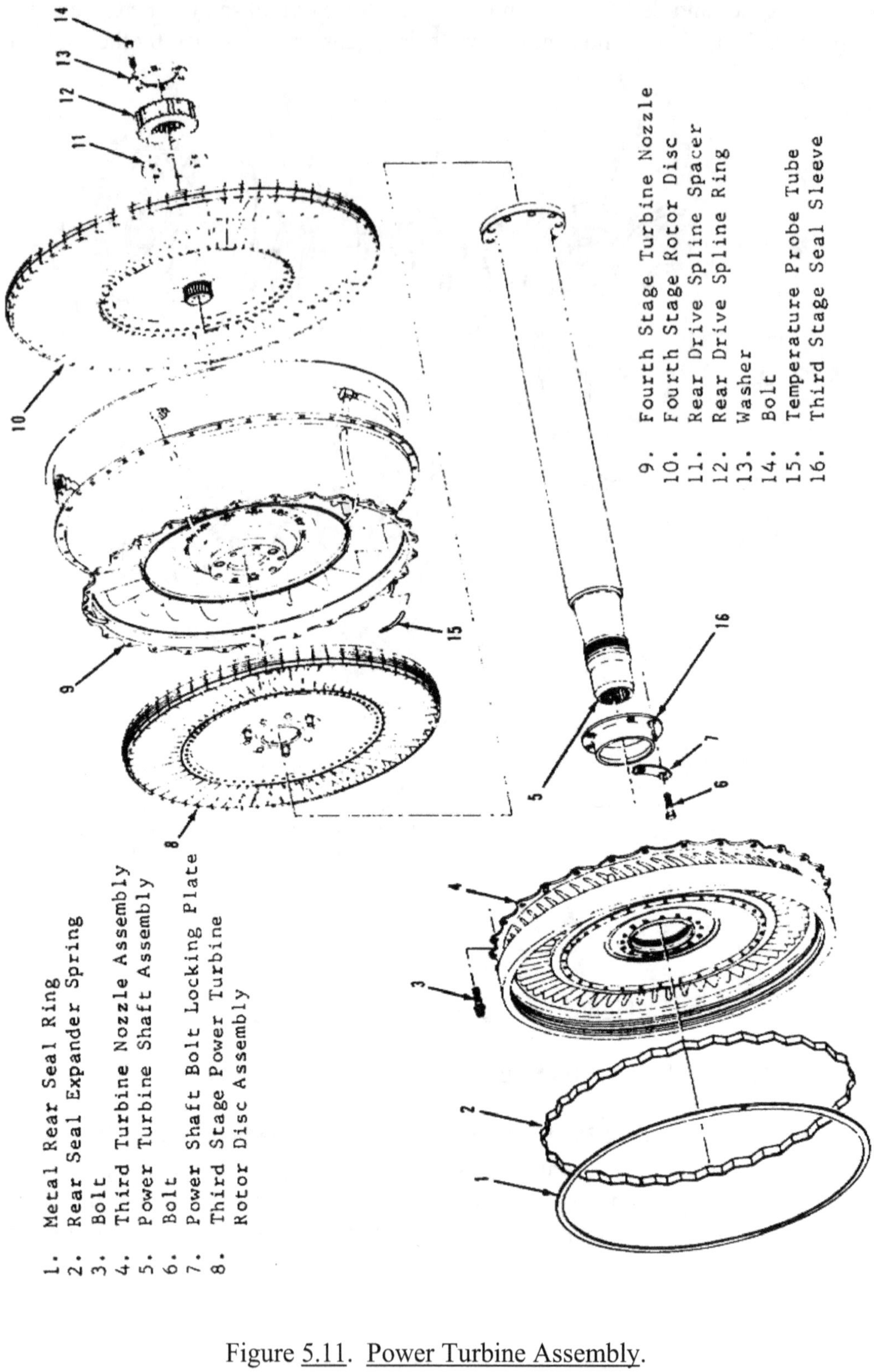

Figure 5.11. Power Turbine Assembly.

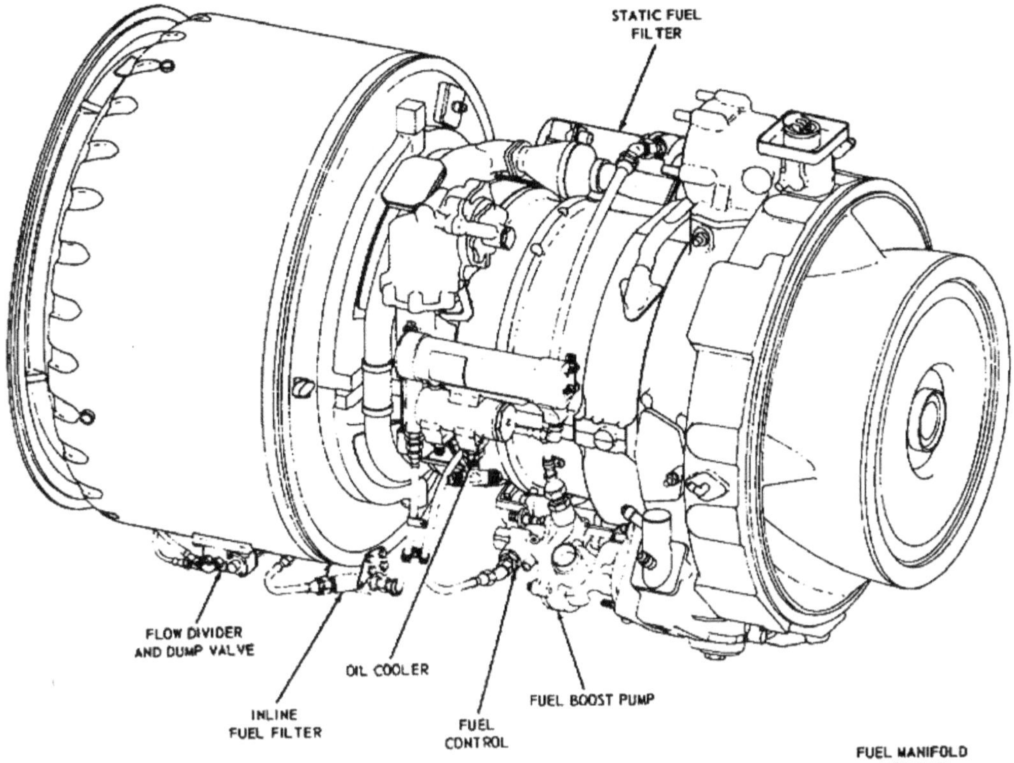

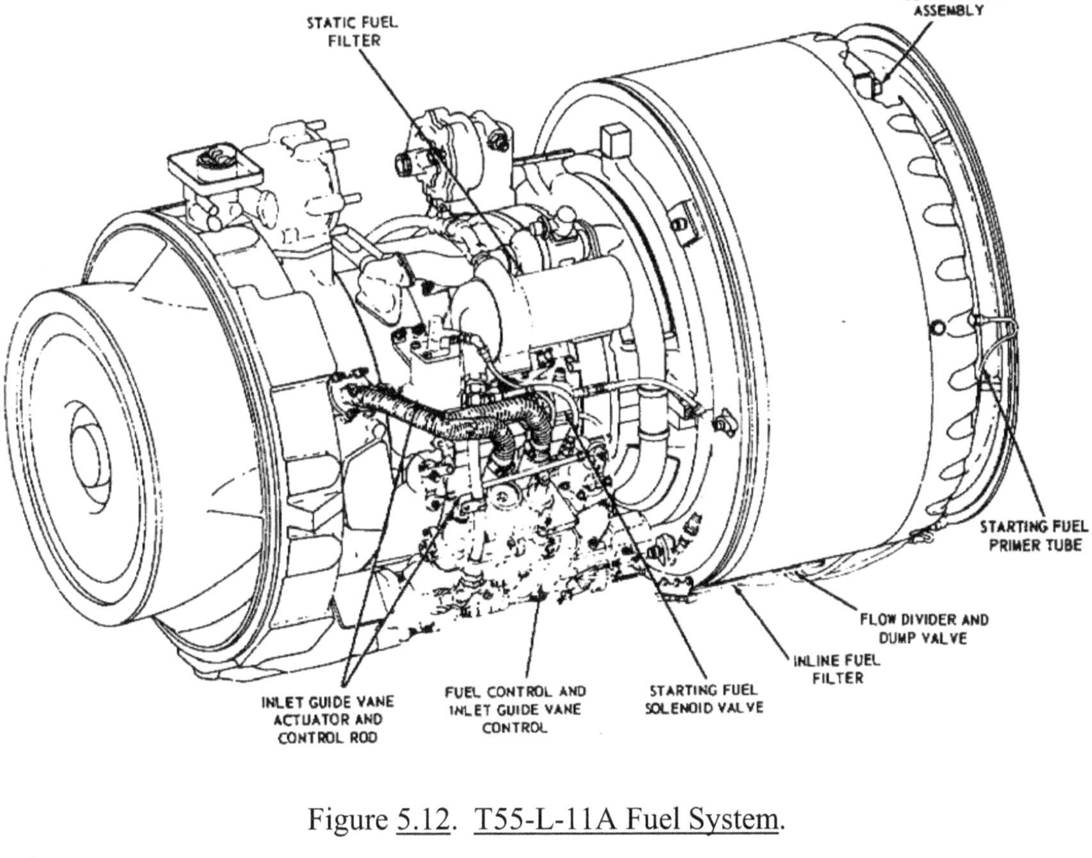

Figure 5.12. T55-L-11A Fuel System.

a. Fuel control. The hydromechanical fuel control contains a dual-element fuel pump, gas producer and power-turbine speed governors, an acceleration and deceleration control, an airbleed signal mechanism, and a fuel shutoff valve. Functionally, the fuel control unit can be divided into two sections: the flow control section and the computer section. A flow control schematic is shown in figure 5.13. The flow control section meters engine fuel flow. The computer section schedules the positioning of the metering valves in the flow control section. The computer section also signals the actuation of the compressor bleed band and the variable inlet vanes systems.

b. Liquid-to-liquid cooler. From the fuel control, fuel passes through the cooler mounted on the engine compressor housing. The cooler uses fuel to aid in cooling the engine oil. The cooler is a counter-flow type with the fuel passing through small aluminum tubes. The heat from the oil is transferred to the fuel. This type of cooler serves two purposes, to cool the oil and heat the fuel to aid in better atomization. A liquid-to-liquid cooler is illustrated in chapter 2, figure 2.12.

c. Start fuel system. The electrically operated, normally closed igniter fuel solenoid valve is located on the compressor housing at the 9 o'clock position. The solenoid valve is actuated by a switch in the cockpit that energizes the valve and allows fuel to flow from the fuel control unit through the starting manifold and to the start fuel nozzles. On T55-L-5 and T55-L-7 engines, two start nozzles are connected to the starting fuel manifold at approximately the 3 and 9 o'clock positions. Starting fuel passes through these nozzles into the combustion chamber where it is ignited by a spark from an igniter plug adjacent to each start fuel nozzle. On T5 5-L-7C engines, four start fuel nozzles are connected to the starting fuel manifold at approximately the 1, 4, 7, and 10 o'clock positions.

d. Main fuel system. The main and start fuel manifold are positioned on the rear surface of the combustion chamber assembly. The main fuel manifold consists of two manifold halves, with attaching points for 14 fuel vaporizer tubes. The main fuel manifolds for the T55-L-7 and L-7C/11A are shown in figure 5.14.

The T55-L-7C and T55-L-11 engines have an improved fuel system. This system illustrated in figure 5.15 consists of 28 dual-orifice fuel nozzles in a two-section main fuel manifold. A flow divider was added to meter fuel to the fuel nozzles at 9 to 10 percent N_1; as the rpm increases to 32 percent N_1, secondary fuel flows to the nozzles.

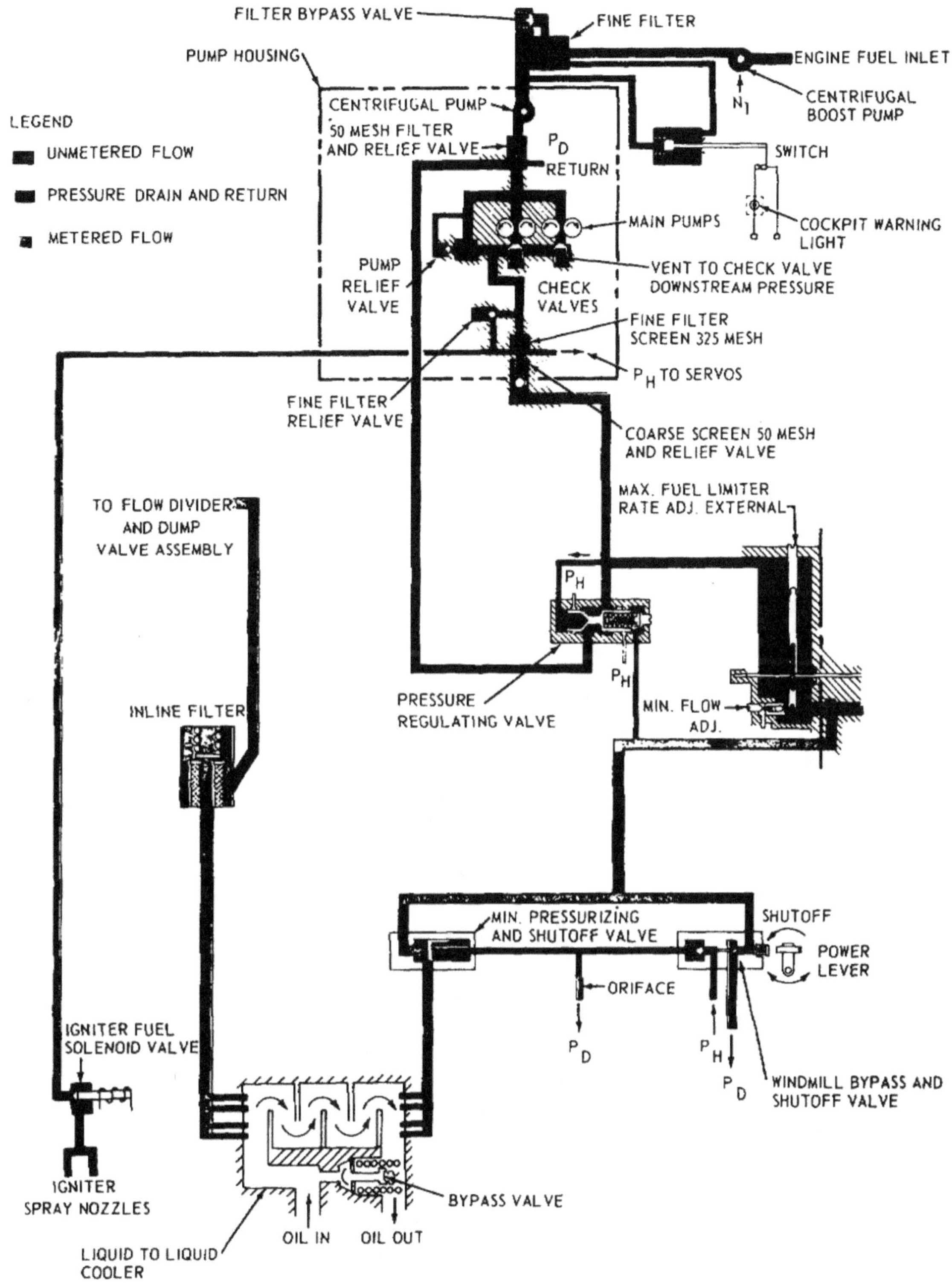

Figure 5.13. Fuel Control Flow Schematic.

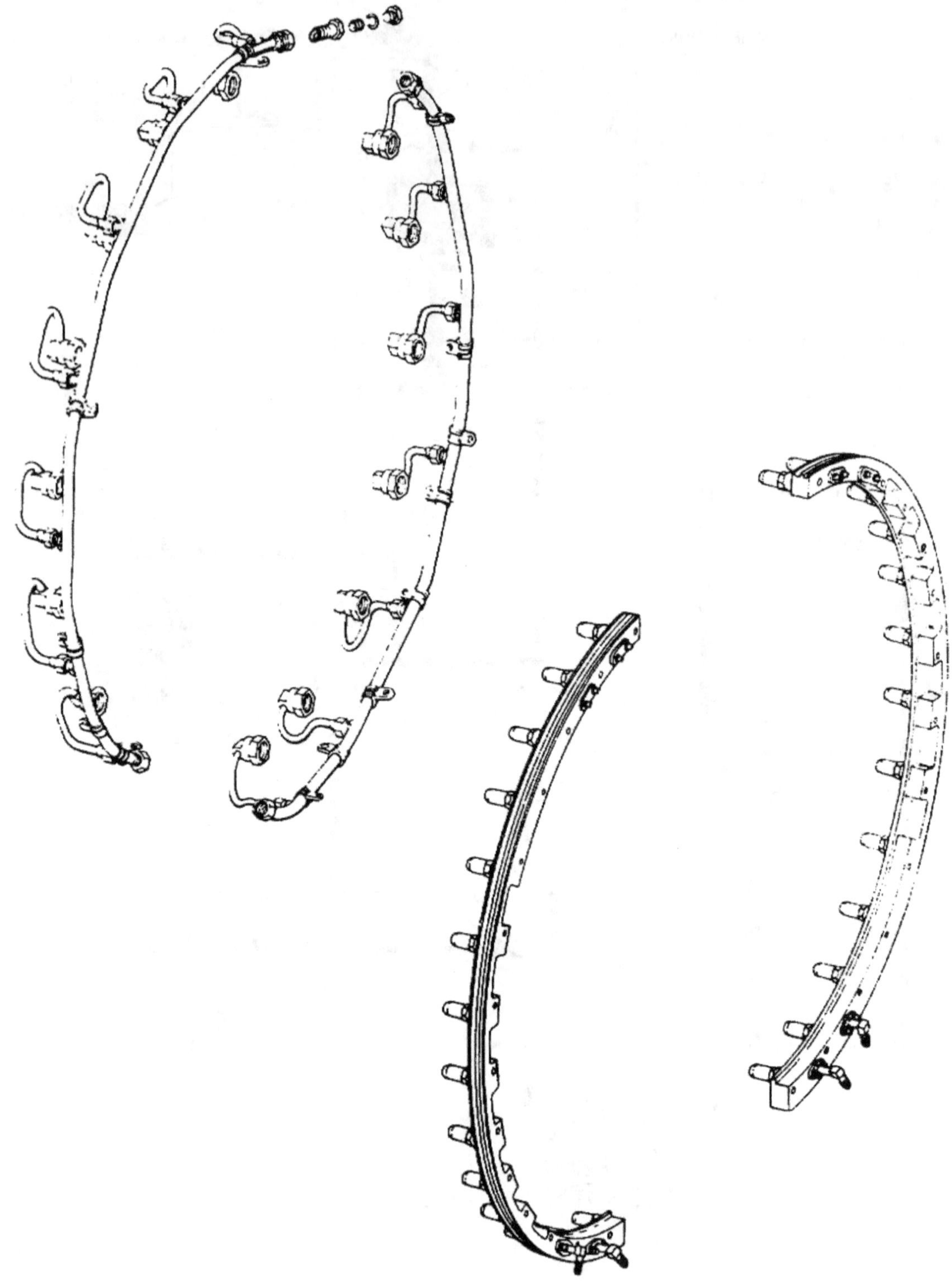

Figure 5.14. Main Fuel Manifolds.

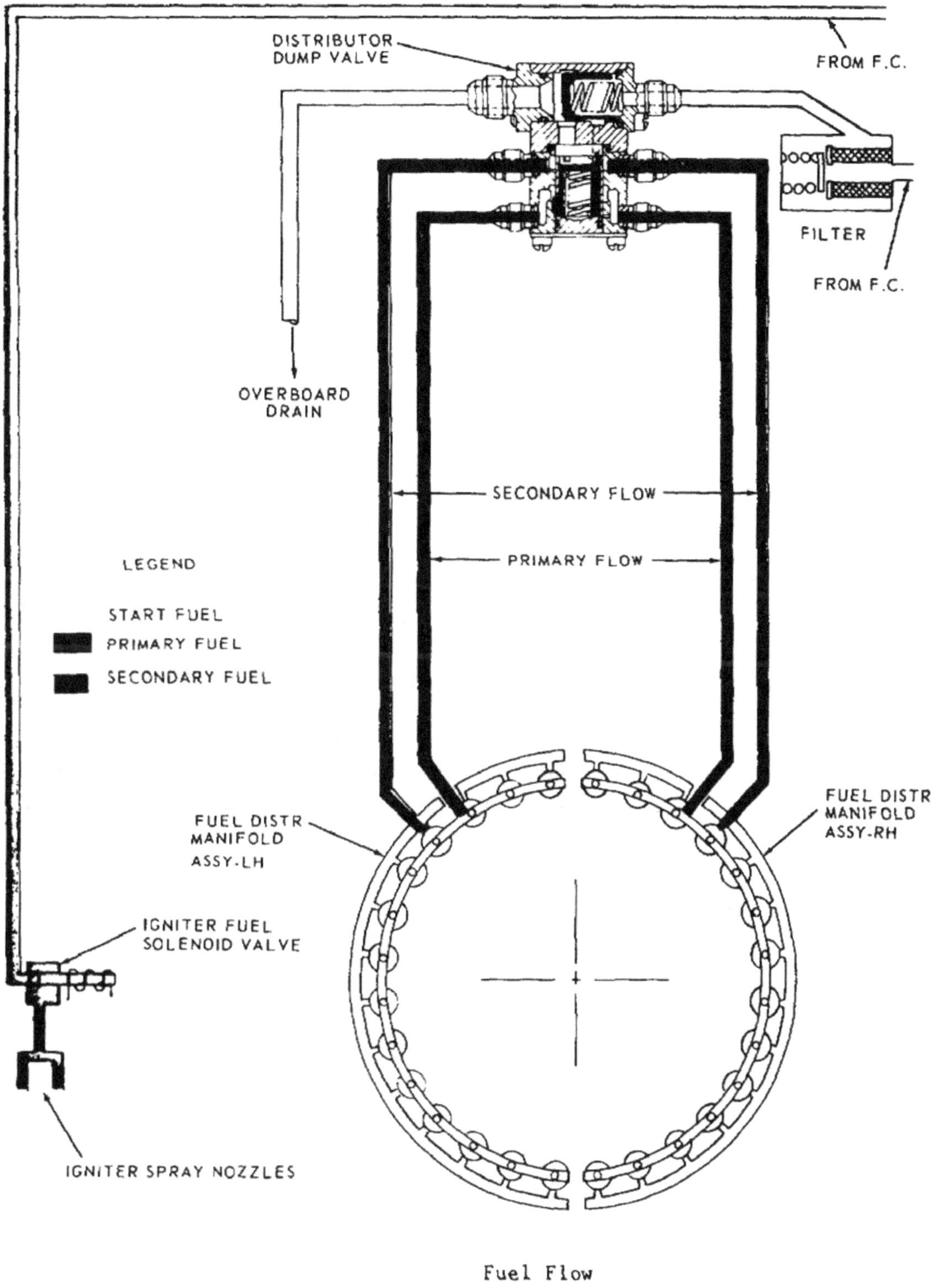

Figure 5.15. T55-L-11 Main and Start Fuel Flow.

5.15. INTERNAL COOLING AND PRESSURIZATION SYSTEM

The internal cooling system cools internal components and ensures extended engine service. A combination of several passages throughout the engine receives air from the main air-flow channel and directs it to cool components located within heat-generating areas. The exits from the cooling passages conduct the heated air to the main exhaust gas flow.

Some of the cooling air is extracted for bearing seal pressurization. This internal airflow is guided to the appropriate bearing seal to protect against oil seepage while the engine is in operation.

5.16. ANTI-ICING SYSTEM

The engine inlet is protected from ice forming on it by the anti-icing system. The walls and struts of the inlet housing have internal passages through which hot, scavenged engine oil circulates. The variable inlet guide vanes are supplied with hot air which is extracted from the centrifugal compressor. The engine hot-air anti-icing system is shown in the schematic in figure 5.16.

This air first passes through a hot-air valve and is distributed by a tube which directs the air into an annulus (circular structure) around the inlet-guide vane assembly. The anti-icing air is routed through the stem of the vanes and is discharged at the base of the leading edge into the main airflow.

5.17. VARIABLE INLET GUIDE VANE SYSTEM

The inlet guide vane assembly, located in front of the first compressor rotor, consists of a series of hollow blades positioned mechanically by a hydraulically operated synchronizing ring. The guide vane control system schedules the positions of the variable inlet guide vanes in response to gas producer speed and compressor inlet temperature. At low N_1 speeds, a high angle of inlet air is required, and the inlet guide vanes are in the closed position of approximately 45.5° of the engine centerline. Guide vane positions are illustrated in figure 5.17.

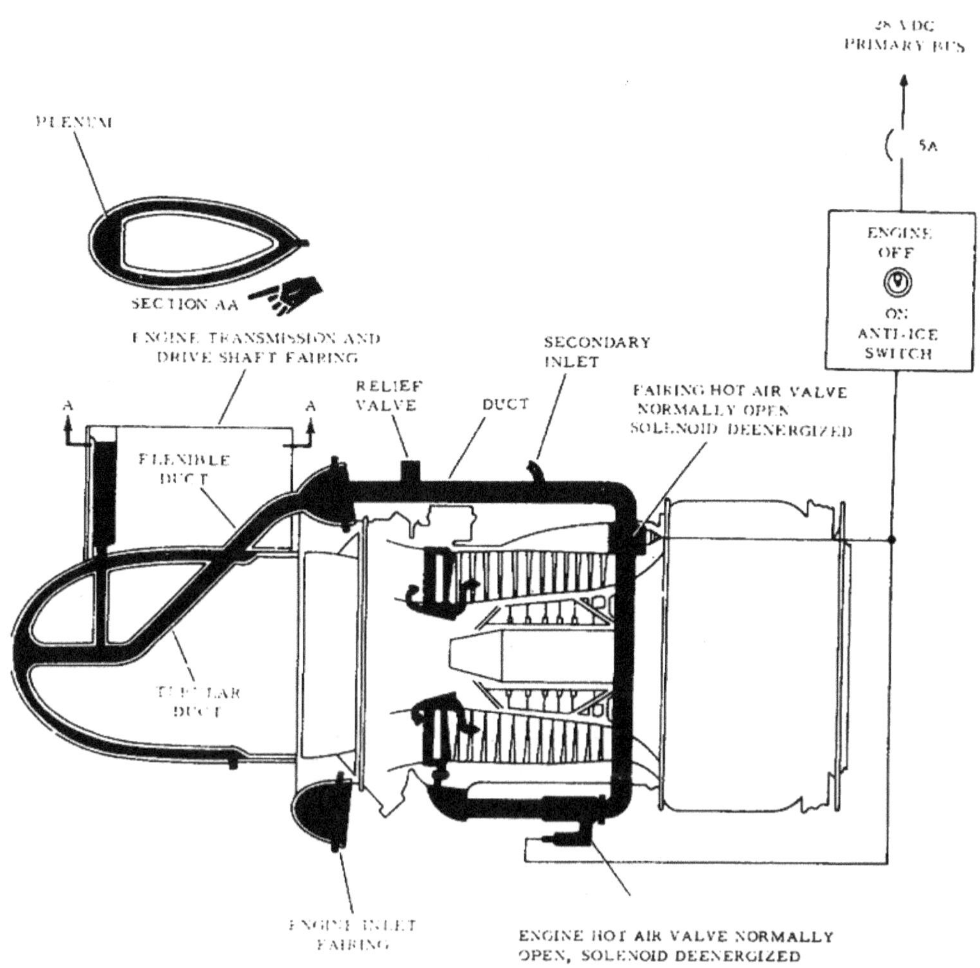

Figure 5.16. Hot-Air Anti-icing System.

When engine speed increases above 65 percent, the inlet guide vanes begin to open. The progression toward the open position is proportional to compressor speed. At 98 percent speed, the vanes are opened to 7° off the centerline, and as speed increases to 100 percent, the vanes cross the centerline to the -4.5° position. At any steady state N_1 speed, between 65 percent and 98 percent, the inlet guide vanes assume a constant position.

5.18. INTERSTAGE AIR BLEED SYSTEM

An interstage air bleed actuator assembly and bleed band, illustrated in figure 4.27, are used to facilitate compressor rotor acceleration and avoid compressor stalls. A series of vent ports around the compressor housing, at the sixth stage of compression,

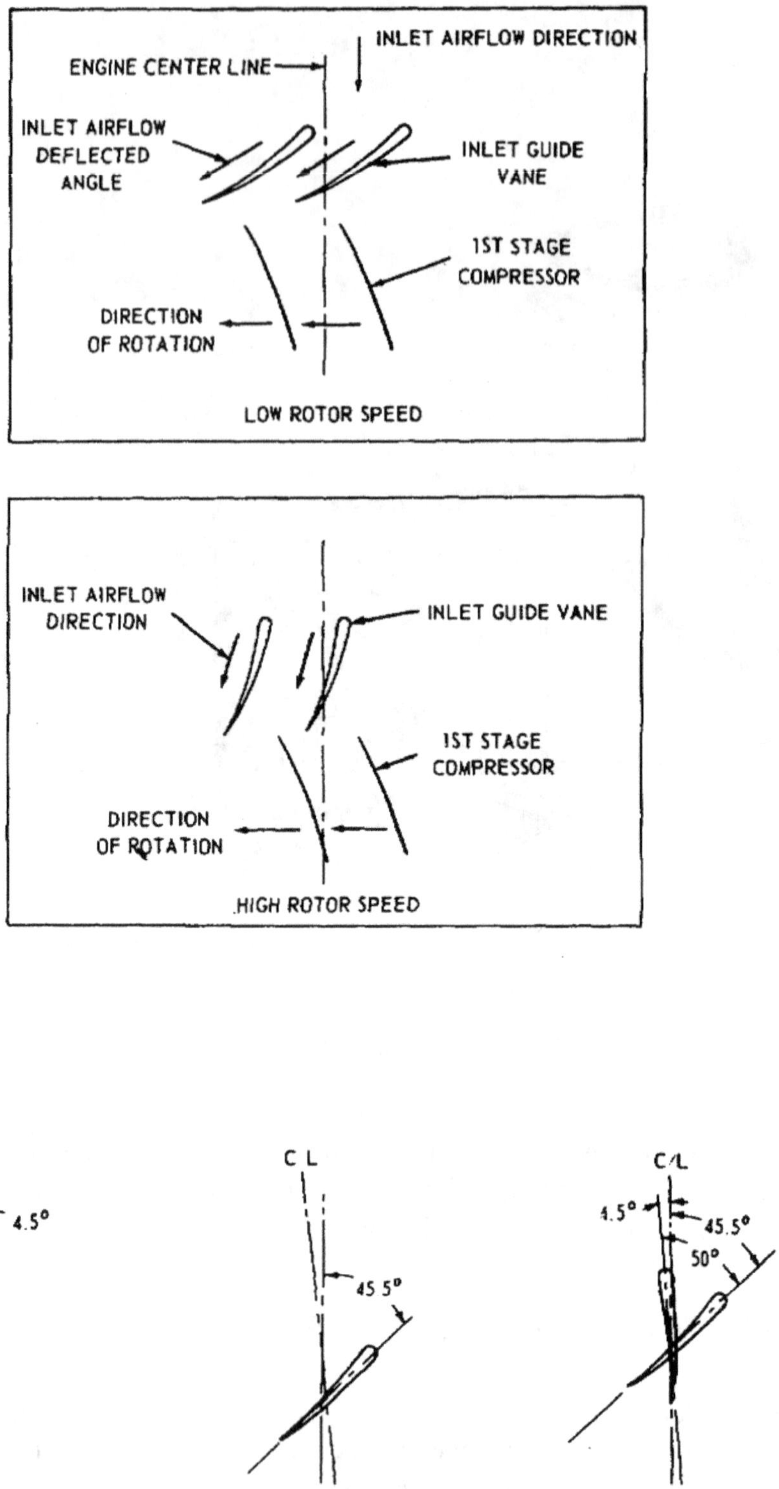

Figure 5.17. Angles of Variable Inlet Guide Vane.

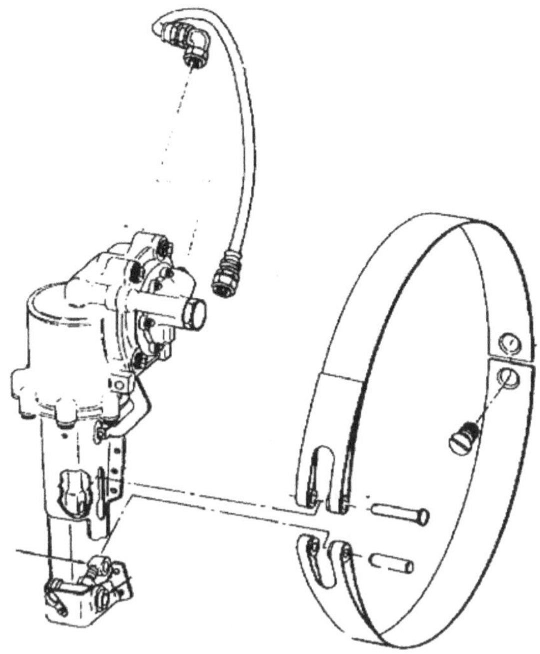

Figure 5.18. Interstage Airbleed Band Assembly.

permits compressor air to bleed off and allows a more rapid acceleration. The pneumatic interstage-actuator assembly controls operation of the air bleed system by tightening or loosening a metal band over the vent holes in the compressor housing. The bleed band and actuator assembly are shown in figure 5.18. The fuel control is equipped with an acceleration air bleed adjustment that sets the compressor rotor speed at which the interstage air bleed closes. This is usually factory set to close the bleed band at approximately 30 percent of normal rated power.

5.19. LUBRICATION SYSTEM

The engine lubrication system has the dual function of lubricating and cooling. The principal components of the system are the integral oil tank, dual element lubrication pump, filter and screen assemblies, oil cooler assembly, oil level indicator assembly, and internal scavenge pumps. The entire lubrication system is self-contained within the engine. A 3.75-gallon oil tank is located in the engine air inlet housing. The tank filler neck is located at the top center of the inlet housing. Oil level indication is taken by means of an externally-mounted mechanical indicator. Two connections, one at the top and one at the bottom of the tank, are for the addition of an external oil tank to increase the oil capacity. The following subparagraphs discuss the lubrication system and components.

a. Oil level indication system. The oil tank has an oil level indicator mounted on the left side of the inlet housing at the 9-o'clock position and can be read from the top or the side. The indicator contains a low-level warning switch for remote indication in the cockpit. Based on using the maximum allowable oil consumption rate, the switch is set to signal when there is a 2-hour supply of usable oil remaining.

b. Oil flow. Oil from the tank flows from the bottom of the engine inlet housing through an external line to the pressure side of the main oil pump. The oil flow may be followed by referring to the schematic shown in figure 5.19.

The main oil pump is located on the aft face of the accessory gearbox. I contains both pressure and scavenge elements. An adjustable relief valve in the pump maintains nearly constant oil pressure during engine operation. Oil goes from the pump and through a filter located in the accessory gearbox. A bypass valve will open if the filter becomes clogged. On the outlet side of the filter, oil temperature is measured by a temperature bulb. Then the oil is routed through the oil cooler into two external low paths. One flow path directs the oil to the rear of the engine. At that location it lubricates number 2, 4, and 5 bearings. The second flow path directs oil to the front bearings and accessory drive gear trains. Oil from the rear bearings is force-scavenged into an external oil-return line by paddle pumps mounted on the power turbine shaft. This scavenge oil is directed back to the accessory gearbox.

A scavenge impeller in the accessory gearbox picks up the scavenge oil and pumps it to the scavenge element of the main oil pump. The oil is then returned to the inlet housing where it is discharged into the oil tank.

c. Chip detectors. On the T55-L-7, the engine magnetic chip detector is located in the scavenge-pump housing on the lower left face of the accessory gearbox. The detector attracts ferrous material, which builds up until it bridges a gap, as shown in the accompanying sketch. This makes it possible to check for the presence of foreign material by checking continuity across the contacts. The detectors are electrically connected to caution lights in the cockpit.

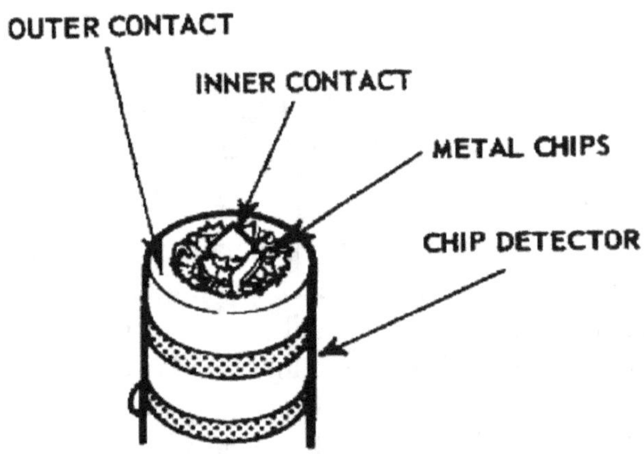

The T55-L-11 engine is equipped with three chip detectors. Each bearing scavenge line dual chip detector is mounted externally. Each detects ferrous and nonferrous metal chips originating in the No. 2, 4, and 5 bearing areas of the engine. Another is located near the gearbox scavenge impeller.

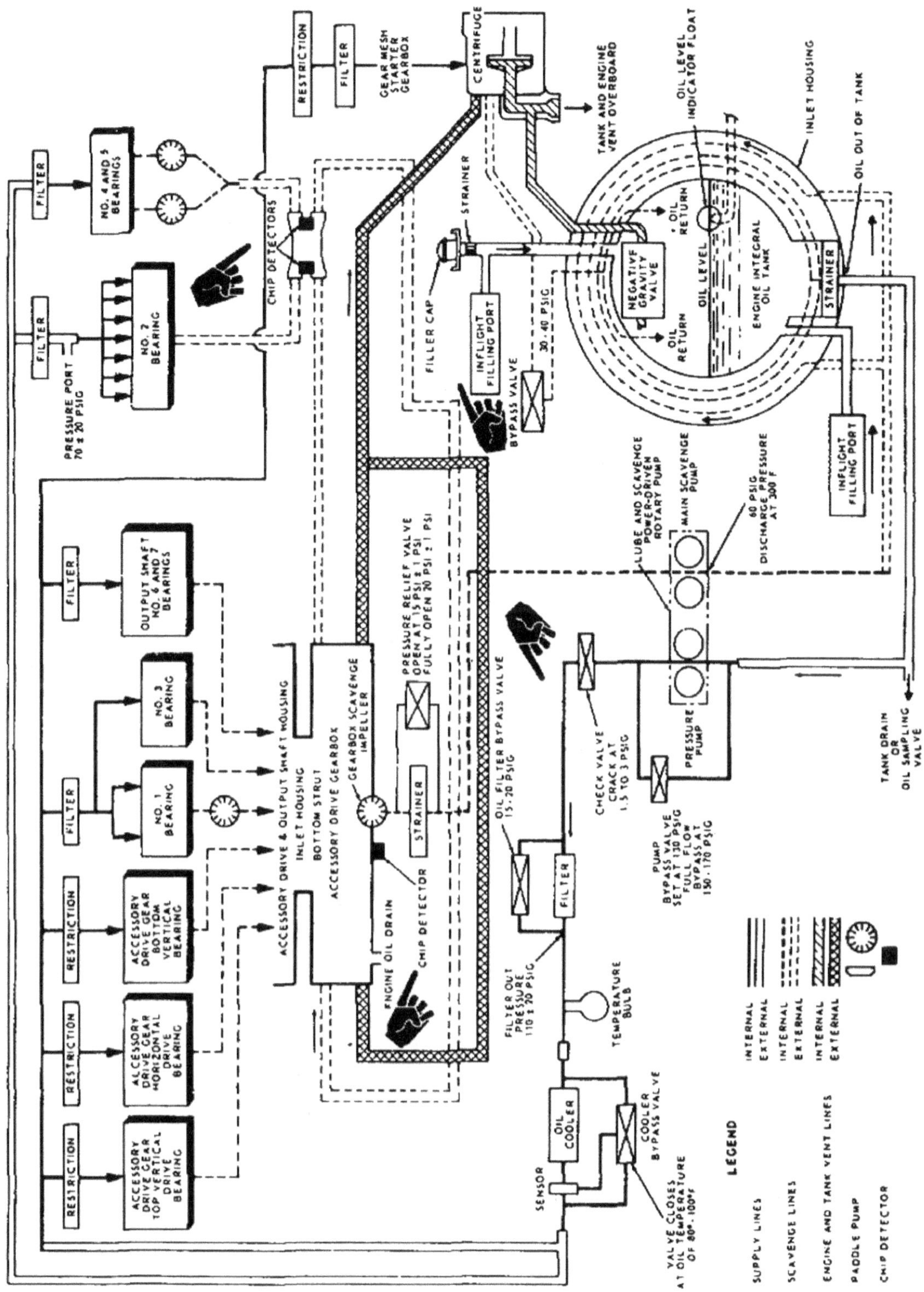

Figure 5.19. Lubrication System Schematic, T55-L-11.

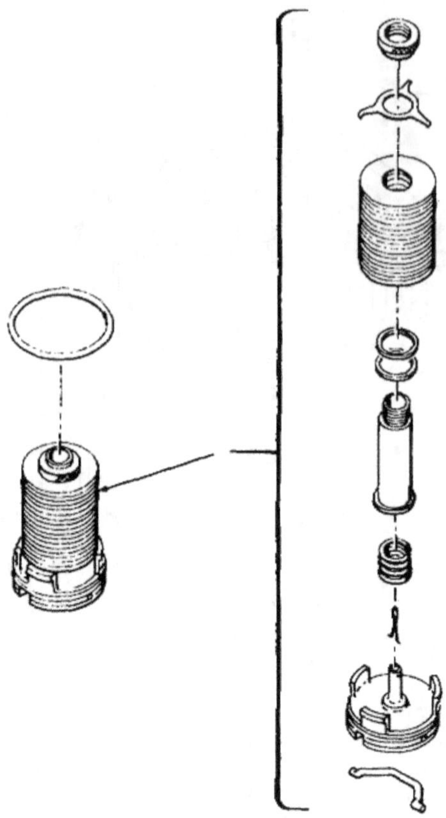

Figure 5.20. Main Oil Filter.

d. Engine oil filters. The oil filter system consists of a main oil filter assembly and five secondary oil filters. Three of these filters are accessible for intermediate level maintenance. These are the No. 2 bearing strainer, located at the 3 o'clock position on the forward face of the diffuser housing assembly, and the No. 4 and 5 bearing strainers located in the supply tube within the exhaust diffuser. The main oil filter assembly, illustrated in figure 5.20, is mounted on the bottom of the accessory gearbox directly in front of the main oil pump, or fuel boost pump.

5.20. TORQUE METER SYSTEM

The Lycoming Electric Torquemeter used on the T55 engine is the same type that is used on the T53-L-701 covered in the previous chapter. To save describing the system in detail again, this paragraph is a brief review. When torque is imposed on the engine power output shaft, tension and compression stresses change the magnetic reluctance of the shaft. This change in magnetic reluctance is transmitted to an indicator in the cockpit and read as percent (%) of torque.

5.21. ELECTRICAL SYSTEM

The engine electrical system includes circuitry to facilitate starting, ignition, anti-icing, and all engine-oriented electrical monitoring devices. The following subparagraphs discuss the operation of the engine electrical system.

a. Main electrical cable assembly. The main electrical cable assembly furnishes all necessary interconnecting wiring between the main disconnect plug and the nine-branch electrical connectors. The nine electrical accessories served by this cable are the gas producer tachometer generator, oil temperature bulb, fuel flow pressure switch, torquemeter system, ignition exciter, anti-icing solenoid valve, starting fuel solenoid valve, oil level indicator, and power turbine tachometer generator. The main disconnect plug mates with an electrical receptacle of the airframe wiring, establishing electrical continuity to the various airframe components.

b. Ignition exciter. The high-potential ignition pulse is developed by the ignition exciter; 24 volts dc is applied to the input of the exciter. Current flows through the primary transformer winding the bias coil and the vibrator points to ground. This generates magnetic lines of force which permeate the transformer core and the core of the bias coil attracting the vibrator reed upward and interrupting the circuit. As current flow ceases, the lines of force collapse and the reed falls back, closing the circuit. This cycle repeats at a rate proportional to the input voltage. The resultant current flows in pulses, causing magnetic lines of force to build up and collapse with each pulsation. These lines induce voltage across the secondary coil which is transformed to a higher potential by an increased number of windings comprising the secondary. The diodes rectify the pulsating current back into direct current to charge the capacitors. The charge on the capacitors continues to build up at a rate proportional to input voltage, until a potential of 2,500 volts exists. The calibrated spark gaps ironize at this voltage, creating an electrical path for the firing pulse. The capacitors discharge through this path into the lead and coil assembly for distribution to each of the spark igniters.

Radio frequency energy is generated within the exciter during normal operation. An inductive capacitive filter has been incorporated at the input to prevent this energy from being fed back into the 24-volt input line. Radio frequency interference on this line could be detrimental to the operation of other electrical accessories. This filter is tuned to radio frequencies and does not offer any appreciable opposition to the flow of 24-volt direct current.

c. Ignition lead and coil assembly. The ignition lead and coil assembly constitutes the high-potential ignition wiring. This assembly incorporates two coils, fed with high voltage from the two outputs of the ignition exciter. The coil assemblies function as spark splitters distributing high voltage to four igniter plugs. Each coil assembly has one input and two outputs with the coil windings forming a transformer having a 1:1 ratio. Any current flowing through either winding will induce a voltage across the other so that even a shorted igniter plug will not short out the high-voltage ignition signal. The entire wiring harness is shielded and grounded at the airframe to suppress radio frequency interference.

d. Exhaust gas temperature harness. The chromel-alumel, thermoelectric measuring system is independent of all other engine electrical wiring. The engine components are ten thermocouple probes to the shielded harness. The aircraft wiring, spool resistor, and indicator complete the system. The ten thermocouple probes protrude into the gas flow of the engine at station number 7. The probes react to variations in power turbine entry temperature by developing a proportional electromotive force across the chromel-alumel junction. This electromotive force results in meter deflection of the cockpit indicator calibrated to read temperature in degrees centigrade.

5.22. SUMMARY

The T55 series gas turbine engine is used to power the CH-47 Chinook helicopter. The CH-47A and B are equipped with the T55-L-7 series engines, and the CH-47C is equipped with the T55-L-7C or T55-L-l1. These are all basically the same except for shaft horsepower ratings and internal details. The fuel system includes starting and main fuel components. The fuel control is a hydromechanical device that automatically meters the proper amount of fuel under varying atmospheric conditions and power requirements. The engine has its own lubrication system with the oil tank contained within the inlet housing. It is equipped with an interstage bleed air system to facilitate acceleration and avoid compressor stalls. The anti-icing system prevents icing by ducting hot air from the diffuser section to the engine inlet. The engine has a starting and ignition system, and its performance is monitored by instruments on the cockpit instrument panel.

Chapter 6

SOLAR T62 AUXILIARY POWER UNIT

6.1. INTRODUCTION

The SOLAR T62 auxiliary power unit (APU) is used in place of ground support equipment to start some helicopter engines. It is also used to operate the helicopter hydraulic and electrical systems when this aircraft is on the ground, to check their performance. The T62 is a component of both the CH-47 and CH-54 helicopters -- part of them, not separate like the ground-support-equipment APU's. On the CH-54, the component is called the auxiliary powerplant rather than the auxiliary power unit, as it is on the CH-47. The two T62's differ slightly.

This chapter describes the T62 APU; explains its operation; discusses the reduction drive, accessory drive, combustion, and turbine assemblies; and describes the fuel, lubrication, and electrical systems.

6.2. DESCRIPTION

The T62 gas turbine engine auxiliary unit consists of a combustor, turbine, reduction drive and accessory drive assemblies, engine accessories, plumbing, and wiring. The engine has a single shaft with the compressor and turbine rotor mounted back-to-back.

The T62 develops approximately 70 shaft horsepower. It has its own fuel control unit, hydraulic starter motor, ignition unit, and reduction gear drive. Operating time for the APU is maintained separately from aircraft engine time by an hour meter mounted on the APU. The T62 used on the CH-47 differs slightly from the one used on the CH-54. The accompanying table IV gives the particulars for each engine.

Both models of the T62 are shown in figures 6.1 and 6.2.

6.3. THEORY OF OPERATION

The T62 gas turbine engine consists of three major sections: the reduction and accessory drives, the combustor, and the turbine sections as shown in figures 6.1 and 6.2. Air is drawn into the inlet of the engine when the hydraulic starter rotates the compressor during the starting cycle. After the engine is started, air continues

Table IV. T62 Engine Leading Particulars.

T-62 Engine Leading Particulars

	T-62T-2A (CH-47)	T-62T-16A (CH-54)
Powerplant		
Weight (dry)	73.00 pounds	70.30 pounds
Weight (wet)	78.81 Pounds	76.10 pounds
Dimensions		
Height	19.03 inches	23.00 inches
Width	18.45 inches	14.50 inches
Length	31.57 inches	26.10 inches
Engine Operating Speed	56,000 rpm	56,700 rpm
Specific Fuel Consumption	1.3 lb/shp-hr	1.3 lb/shp-hr
Exhaust Gas Temperature (Max)	1,090°F	1,090°F
Starting Limits		
Sea Level	-65°F to +125°F	-65°F to +125°F
6000 feet	-65°F to +95°F	-65°F to +95°F
15,000 feet	-65°F to +45°F	-65°F to +45°F
Operating Limits		
Sea Level	-65°F to +125°F	-65°F to +125°F
6000 feet	-65°F to +125°F	-65°F to +125°F
Reduction Gear and Accessories		
Input Speed (rated)	56,000 rpm	56,700 rpm
Output Speed (at rated Input)	6,000 rpm	8,000 rpm
Fuel Control Assembly	4,200 rpm	4,118 rpm
Speed Switch	4,200 rpm	4,118 rpm
Starter Motor/2nd Stage Pump	6,000 rpm	8,000 rpm
Fuel and Lubricants		
Fuel	JP-4, JP-5, & AV/Gas	JP-4, JP-5, & AV/Gas
Lubricating Oil	MIL-L-23699, MIL-L-7808	MIL-L-23699, MIL-L-7808
Oil Capacity	3 US quarts	3 US quarts
Fuel Consumption		
No Load	49 phr	49 phr
Normal (Full Load)	85 phr	85 phr
Oil Consumption (Max)	0.10 phr (1.7 FL. OZ)	0.10 phr (1.7 FL. OZ)
Instrumentation		
Engine Operating Time	Hour Meter	Hour Meter
Exhaust Gas Temperature	None	Pyrometer

Figure 6.1. T62T-2A Used on the CH-47.

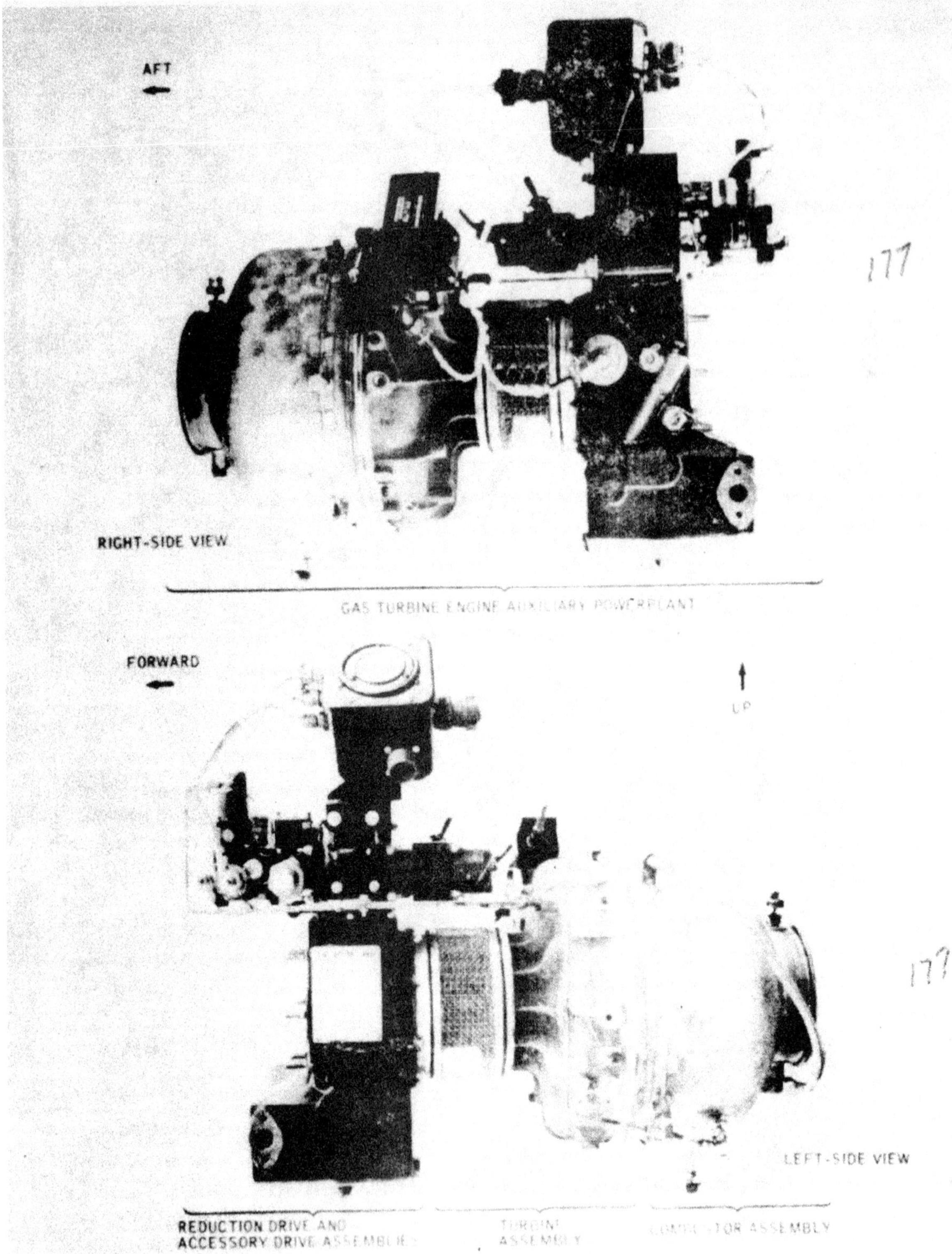

Figure 6.2. T62T-16A Used on the CH-54.

to be drawn into the compressor by the power produced by the engine. The air is compressed and directed into the combustor; fuel is introduced through six vaporizer tubes and is burned. During the starting cycle, fuel from the start nozzle is ignited by a spark plug. When the APU reaches 90 percent speed, a speed switch opens which closes the start fuel solenoid valve, shutting off fuel flow to the start fuel nozzle. The hot expanding gas flows through a turbine nozzle and to the turbine assembly. Power is extracted by the turbine rotor and is then transmitted to the reduction drive assembly.

6.4. PROTECTIVE DEVICES FOR THE APU

The T62 is equipped with protective devices that shut the APU down if any of the operating limitations are exceeded. If the oil pressure drops below approximately 6 psi, the low oil pressure switch shuts off the fuel solenoid valve, which stops the flow of fuel to the engine. When this occurs the light marked "Low Oil Press" will be illuminated on the instrument panel in the cockpit.

Overspeeding the APU is prevented by an overspeed switch, set at 110 percent. When this limit is exceeded, the overspeed switch opens and shuts off the fuel flow to the engine. This causes the overspeed (OVSP) light to illuminate on the instrument panel.

Hot starting of the APU is prevented by the high exhaust-temperature switch, which shuts the engine off if the engine is too hot. Like the low oil pressure and overspeed switches, it stops the fuel flow to the engine. When the operating temperature is exceeded, the high exhaust temperature light, marked "HIGH EXH TEMP," in the cockpit will be illuminated.

If any of these conditions occur and the APU shuts down, the control switch in the cockpit must be moved to the STOP position before attempting to restart the engine.

6.5. REDUCTION DRIVE AND ACCESSORY DRIVE ASSEMBLIES

The major difference between the T62T-2A and the T62T-16A is the accessory drive housing. A comparison can be made by looking at item 31 in figures 6.3 and 6.4. Reduction of high engine rpm is accomplished by using a single stage planetary gear reduction. Three fixed-center planetary gears are driven by the externally splined pinion of the rotor shaft, item 7. An internally splined ring gear transmits the drive from the planetary gears to the output shaft. External gear teeth on the output portion of the shaft drive an upper intermediate gear and oil pump gear. The intermediate

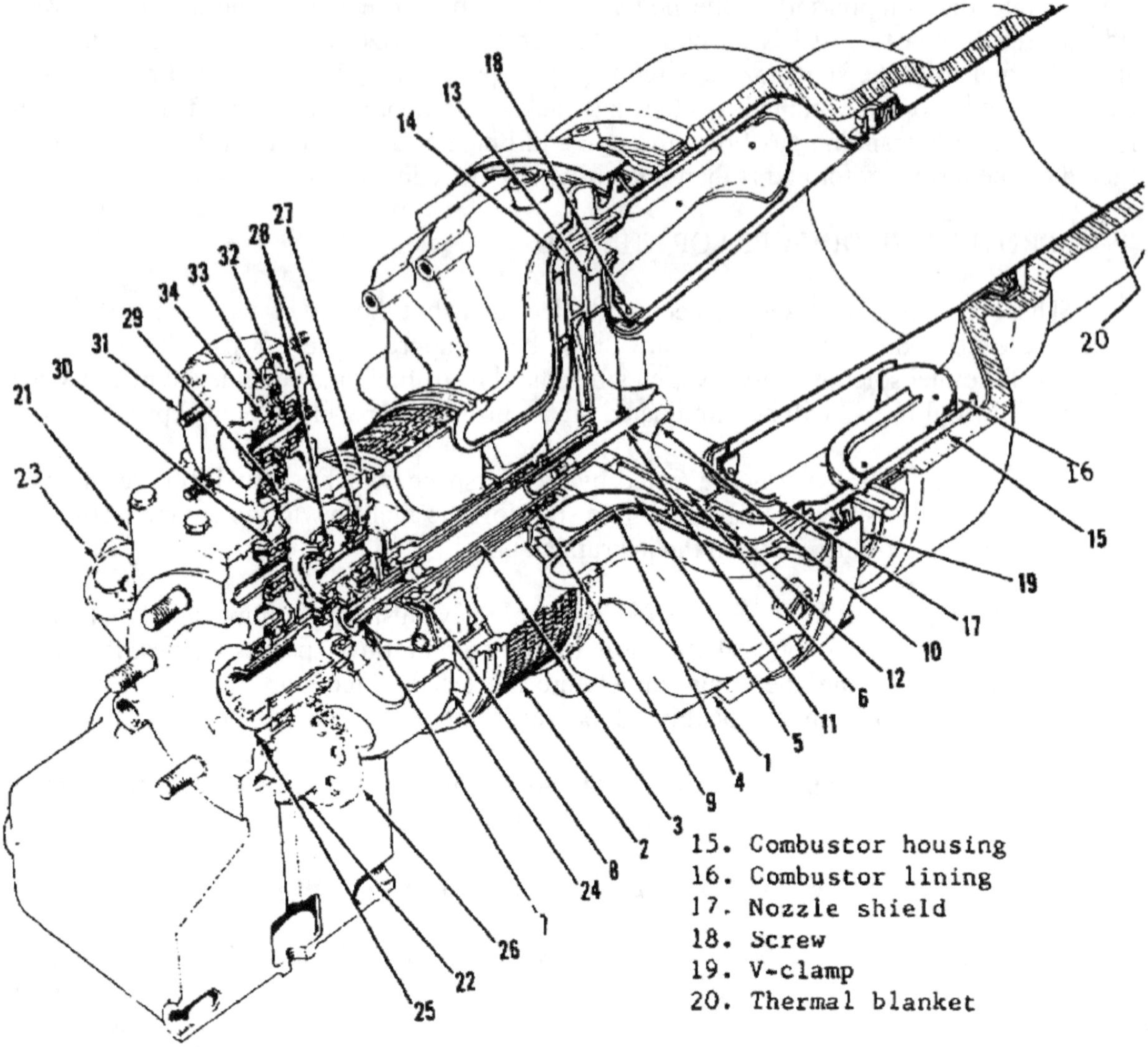

15. Combustor housing
16. Combustor lining
17. Nozzle shield
18. Screw
19. V-clamp
20. Thermal blanket

1. Air inlet housing
2. Wire-mesh screen
3. Rotor shaft
4. Inducer
5. Compressor wheel
6. Turbine wheel
7. Input pinion
8. Forward ball bearing
9. Aft roller bearing
10. Exducer
11. Bolt
12. Compressor-to-turbine seal
13. Turbine nozzle
14. Diffuser
21. Reduction drive housing
22. Oil pump
23. Oil filter
24. Ring gear
25. Output shaft
26. Oil pump drive gear
27. Bearing carrier housing
28. Ball bearing
29. Intermediate accessory drive gear
30. Double row ball bearing
31. Accessory drive housing
32. Accessory drive gear
33. Oil separator plates
34. Ball bearings

Figure 6.3. T62T-2A Cutaway Engine Assembly.

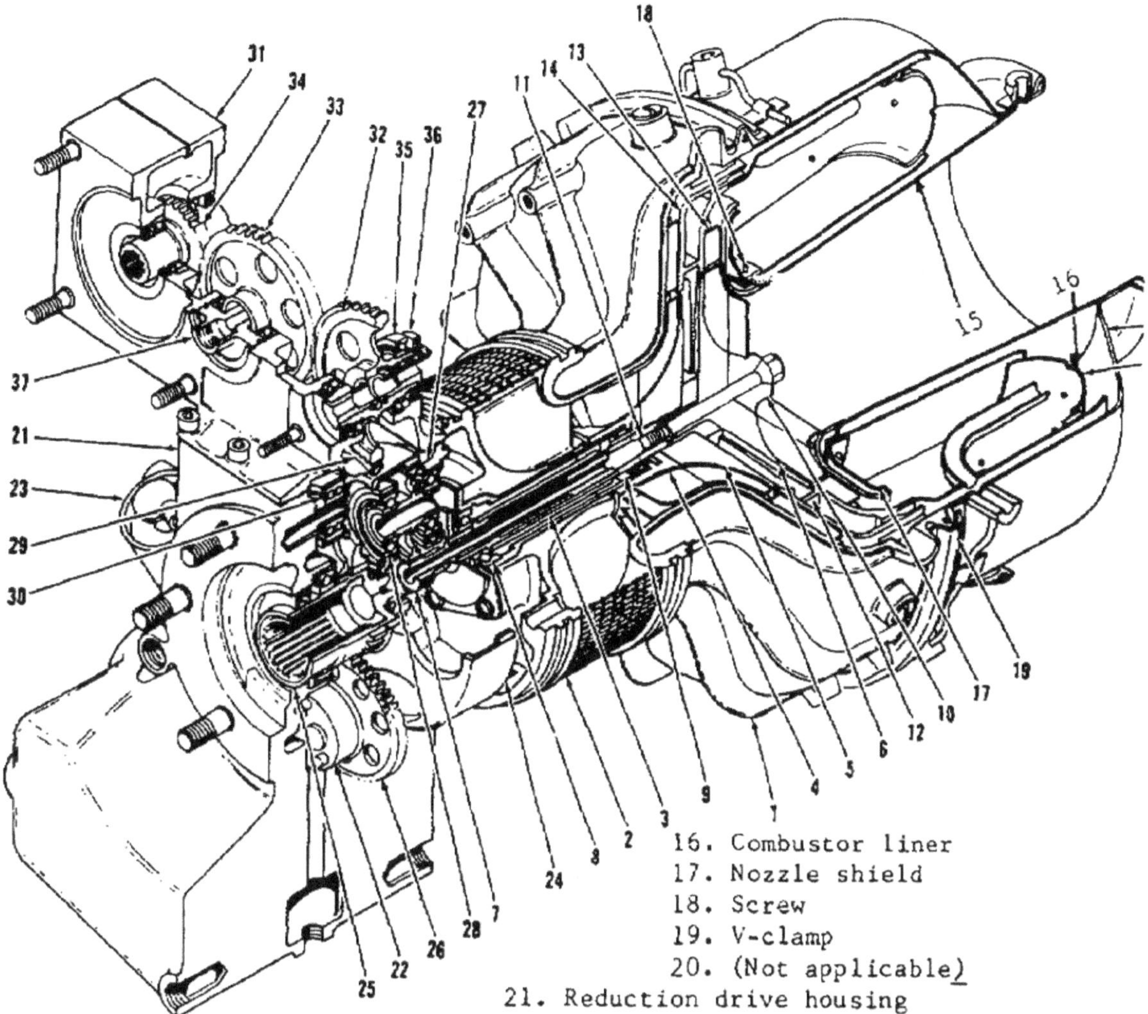

1. Air inlet housing
2. Wire-mesh screen
3. Rotor shaft
4. Inducer
5. Compressor wheel
6. Turbine wheel
7. Input pinion
8. Forward ball bearing
9. Aft roller bearing
10. Exducer
11. Turbine bolt
12. Compressor-to-turbine seal
13. Turbine nozzle
14. Diffuser
15. Combustor housing
16. Combustor liner
17. Nozzle shield
18. Screw
19. V-clamp
20. (Not applicable)
21. Reduction drive housing
22. Oil pump
23. Oil filter
24. Ring gear
25. Output shaft
26. Oil pump drive gear
27. Bearing carrier housing
28. Ball bearing
29. Intermediate accessory drive gear
30. Double row ball bearing
31. Accessory drive housing
32. Accessory drive gear
33. Idler gear
34. Starter gear
35. Ball bearing
36. Seal
37. Oil separator closure

Figure 6.4. T62T-16A Cutaway Engine Assembly.

gear drives the accessory section. The accessory drive assembly consists of accessory gears and the intermediate gear. These gears drive the hydraulic starter, the fuel control, and a speed switch and tachometer generator.

6.6. COMBUSTOR ASSEMBLY

The annular combustor assembly consists of a housing, liner, and nozzle shield, items 15 through 17 in figure 6.3. The assembly is secured to the turbine housing by a V-clamp, item 19 in the figure. Six self-tapping screws support and center the liner in the housing. Fuel for the combustor is supplied by six equally spaced fuel vaporizers connected by a circular manifold on the combustor housing. A combustion chamber drain valve is installed at the 6 o'clock position on the combustion housing.

6.7. TURBINE ASSEMBLY

The turbine assembly consists of the rotor, air inlet, diffuser, and nozzle assemblies. These items are shown in the airflow diagram in figure 6.5.

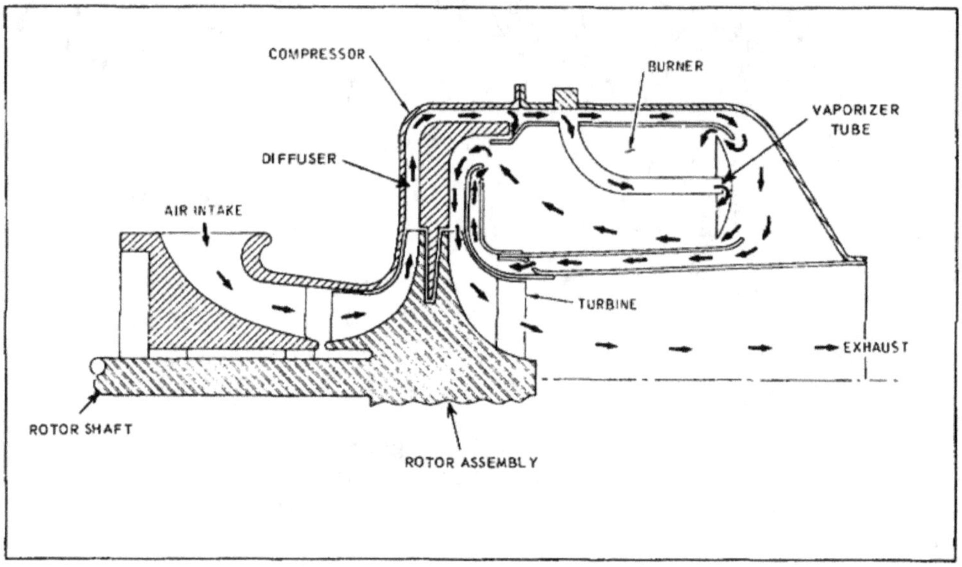

Figure 6.5. Airflow Diagram.

The centrifugal compressor rotor and radial inflow turbine rotor are bolted to the aft end of the rotor shaft. A single forward ball bearing and an aft roller bearing support the shaft. These are items 8 and 9 in figure 6.3 and 6.4. The air inlet assembly serves as a structural support between the combustor assembly and the reduction drive. The cylindrical, contoured casting is open at both ends and is equipped with a wire mesh screen to cover the intake duct. Intake air passes through the inlet assembly and into the impeller. The air then passes through the diffuser to the combustion section where it flows between the outer walls of the housing and combustion liner. At the end of the liner, the flow reverses direction and enters the combustion chamber. After combustion, the gas exits at the forward end of the combustion liner through a nozzle and the turbine wheel.

6.8. FUEL SYSTEM

The T62T-2A and T62T-16A have identical fuel systems. The fuel system is illustrated in figure 6.6. The system consists of an inlet filter, fuel control, six main fuel injectors, a start fuel nozzle, main and start fuel solenoid valve, a fuel pressure switch, and the necessary plumbing. Fuel for the APU is supplied from the same source that supplies the engines.

6.9. FUEL SYSTEM OPERATION

The accessory drive gear drives both the fuel pump and the acceleration control assembly to deliver fuel to the engine at approximately 300 psi. Fuel is directed through the inlet filter, in the fuel pump, through the outlet filter with the fuel pump, and into the governor housing. A relief valve installed in the governor housing bypasses excess fuel from the pump outlet to the pump inlet.

During engine starting, fuel is forced through internal passages to the fuel pressure switch. Normally closed, the fuel pressure switch opens on increasing fuel pressure at 100 to 120 psi. When open it actuates the start fuel solenoid valve and the main fuel solenoid valve to the open position, and completes the ignition circuit. Combustion takes place in the combustor when atomized fuel from the start fuel nozzle is ignited by the spark plug.

At 90 percent speed, fuel flow to the start fuel nozzle is cut off. The main fuel injectors continue to supply fuel to the combustor. Fuel flow to the main fuel injectors is increased in direct proportion to increasing air pressure, thus eliminating the possibility of over temperature or compressor surge during acceleration.

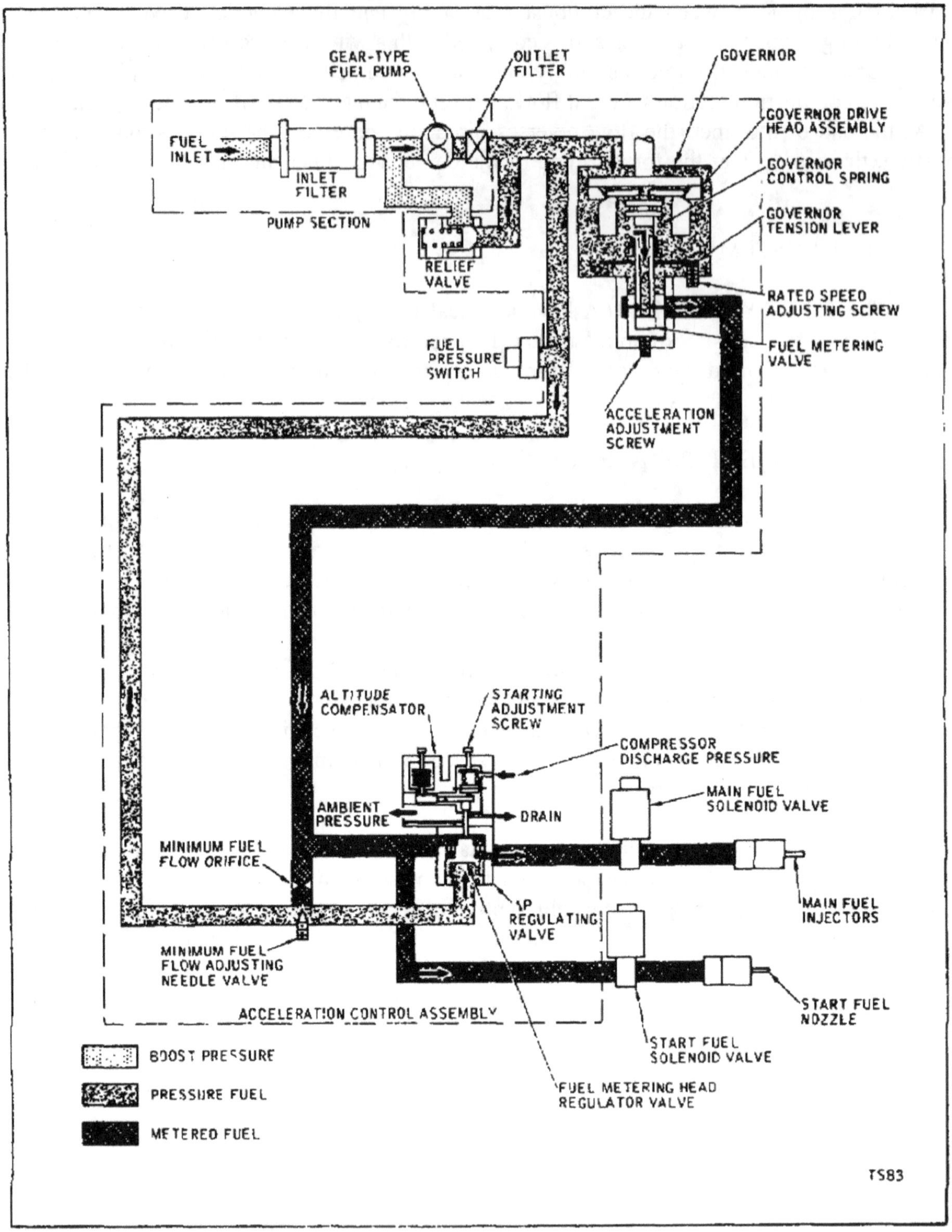

Figure 6.6. Fuel Control Schematic.

Ambient air pressure varies inversely with altitude. The altitude compensator has an aneroid bellows assembly that reacts to changes in ambient air pressure. It thereby reduces fuel flow to the fuel injectors during acceleration if the engine is operated above sea level, up to a maximum altitude of 15,000 feet.

A minimum fuel flow needle, installed in the fuel passage that bypasses the governor, allows a small amount of fuel under pump pressure to flow to the fuel injectors when fuel flow is reduced by the governor. The engine speed is controlled throughout the operating range of 100 to 102 or 105 percent rpm by the flyweight assembly, in the governor housing.

The combustor drain valve is mounted on the bottom of the combustor housing. This valve drains unburned fuel from the combustor during engine shutdown after a false or aborted start.

6.10. LUBRICATION SYSTEM

Figure 6.7 shows a schematic of the lubrication system. The pump draws oil out of the sump through an oil passage and into the housing. Oil under pump pressure enters the bottom of the filter housing, passes through the filter element, and flows out of the housing through a passage in the filter cap. A relief valve in the filter assembly opens at a differential pressure of 15-to 25 psi. This allows oil to flow from outside the filter element, through a passage in the filter element cap, to the filter outlet passage. If the filter element becomes clogged, this valve will open and allow oil to bypass the filter.

From the filter, oil is forced into a passage to the pressure relief valve and to four oil jets. The oil jet ring which encircles the high speed input pinion contains three of these jets, and sprays oil to the points where the high speed input pinion meshes with the three planetary gears. One jet directs a spray between the end of the output shaft and the high speed pinion to create a mist to lubricate the rotor shaft bearings. The remaining gears and bearings are lubricated by air-oil mist created when oil strikes the planetary gears and high speed pinion.

System pressure is maintained at 15 to 25 psi by a pressure relief valve. The valve regulates pressure by bypassing excessive pressure directly into the reduction drive housing. The bypassed oil strikes the inside surface of the air inlet housing, thus aiding in cooling the oil. Bypassed oil returns to the sump by gravity flow through an opening in the bottom of the planetary gear carrier.

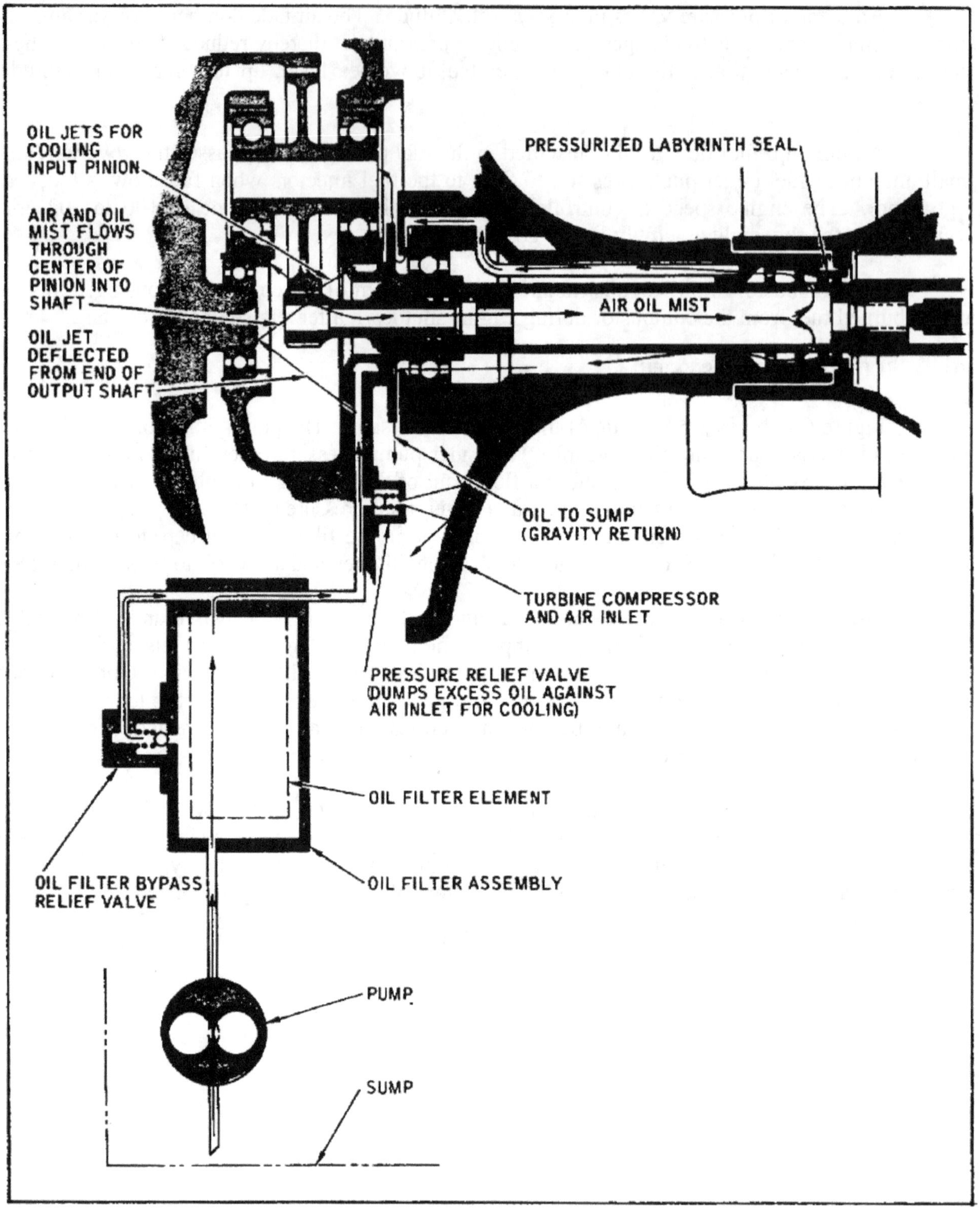

Figure 6.7. Lubrication System Schematic.

The low oil pressure switch, normally open, closes on increasing oil pressure to 5 to 7 psi. When the switch contacts close, the low oil-pressure circuit is deenergized. At rated engine speed, a drop in oil pressure below 5-7 psi opens the low oil-pressure switch contacts, and, through electrical circuitry, closes the main fuel solenoid valve and shuts down the engine.

6.11. ELECTRICAL SYSTEM

Circuitry for ignition and engine electrical accessories is included in the electrical system. The system includes the ignition exciter, spark plug, speed switch, tachometer generator, thermocouple, and hour meter. The engine harness-assembly connectors mate with receptacles on the junction box, oil pressure switch, ignition exciter, speed switch, and tachometer-generator. Descriptions of the components in the electrical system are given in the following subparagraphs.

a. Ignition exciter. The input voltage to the ignition exciter is 10 to 29v dc. Input voltage is converted to an intermittent high-energy current which is directed to the spark plug for ignition. Minimum spark rate is two per second at 14 volts.

b. Spark plug. The shunted surface-gap type spark plug is threaded into a boss in the end of the combustor. The plug furnishes the spark necessary for initial ignition of fuel during the starting cycle. Ignition is terminated by the No. 1 switch in the speed switch when the engine attains 90 percent rated speed.

c. Speed switch. Mounted on the rear accessory drive pad in tandem with the tachometer is the speed switch. Inside the switch housing, two flyweights are mounted to the speed switch drive shaft. The flyweights move an actuating plate held in position by two springs of unequal strength. The speed switch is two switches in one housing. The No. 1 switch is set at 90 percent rated speed, and the No. 2 switch is set at 110 percent rated speed. As rpm is increased, the force of the lighter and then the heavier spring is overcome, allowing the actuating plate to actuate the No. 1 switch (90 percent speed). When the No. 1 switch is actuated, the start fuel and ignition systems are shut off. When the No. 2 switch is actuated at 110 percent speed, the main fuel solenoid valve is closed, and this shuts off fuel to the engine.

d. Turbine exhaust thermal switch. A thermocouple projects into the exhaust gas stream at the aft end of the combustor. If an over temperature occurs, the contacts normally closed in the turbine exhaust thermal switch will open, close the main fuel solenoid valve, and shut the engine down.

e. Tachometer generator. The APU tachometer generator is mounted in tandem with the speed switch on the accessory drive assembly. A synchronous rotor in the tachometer generator produces three-phase ac voltage proportional to the speed at which the rotor is turning. This voltage is transmitted to a tachometer indicator in the cockpit which indicates the engine speed in percent of rated rpm.

f. Hour meter. Engine operating time is recorded by the hour meter attached to the engine. The meter operates on 24v dc supplied by the aircraft bus.

6.12. SUMMARY

The T62 is used on the CH-47 and CH-54 helicopters as an auxiliary power unit. The engine consists of a combustor, turbine, and reduction and accessory drive assemblies. Both the compressor and turbine rotor are mounted on the same shaft. A fuel control, hydraulic starter motor, and ignition unit are mounted on the engine. Operating time for the APU is maintained separately from the aircraft engine time by an hour meter mounted on the APU.

The T62 is equipped with protective devices that shut the APU down if any of the operating limitations are exceeded. If the oil pressure, rpm, or exhaust temperature limits are exceeded, a light on the instrument panel will be illuminated and the APU shut down.

Chapter 7

ALLISON T63

7.1. INTRODUCTION

The T63 series turboshaft engine is manufactured by the Allison Division of General Motors Corporation. The T63-A-5A is used to power the OH-6A, and the T63-A-700 is in the OH-58A light observation helicopter.

Although the engine dash numbers are not the same for each of these, the engines are basically the same. As shown in figure 7.1, the engine consists of four major components: the compressor, accessory gearbox, combustor, and turbine sections. This chapter explains the major sections and related systems.

7.2. OPERATIONAL DESCRIPTION

The T63 turboshaft engine, being an internal combustion engine, requires intake, compression, combustion, and exhaust, as does a reciprocating engine.

Figure 7.2 is a cutaway view of the T63 engine. By looking at the letters in the arrows in figure 7.2, you will be able to follow the air's path through the engine.

Air is routed to the compressor assembly by intake ducts on the aircraft. The (A) compressor assembly "squeezes" incoming air to high pressures. This compressed air is discharged through twin air-transfer tubes (B) into the combustion chamber (C) and mixed with fuel which is then burned. The exhaust gases expand through the compressor turbine (D) which extracts energy from these hot gases to drive the compressor (A). The power turbine (E) takes the remaining energy to drive the power-output shaft (F). The gases are then released through exhaust ducts (G).

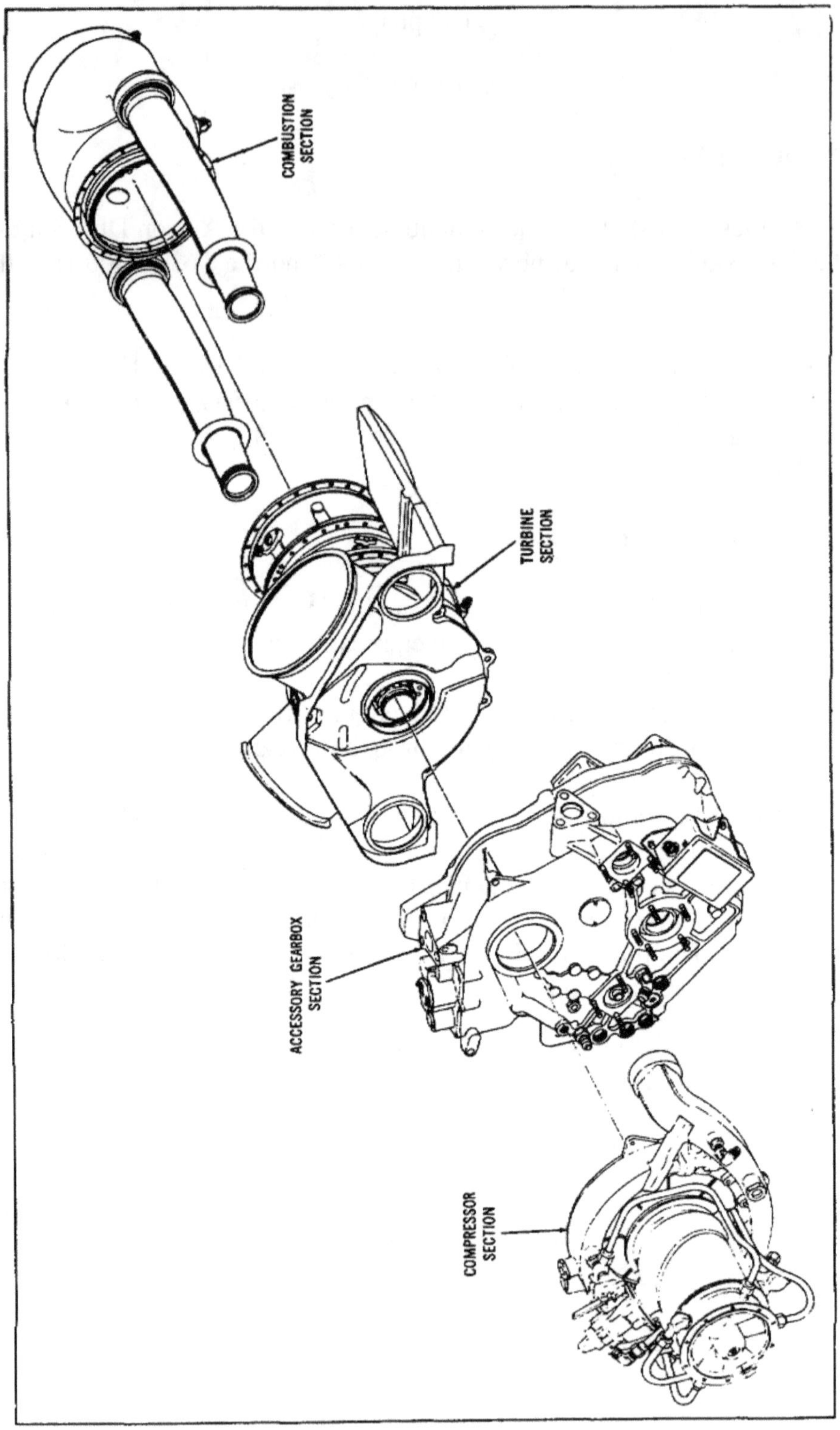

Figure 7.1. Sections of the T63 Engine.

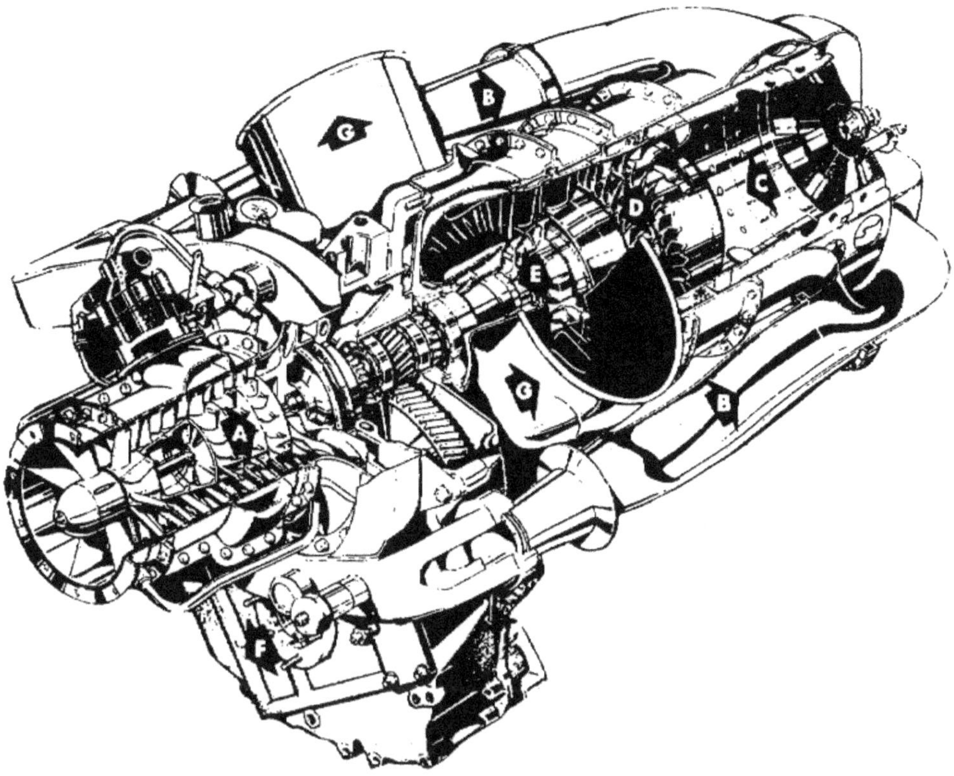

Figure 7.2. Airflow Diagram.

7.3. ALLISON T63 SPECIFICATIONS

Specifications for the T63-A-5A and 700 engines used in Army aircraft are summarized in the following chart.

Allison T63-A-5, -A-700 Gas Turbine Engines.	
MILITARY DESIGNATION	T63-A-5, T63-A-700.
ALLISON DESIGNATION	250-C18
COMPRESSOR STAGES	6 AXIAL, 1 CENTRIFUGAL
COMBUSTOR	REVERSE FLOW ANNULAR TYPE AT REAR OF ENGINE, WITH FORWARD GAS FLOW
TURBINE (GP)	2 STAGE

TURBINE (PT)	2 STAGE
POWER OUTLET	6,000 RPM
FUEL CONTROL	HYDRO-MECHANICAL
IGNITION SYSTEM	HIGH-ENERGY, 1 IGNITER
LUBRICATION SYSTEM	EXTERNAL OIL TANK, COOLER
WIDTH, HEIGHT, LENGTH	19.0 IN., 22.5 IN., 40.3 IN
WEIGHT	155 lb
FUEL	MIL-L5624/JP-4
OIL	MIL-T-23699
RATING (MILITARY)	317 SHP
(MAX CONTINUOUS)	270 SHP
APPLICATION	OH-6A, OH-58A

7.4. COMPRESSOR SECTION

The compressor assembly consists of the following: compressor front support, compressor case, diffuser scroll, front diffuser, rear diffuser, rotor, rotor bearing, and oil seals. An exploded view of the compressor assembly is shown in figure 7.3.

The compressor is a combination axial-centrifugal type with six stages of axial compression and one stage of centrifugal compression. The rotor hub and blade assemblies and the impeller are made from stainless steel. The compressor rotor front bearing (No. 1) is housed in the compressor front support, and the compressor rotor rear (No. 2) bearing is housed in the compressor rear diffuser. The No. 2 bearing is the thrust bearing for the compressor rotor assembly.

The compressor case assembly consists of upper and lower halves and is made of stainless steel. Thermal-setting plastic is centrifugally cast to the inside surface of the case halves and vane outer bands.

To achieve maximum compressor efficiency, a minimum clearance is necessary between the compressor-blade tips and the case. The first time an engine is started, the blade tips cut their own tip clearance in the plastic coating. The inside of the compressor must be kept free of dirt accumulation. A dirty compressor can cause high turbine outlet temperatures, low engine power, and

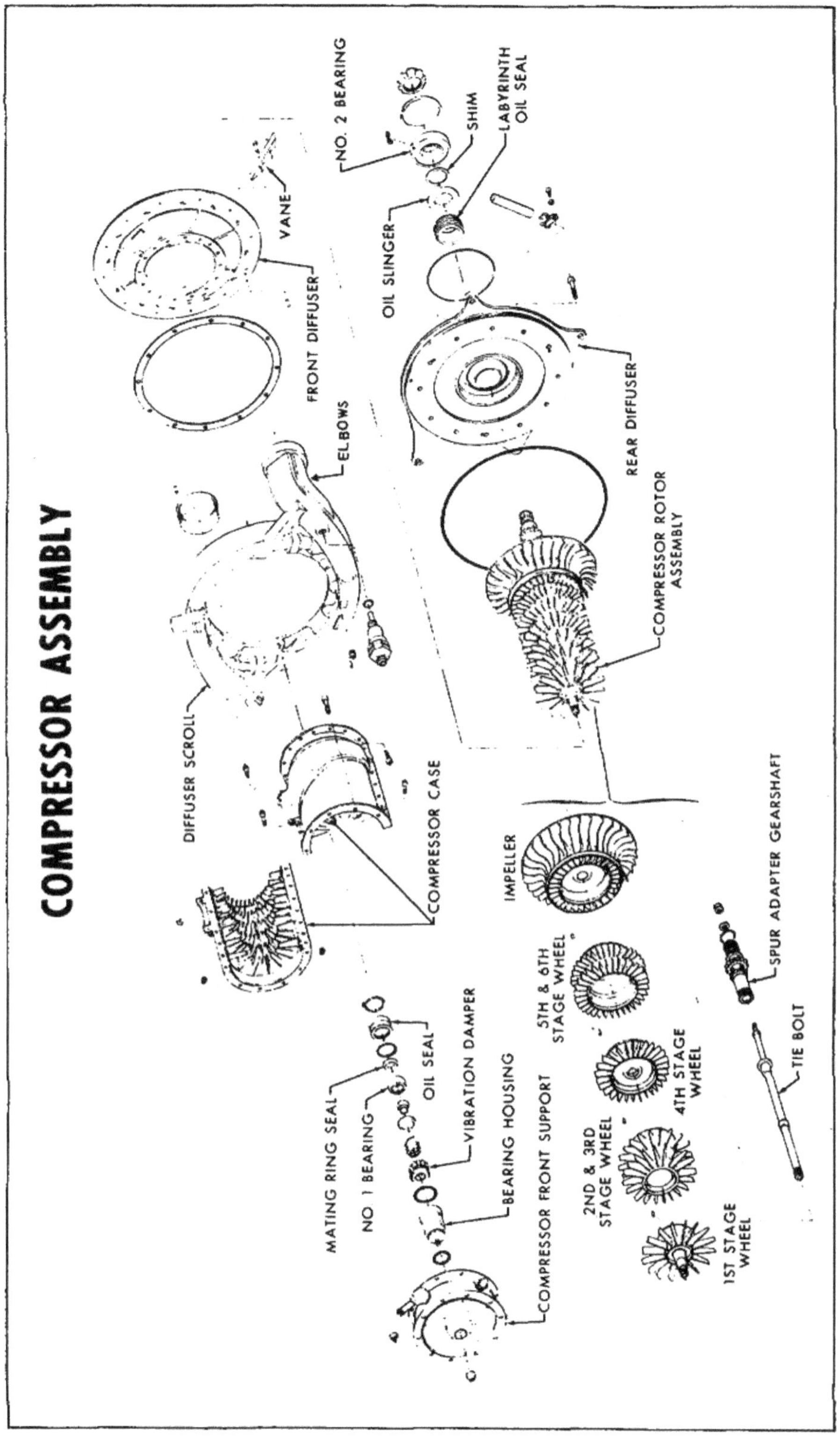

Figure 7.3. Compressor Assembly.

eventual engine failure. Engine cleaning procedures can be found in the appropriate aircraft or engine maintenance manual. The compressor of the T63 must never be cleaned with an ordinary cleaning solvent, because this will dissolve the plastic coating on the inside of the compressor case and cause engine failure.

A control valve is mounted on the compressor case assembly to bleed air off the 5th stage of the compressor during starting and all engine operations at low pressure ratios.

The compressor diffuser assembly consists of stainless steel front and rear diffusers and a magnesium alloy scroll. The scroll collects the air and delivers it to two elbows. Compressor-discharge air tubes deliver compressed air from the outlet of the elbows to the combustion outer case. The diffuser scroll has five ports from which air can be bled or compressor discharge air pressure sensed. Two of these ports are customer bleed air ports, and the remaining ports are used by the anti-icing valve, fuel control pressure sensing, and bleed air pressure sensing. Customer air ports are used by the airframe manufacturer to tap bleed air to run pumps, heaters, and so forth.

7.5. ACCESSORY GEARBOX SECTION

Shown in figure 7.4, the accessory gearbox is the primary structural member of the engine. All engine components, including

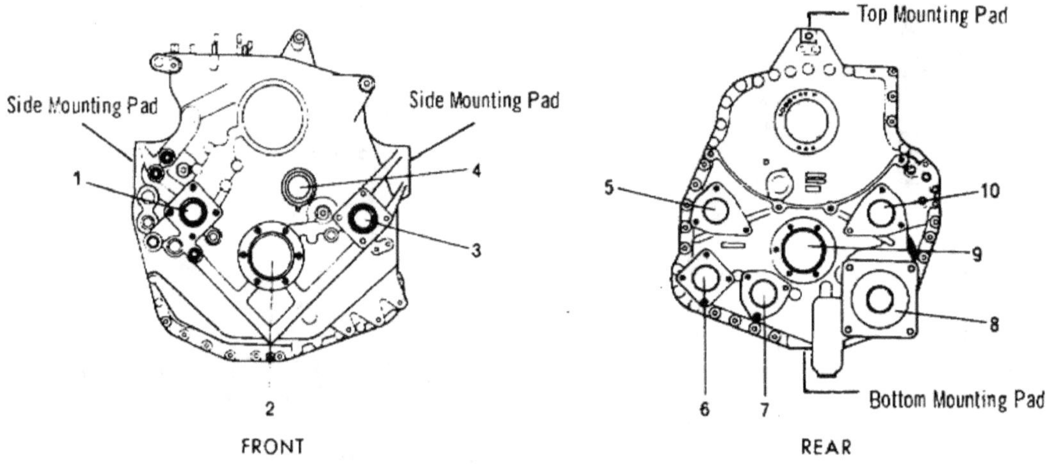

1. Gas producer tachometer-generator
2. Power output
3. Power turbine tachometer generator
4. Torquemeter spanner nut
5. Power turbine fuel governor
6. Spare (not used)
7. Fuel pump
8. Starter-generator
9. Power takeoff (not used)
10. Gas producer fuel control

Figure 7.4. Accessory Gearbox Case.

the engine mounted accessories, are attached to the case, which has four mounting pads, one on each side, one on the top, and one on the bottom. The side pads must be used, and the helicopter manufacturer has his choice of using the top and bottom pads. The accessory gearbox contains most of the lubrication system components and houses the power turbine and gas producer gear trains. The power turbine gear train, shown in figure 7.5, reduces engine N_2 speed from 35,000 to 6,000 rpm. The power turbine gear train drives the torquemeter, N_2 tachometer-generator, and N_2 governor. The gas producer gear train, illustrated in figure 7.6, drives the oil pumps, fuel pump, gas producer fuel control, N_1 tachometer generator, and starter-generator. During starting, the starter-generator cranks the engine through the gas producer gear train.

7.6. TURBINE SECTION

As shown in figure 7.1, the turbine section is mounted between the combustion section and the power and accessory gearbox. The turbine section consists of a two-stage gas producer turbine, shown in figure 7.7, and a two-stage power turbine, shown in figure 7.8. Power to drive the compressor rotor is furnished by the gas producer turbine rotor through a direct drive. The power turbine rotor converts the remaining gas energy into power which drives the power output shaft. The exhaust gases are directed into the exhaust collector support and through the twin exhaust ducts on the top of engine.

The power turbine is a free turbine, because it is free to rotate at a different speed than the gas producer turbine rotors.

The gas producer turbine nozzles (1st and 2nd) are housed in the gas producer turbine support. The power turbine (3rd and 4th stage) nozzles are housed in the power turbine support and the aft end of the exhaust collector support. These nozzles may be seen by looking at figures 7.7 and 7.8. These nozzles increase the velocity of the expanding gas and direct the gas at the proper angle onto the turbine rotors.

7.7. LUBRICATION SYSTEM

The oil system is a circulating dry-sump type with an external oil tank and cooler. The engine oil-system schematic is illustrated in figure 7.9. This system is designed to supply oil for lubrication, scavenging, and cooling as needed.

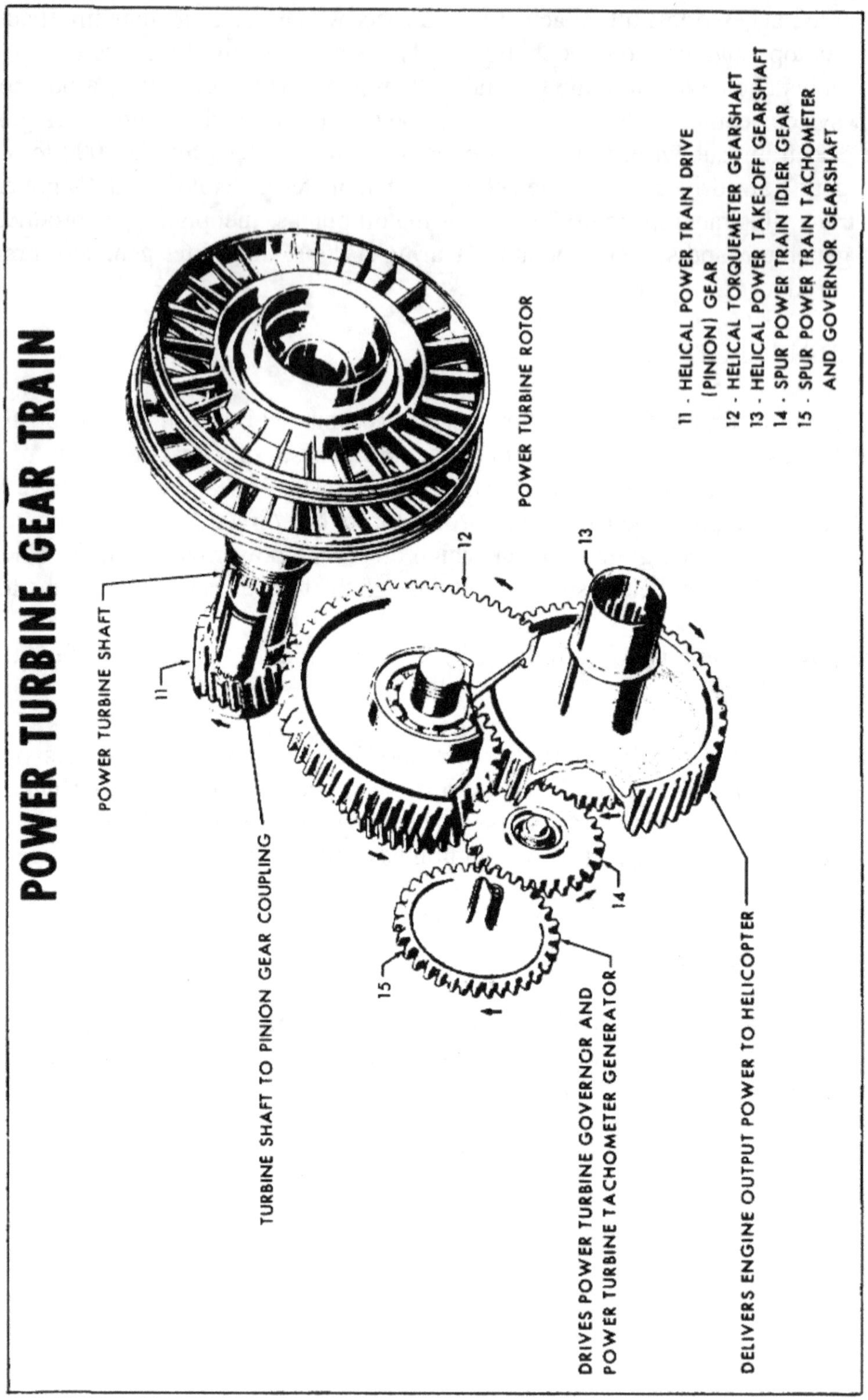

Figure 7.5. Power Turbine Gear Train.

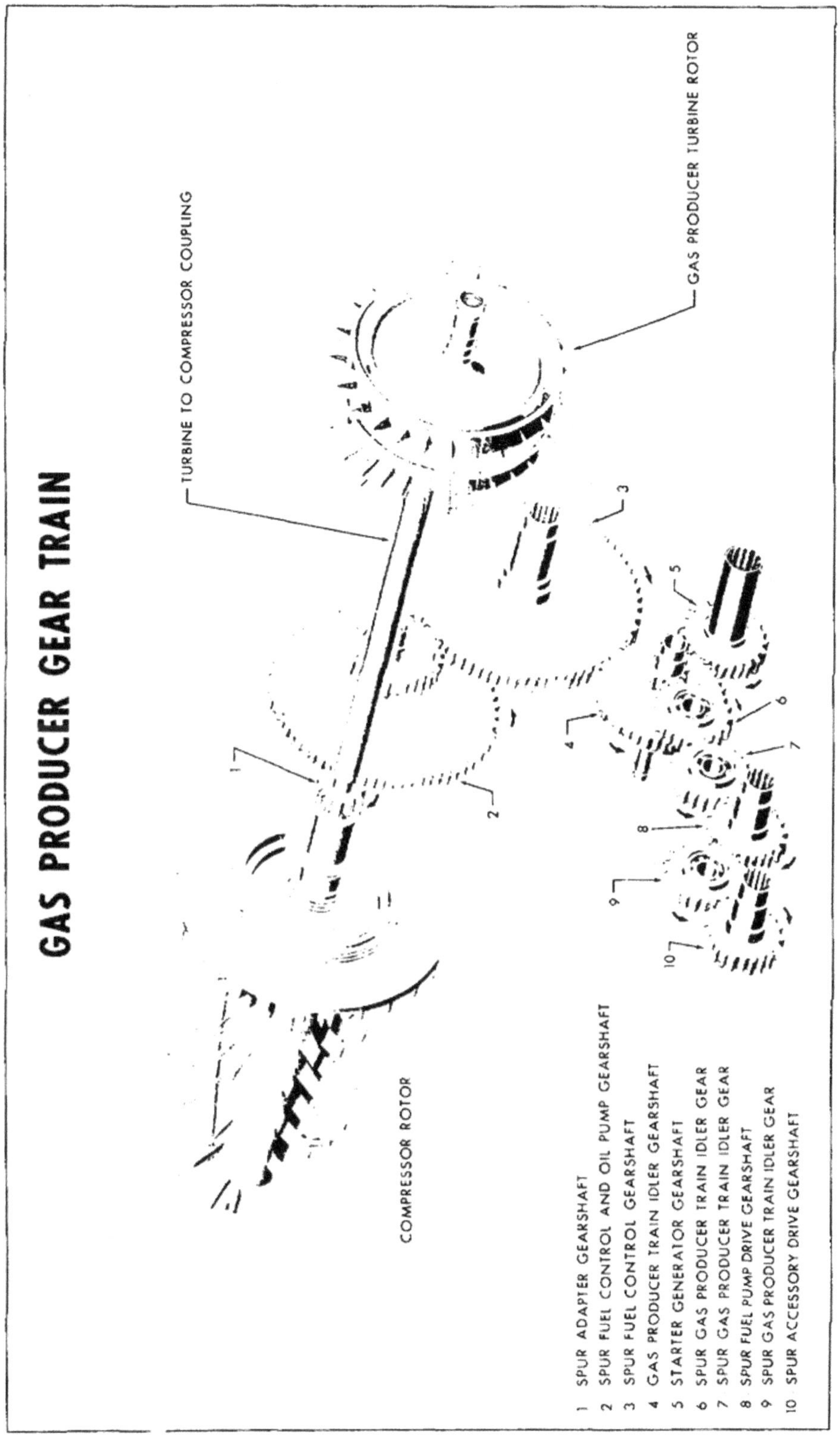

Figure 7.6. Gas Producer Gear Train.

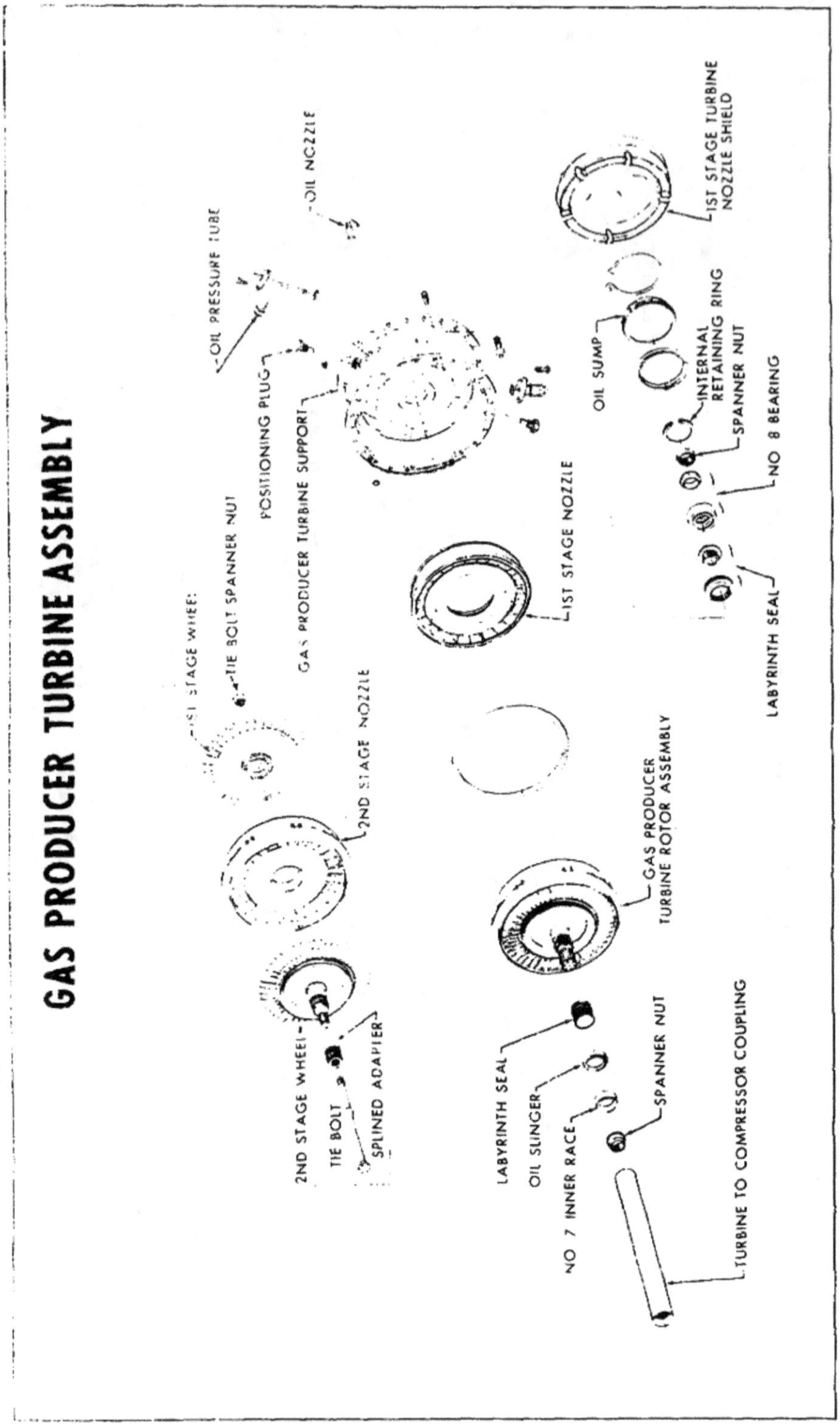

Figure 7.7. Gas Producer Turbine Assembly.

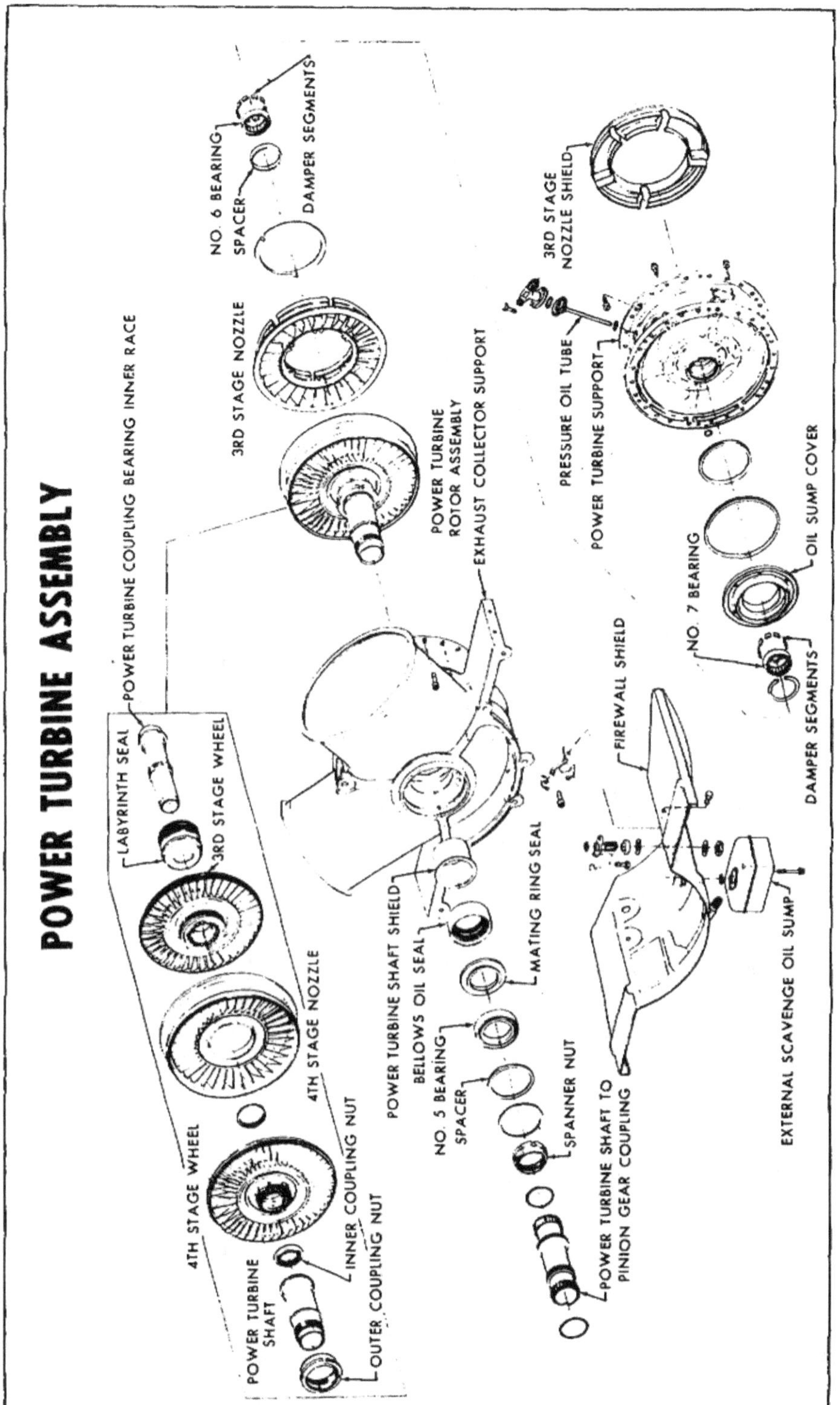

Figure 7.8. Power Turbine Assembly.

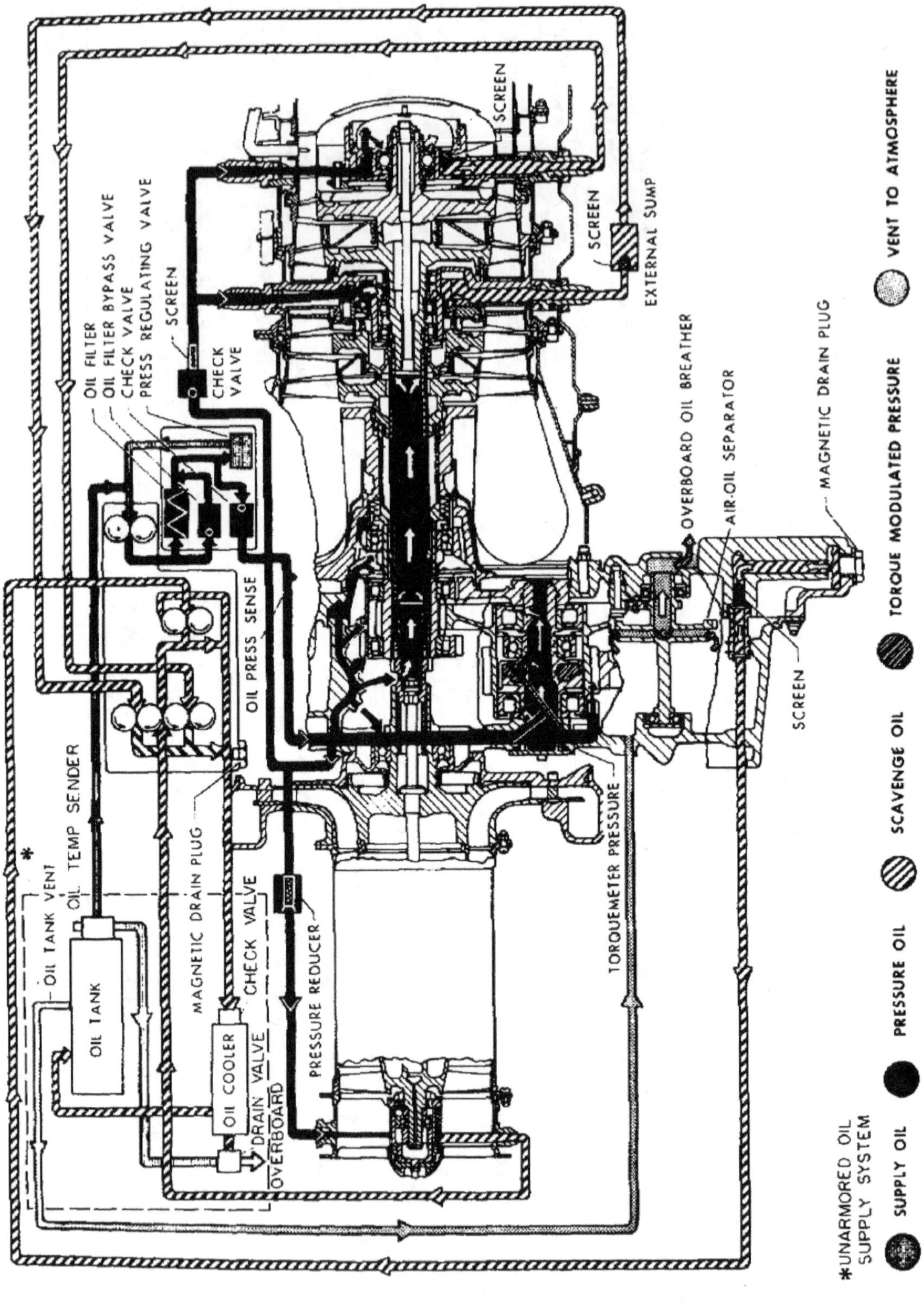

Figure 7.9. Engine Lubrication Schematic.

Spray jet lubrication is used on all compressor, gas producer turbine, and power turbine rotor bearings, and to bearings and gear meshes of the power turbine gear train, with the exception of the power output shaft bearings. The power output shaft bearings and all other gears and bearings are lubricated by oil mist.

A spur-gear oil pump assembly, illustrated in figure 7.10, consisting of one pressure element and four scavenge elements, is mounted within the accessory gearbox.

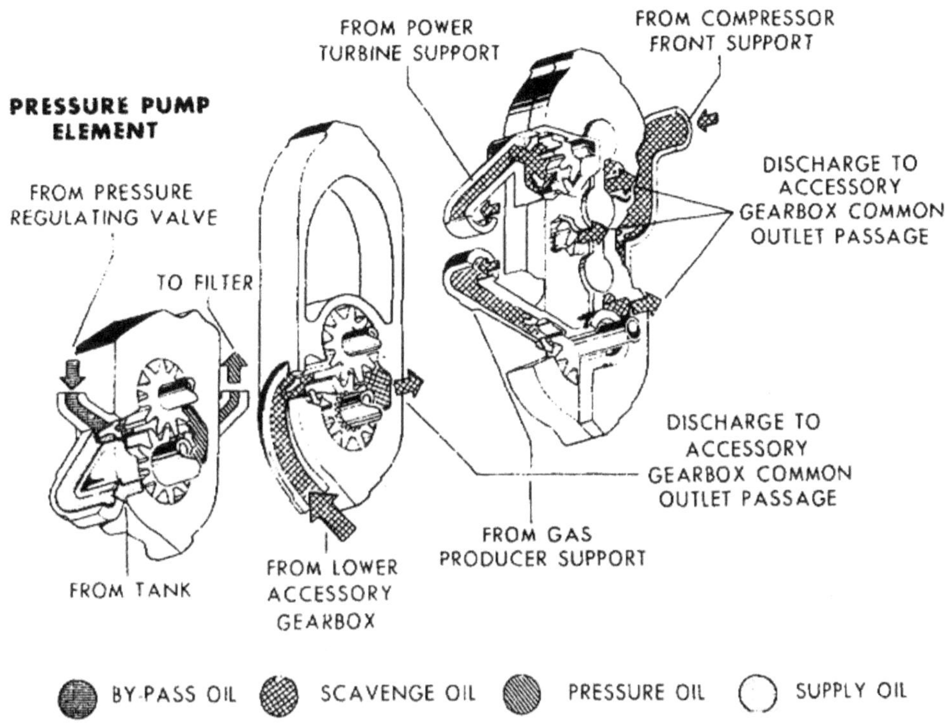

Figure 7.10. Oil Pump.

The oil filter assembly, located in the upper right-hand side of the accessory gearbox, consists of a filter bypass valve and a pressure regulating valve. Oil from the tank is delivered to the pressure pump which pumps oil through the filter and then to the points of lubrication. The oil pressure is regulated 115-130 psi by the pressure regulating valve.

The engine lubrication system incorporates four screens, two magnetic chip detector plugs, and two check valves. The check valve in the oil filter outlet passage prevents oil in the tank from draining into the engine when it is not in operation.

To further prevent internal oil leakage at engine shutdown, an external sump is connected to the scavenge oil line at the power turbine support. One magnetic chip detector plug is in the accessory gearbox sump, and the other one is in the scavenge oil pressure line which delivers oil to the oil outlet port on the gearbox.

7.8. TORQUEMETER

A torquemeter is located on the instrument panel and is connected to a transmitter which is part of the engine oil system. The torque indicating system converts the pressure sensed at the torquemeter pressure port, on the front side of the accessory gearbox, into an indication in psi of engine torque output.

7.9. FUEL CONTROL SYSTEM

The fuel control system meters fuel during starting, acceleration, constant speed, and deceleration. This control system consists of governor, accumulator and check valve, and fuel nozzle. These components are illustrated in figure 7.11. The following subparagraphs cover each component in the fuel system.

a. The fuel pump assembly consists of two spur-gear pumps, filter, filter bypass valve, regulator valve, and two discharge check valves. As shown in figure 7.12, the fuel pumps are parallel, that is, fuel entering the inlet port can be pumped by either pump to the outlet port. Each pump has a separate shear point. In the event of a pump failure, the discharge check valve of the failed pump closes to prevent reverse flow through the pump. Failure of one pump will not affect engine operation.

b. The fuel control (N_1) is located at the 3 o'clock position on the accessory gearbox case. The fuel control delivers metered fuel (P_2 in figure 7.11) to the fuel nozzle. The gas producer fuel control, driven by the N_1 gear train, senses compressor discharge pressure (P_C). The N_1 fuel control lever, positioned by the twist grip throttle, is mechanically linked to a pointer and the fuel cutoff valve. A quadrant, shown in figure 7.13, with 0, 5, 30, and 90 degree indications is located adjacent to the pointer. The pilot's twist grip throttle has three basic positions: cutoff is 00 to 50; ground idle is at 300, and full open at 900. When the twist grip is moved from cutoff to ground idle during engine start, the fuel control automatically meters fuel in response to N_1 rpm. The engine starts, accelerates, and stabilizes at ground idle rpm. Fuel flow during this phase is metered entirely by the fuel control.

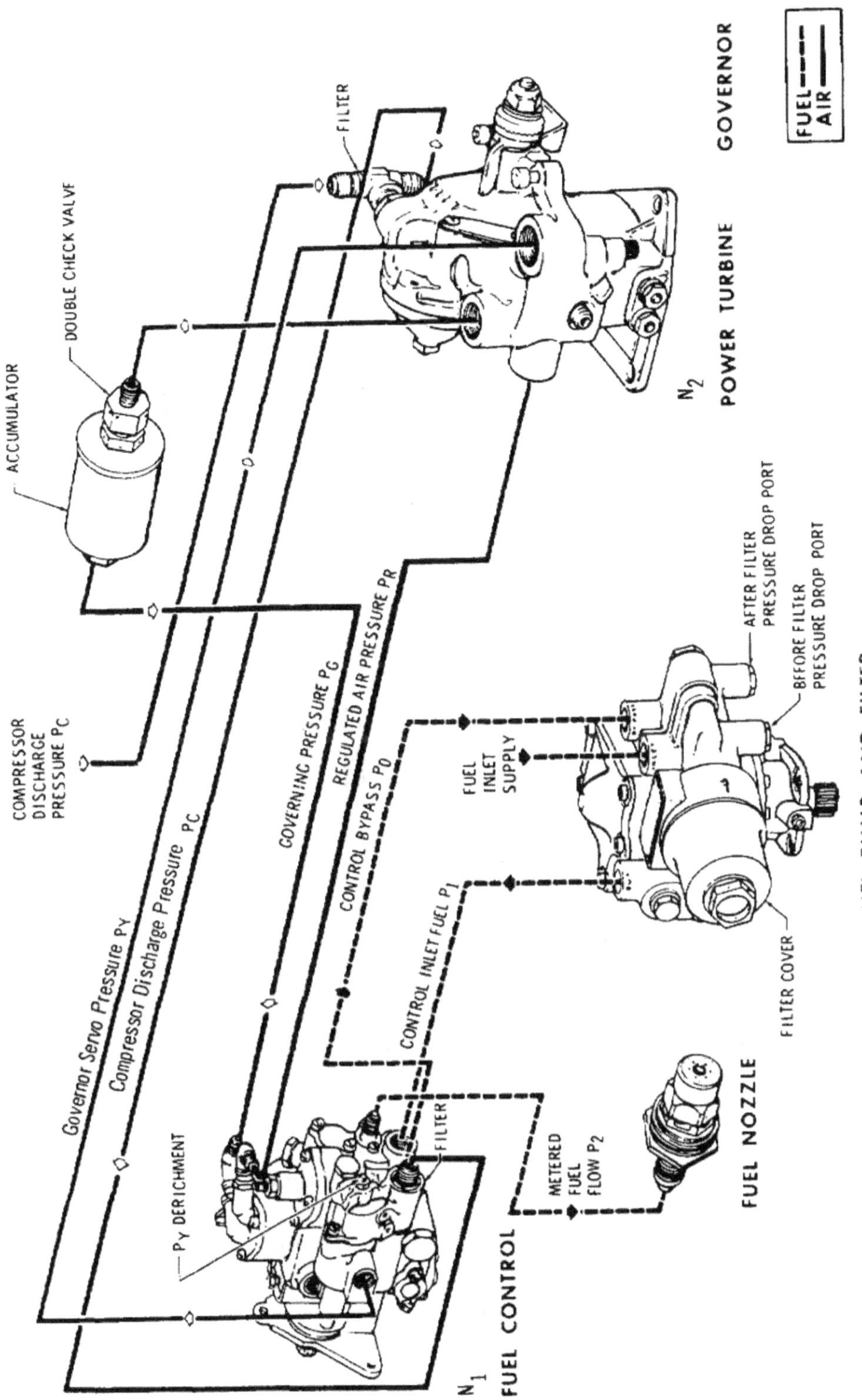

Figure 7.11. Fuel Control System.

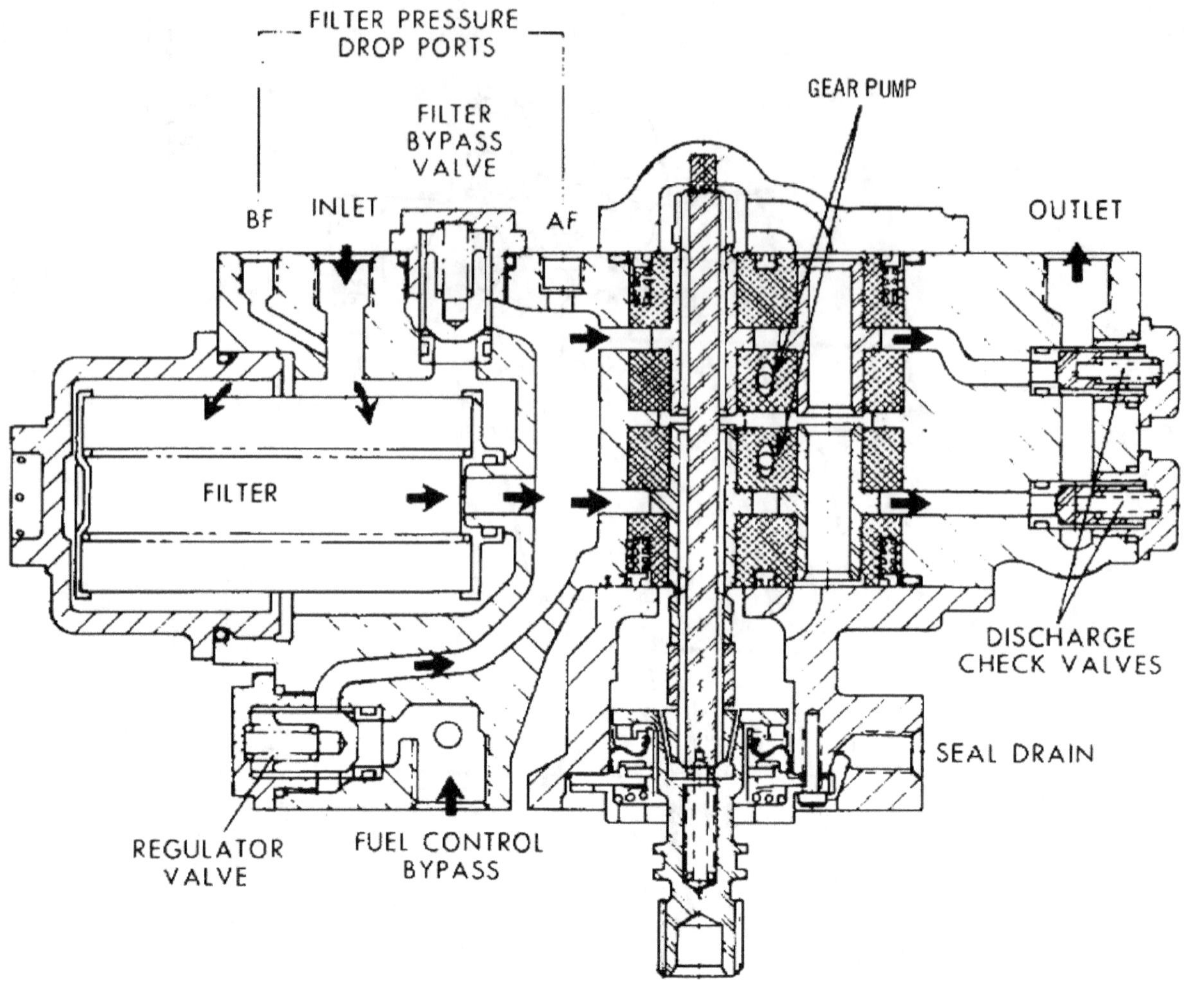

Figure 7.12. Fuel Pump and Filter Assembly.

c. The power-turbine governor (N_2) is not required for starting or ground idle operation, but it is required for speed governing of the power turbine rotor. The gas-producer fuel control and the power-turbine governor are connected together by two pneumatic lines, labeled P_G and P_R in figure 7.11. The power-turbine governor, driven by the power-turbine gear train, senses compressor discharge pressure (P_C). The power-turbine governor lever, as shown in figure 7.14, is positioned by the helicopter droop compensator or the N_2 actuator. The power-turbine governor is required for speed governing of the power turbine rotor (N_2). The N_2 rpm at which the power-turbine governor will govern is under the control of the N_2 actuator. Normally, the power-turbine governor is operated at 100 percent N_2. However, the pilot can select a setting of the governor either higher or lower than 100 percent N_2.

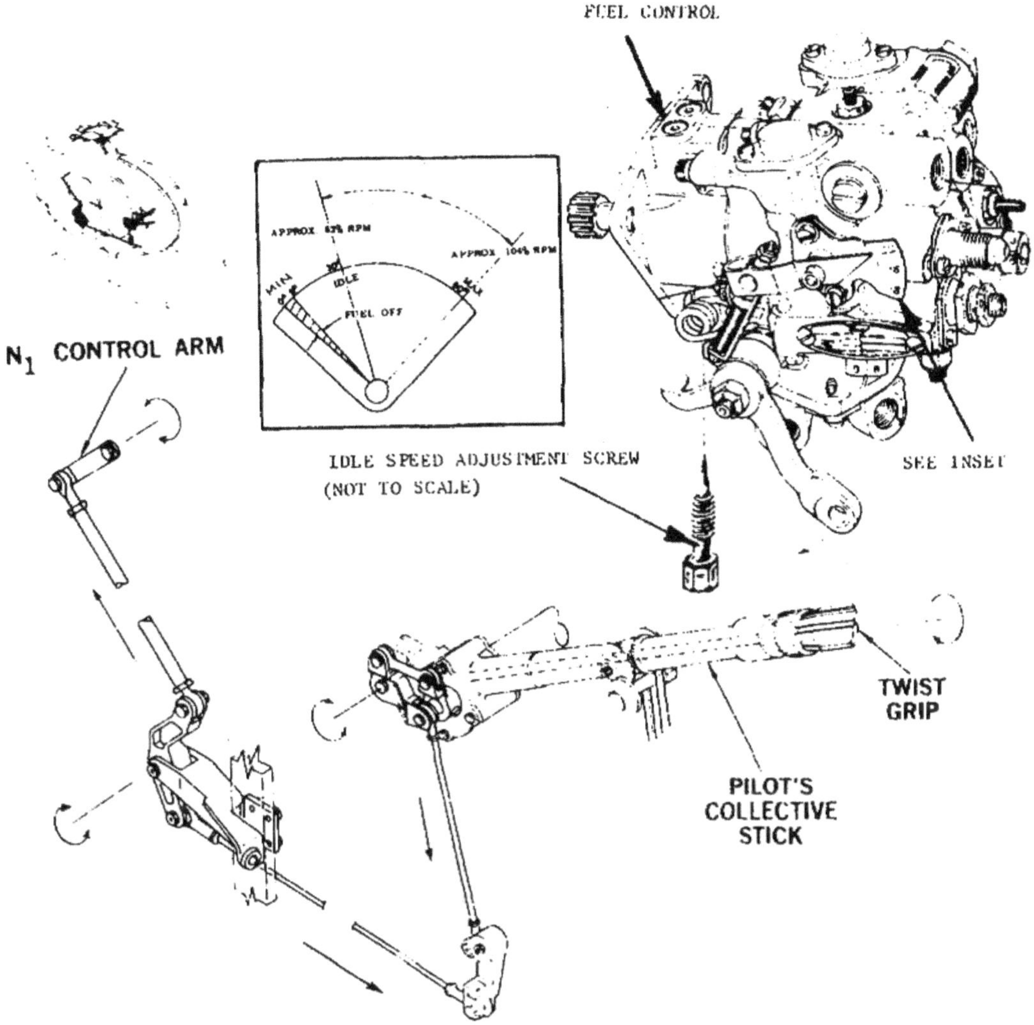

Figure 7.13. Fuel Control System.

The droop compensator moves the governor Lever any time the pilot's collective stick is increased or decreased. This is necessary to maintain the N_2 rpm the pilot has set with the N_2 actuator. When collective pitch is increased, the power requirements are increased, and N_2 rpm droops slightly. The governor senses the N_2 droop and it signals the gas-producer fuel control to meter more fuel. As fuel flow increases, N_2 returns to the setting of the power-turbine governor. When collective pitch is decreased, the power requirements are decreased, and N_2 increases slightly. The governor senses the N_2 increase and signals the fuel control to meter less fuel. As fuel flow decreases, the N_2 decreases to the setting of the power-turbine governor.

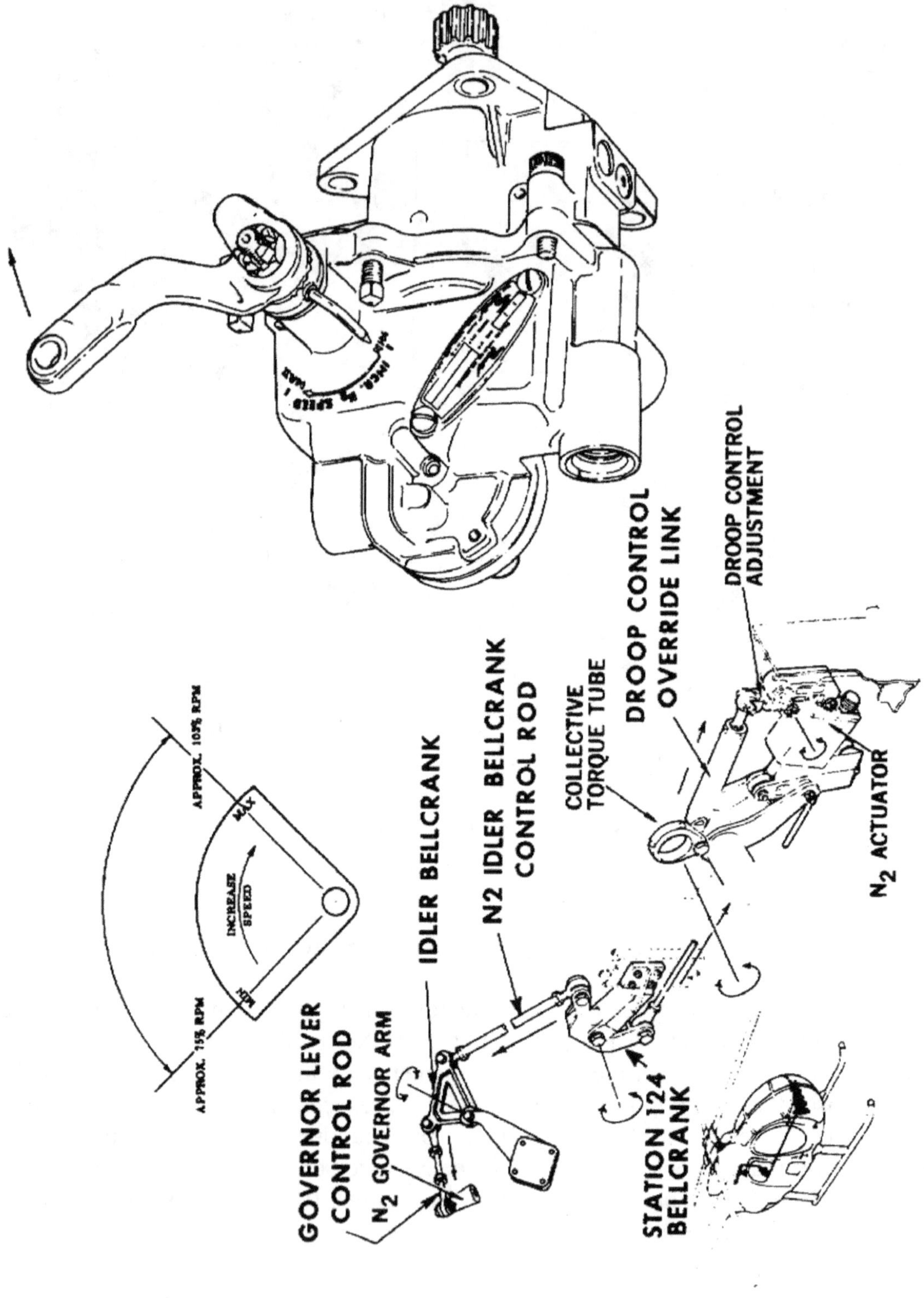

Figure 7.14. Power Turbine Governor Control System.

d. The check valve assembly and accumulator are installed to dampen torsional vibrations encountered in helicopter rotor systems. Because of wind gusts and turbulent air conditions, the rotor rpm will fluctuate slightly. The power turbine governor senses this rpm change and causes the fuel control to vary fuel flow. By installing an accumulator and damping out the governing pressure (P_G) to the gas producer fuel control, the design prevents the engine from responding to torsional vibrations.

e. The fuel nozzle is a single-entry, dual-orifice type nozzle. It threads into the combustor outer case and extends into the aft end of the combustor liner. The fuel control delivers fuel to the nozzle which atomizes and injects fuel into the combustion chamber, where it is mixed with air and burned.

The fuel nozzle must properly atomize and inject the fuel in all ranges of fuel flow from starting to maximum power. This is accomplished by means of a dual-orifice design. The primary orifice has fuel delivered to it whenever the engine is operating, but the secondary orifice receives fuel only when the fuel pressure to the fuel nozzle exceeds 150 psi.

7.10. AIR BLEED AND ANTI-ICING SYSTEMS

The compressor air bleed system permits rapid engine response by relieving compressor pressure during engine acceleration. A bleed air control valve is mounted to the bleed air manifold on the compressor case. The compressor bleed air and anti-icing systems are illustrated in figure 7.15.

Elongated slots between every other vane at the compressor fifth stage bleed compressor air into a manifold on the compressor case. The air bleed control valve is open during starting and ground-idle operation, and it remains open until a predetermined pressure ratio is obtained, at which time the valve begins to move from the open to the closed position.

The engine is equipped with an anti-icing system that conducts hot air to the compressor front-support struts to prevent ice forming on the struts. The system is entirely separate and independent of any other bleed air system. The engine anti-icing system must be turned on by the pilot. As air passes through the compressor, it is compressed. As a result of this compression, the air is heated and is a source of hot air required by the engine anti-icing system.

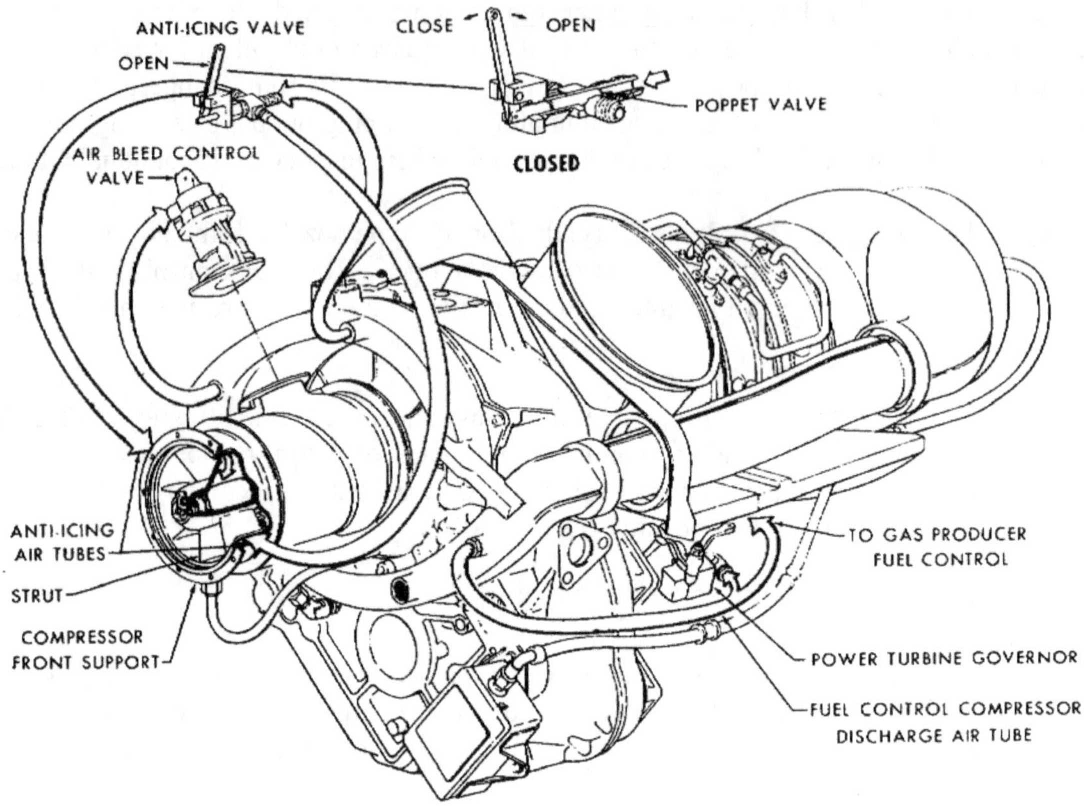

Figure 7.15. Air Bleed and Anti-Icing System.

The anti-icing system consists of an anti-icing valve, two tubes, and passages within the compressor front support. During operation of the system, the anti-icing tubes deliver hot air from the anti-icing valve to the two ports on the compressor front support. These ports deliver the hot air into an annular passage formed between the dual-walled housing. Hot air flows from the annular passage through the hollow inlet guide vanes into the front bearing support hub. Some of the air flowing through the struts is exhausted out of slots on the trailing edge of the guide vanes, and the remaining air is exhausted out of holes in the front bearing support hub. During anti-icing operation, all surfaces of the compressor front support which come in contact with compressor inlet air are heated and ice cannot form.

7.11. IGNITION AND TURBINE OUTLET TEMPERATURE MEASUREMENT SYSTEMS

The ignition system is composed of three components: a low-tension capacitor exciter-assembly, a spark igniter lead, and a shunted surface gap spark igniter. The system is powered by the aircraft 28-volt dc electrical system. This ignition system is only required during starting, because continuous combustion takes place after the engine is started. Components of the ignition and turbine outlet temperature (TOT) systems are illustrated in figure 7.16.

The TOT thermocouple harness contains four probes used to sense the temperature of the gases on the outlet side of the gas-producer turbine rotor. Each thermocouple probe generates a dc millivoltage which is directly proportional to the gas temperature it senses. The thermocouple harness averages the four voltages produced and indicates TOT on a gage in the cockpit.

7.12. SUMMARY

The Allison T63 is a free-power gas turbine engine which has four major sections: the compressor assembly, power and accessory gearbox, turbine assembly, and combustion assembly. The power-turbine governor senses power turbine speed and relays this to the fuel control that controls the compressor speed. The fuel control sends fuel to the nozzle located in the combustion section, and the nozzle sprays fuel into the combustion liner.

The engine is lubricated by a dry-sump pressure system. Ignition for engine starting comes from an ignition exciter and spark igniter located next to the fuel nozzle in the engine combustion section.

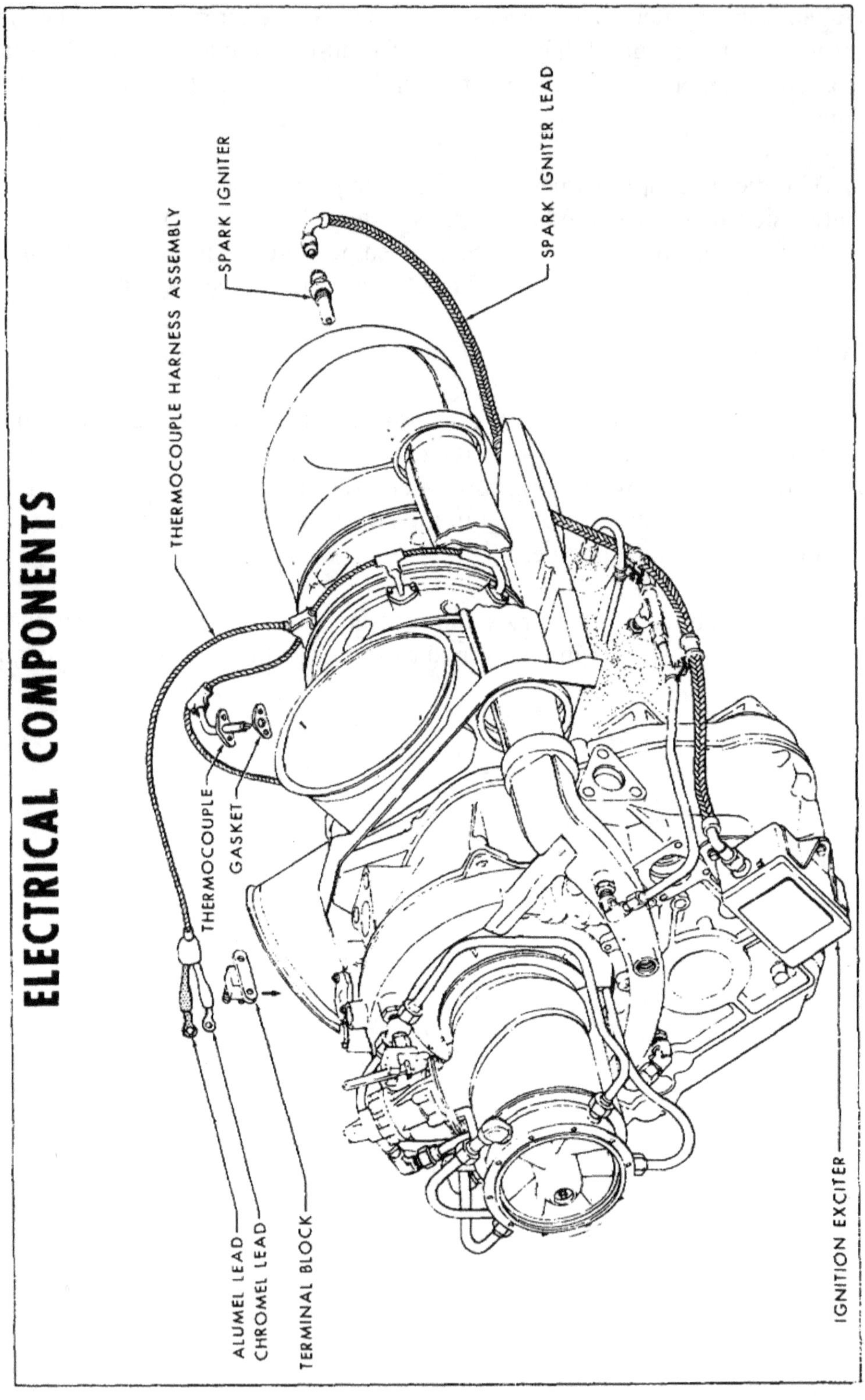

Figure 7.16. Electrical Components.

Chapter 8

PRATT AND WHITNEY T73

8.1. INTRODUCTION

The Pratt and Whitney T73-P-1 and T73-P-700 are the most powerful engines used in Army aircraft. Two of these engines are used to power the CH-54 flying crane helicopter. The T73 design differs in two ways from any of the engines covered previously. The airflow is axial through the engine; it does not make any reversing turns as the airflow of the previous engines did, and the power output shaft extends from the exhaust end.

Chapter 8 describes and discusses the engine sections and systems. Constant reference to the illustrations in this chapter will help you understand the discussion.

8.2. GENERAL DESCRIPTION

The T73-P-1 and the T73-P-700 engines are straight-flow, free-turbine power plants using a two-stage turbine to drive a nine-stage, axial-flow compressor. The free turbine uses the exhaust gas to drive a two-stage free turbine rotor. External views of the T73 gas turbine engine are shown in figure 8.1. A cross-sectional view is shown in figure 8.2.

The axial flow compressor consists of a nine-stage rotor and nine stator stages. The gas path of the compressor is so designed that it forms a convergent duct. The compressor has a moderate compression ratio.

An automatically-controlled interstage airbleed is used for starting and low power operation of the engine. The engine's anti-icing system prevents dangerous accumulations of ice on compressor-inlet surfaces by directing compressor-discharge air into the hollow compressor inlet guide vanes.

To the rear of the compressor is the diffuser section which reduces air velocity and increases air pressure for entry into the combustion chambers.

The combustion section houses the canannular combustion chambers and the fuel. manifolds. The eight separate combustion chambers, arranged annularly, are connected by flame tubes.

Figure 8-1. T73-P-700 Gas Turbine Engine.

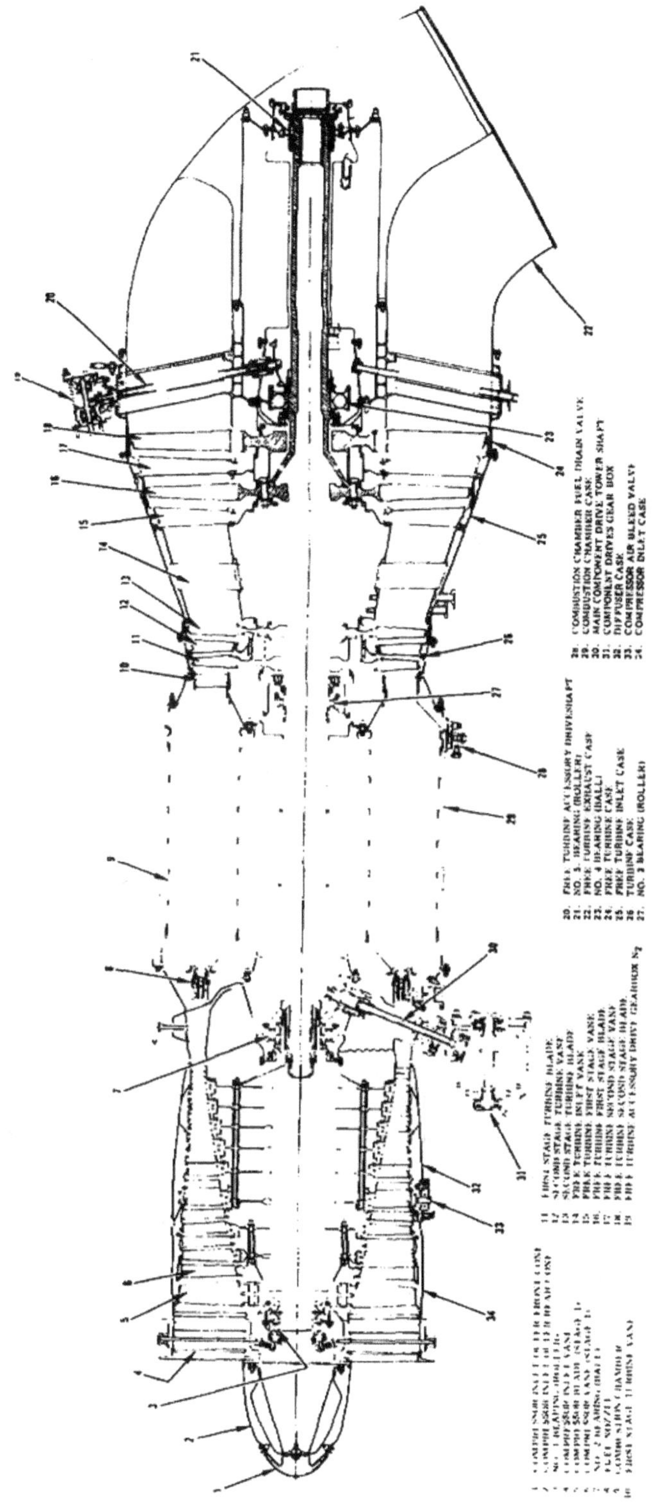

Figure 8.2. Cross-Section View of T73-P-700.

Igniter plugs are installed in number three and six combustion chambers. Right and left, clockwise and counterclockwise, upper and lower, and similar directional references apply to the engine as viewed from the rear, with the accessory section at the bottom. Table V gives the leading particulars for the T73-P-700 and T73-P-1 engines.

8.3. COMPRESSOR SECTION

The compressor inlet case, illustrated in figure 8.3, consists of the outer and inner inlet cases, outer and inner inlet vane shrouds, and hollow inlet vanes. Bosses and pads are located on the outer case for the No. 1 bearing compartment breather, scavenge, and pressure connections, inlet air pressure and temperature sensing connections, and anti-icing a air-icing air ports. The No. 1 bearing housing is bolted to the inner-inlet-case rear flange.

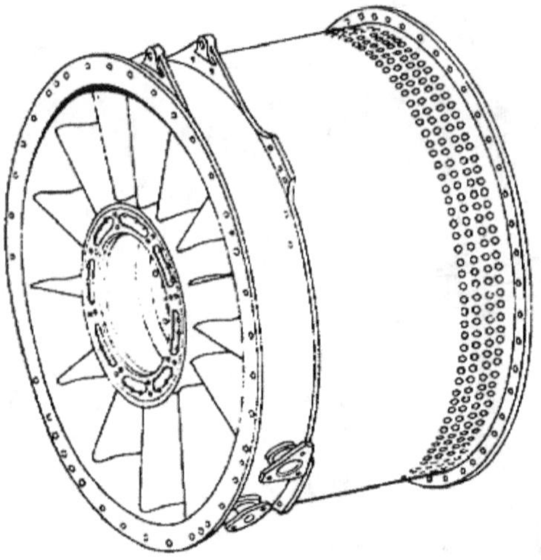

Figure 8.3. Compressor Inlet Case.

The compressor inlet case houses the first four compressor rotor and compressor vane and shroud stages. Ports in the rear of the case allow bleed air to be discharged overboard by a bleed band assembly to aid engine starting and acceleration.

The compressor rotor, shown in figure 8.4, consists of a forward hub assembly, a nine-stage compressor rotor, and an aft hub assembly. The No. 1 bearing is installed on the forward hub assembly of the compressor rotor shaft.

The first stage compressor blades are attached directly to the forward hub while all other stages of compressor blades are attached to individual disks. The blades of the first and second stages are pin mounted to allow the blades to flex. Stages three through nine have a single dovetail and are secured in broached slots in the disk by expend able blade locks.

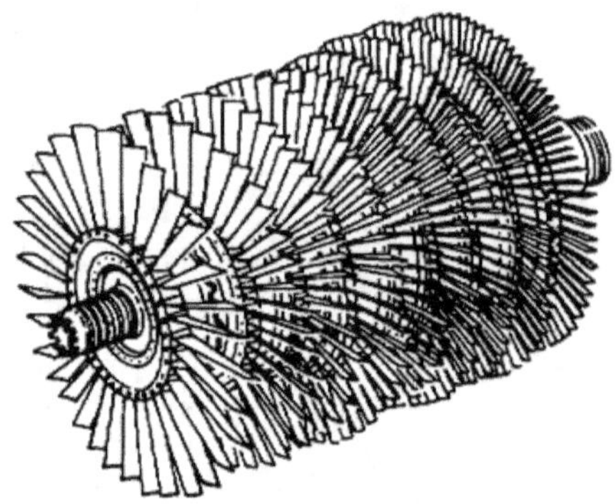

Figure 8.4. Compressor Rotor.

TABLE V. LEADING ENGINE PARTICULARS

POWER PLANT

A free turbine drive engine consisting of a multistage axial flow compressor driven by a two-stage reaction turbine. Another two-stage reaction turbine, which has no mechanical connection to the first turbine, drives the power shaft directly.

MODELS	(T73-P-1) (T73-P-700)

COMPRESSOR

Type	Axial, single spool
Stages	Nine
Direction of rotation	Clockwise

TURBINE

Type	Reaction
Stages	Two
Direction of rotation	Clockwise

FREE TURBINE

Type	Reaction
Stages	Two
Direction of rotation	Counterclockwise

COMBUSTION CHAMBERS

Type	Can-Annular, straight flow
Number	Eight

ENGINE DIMENSIONS (at room temperature)

Length (maximum)	108 inches
Diameter	30 inches

FUEL AND LUBRICANTS

Fuel specification	MIL-T-5624/JP4
Lubrication specification	MIL-L-23699
Oil consumption, maximum (as measured over a ten hour period)	0.11 gal/hr.

ENGINE DRY WEIGHT*

T73-P-1 (approximate)	920 pounds
T73-P-700 (approximate)	935 pounds

ADDITIONAL EQUIPMENT WEIGHT

Oil tank	14.0 pounds
Fuel oil cooler	15.0 pounds
Fuel heater	14.0 pounds
Inlet bullet nose cone	3.0 pounds

*Not including oil tank and associated hardware and plumbing, fuel oil cooler and associated plumbing, flowmeter fireseal brackets, ground handling brackets inlet nose cone, combustion chamber case rear mount bracket, instrumentation provisions and bosses, fuel heater, aircraft equipment brackets, shipping covers, and enclosures.

The diffuser case, illustrated in figure 8.5, is a welded steel assembly located between the compressor inlet case and the combustion chamber case. It consists of an inner and outer case and 8 hollow struts. It houses stages 5 through 9 of the compressor rotor shroud and vane assembly, the 9th stage exit guide vanes, and No. 2 main bearing and seal.

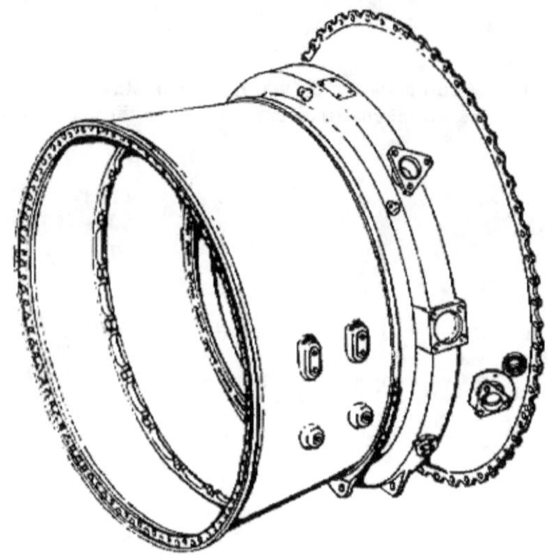

Figure 8.5. Diffuser Case.

The diffuser assembly reduces air velocity and increases air pressure for entry into the combustion chambers. High pressure air from this case is bled off for anti-icing, fuel heating, and bearing pressurizing.

On the outer case at the 3 and 9 o'clock positions are the engine mount pads. At the bottom of the case are bosses for installing the gearbox and the fuel-pressurizing dump valve.

8.4. COMBUSTION SECTION

The combustion chamber case is bolted between the diffuser case and the free turbine case. This combustion chamber outer case, shown in figure 8.6, forms the outer rigid support member of the engine and must be removed for completing a hot section inspection. Located within the combustion outer case are eight combustion chambers (cans), the combustion inner case, fuel manifolds, and the combustion chamber outlet duct. The combustion chambers are attached by clamps to the combustion chamber outlet duct in the aft end of the combustion chamber case. Flame tubes interconnect all combustion chambers. Chambers three and six have igniter plug cutouts. Each combustion chamber, illustrated in

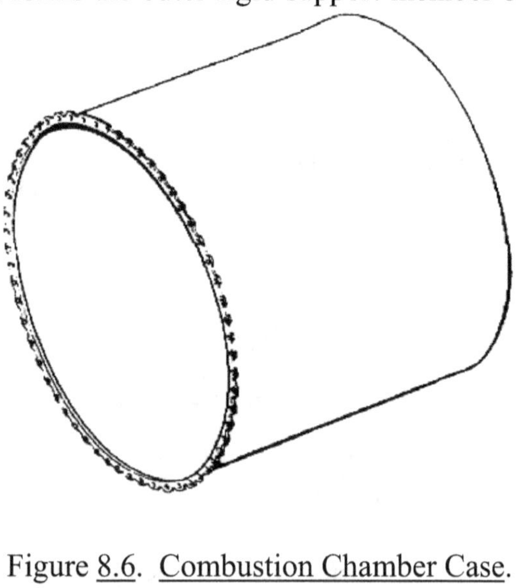

Figure 8.6. Combustion Chamber Case.

figure 8.7, consists of six liners, a fuel nozzle cap, a fuel nozzle cup adapter, and male or female flame tubes.

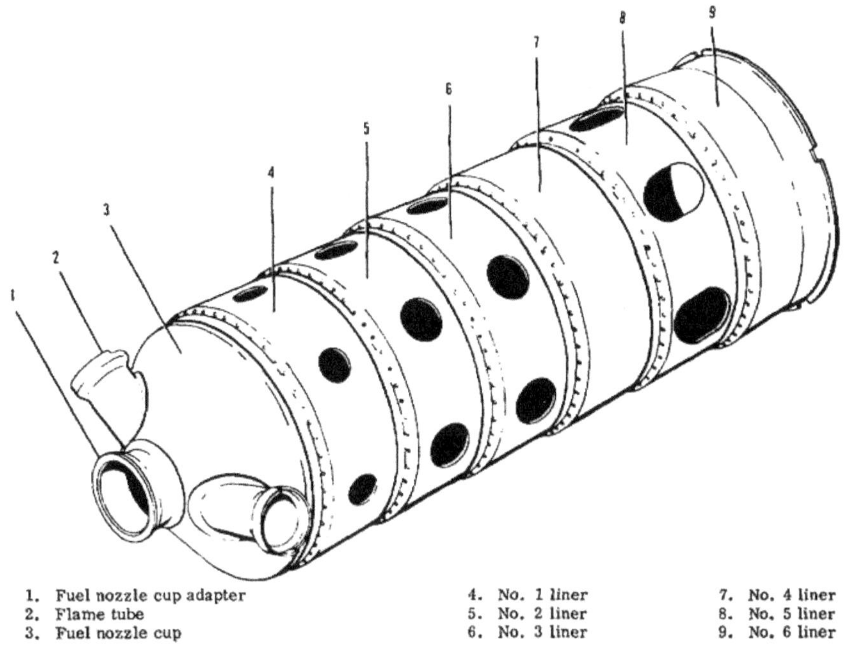

Figure 8.7. Combustion Chamber.

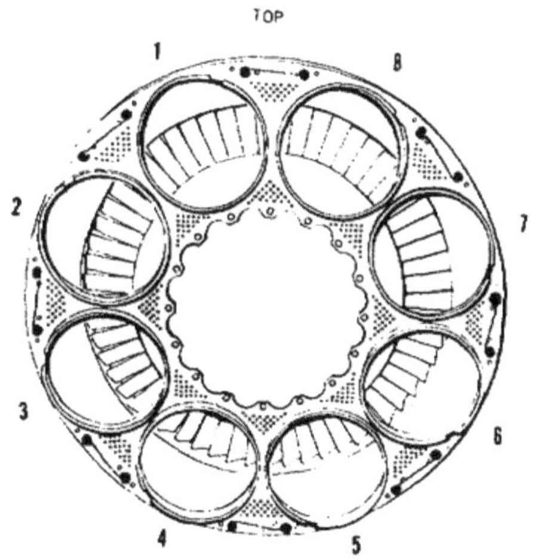

Figure 8.8. Combustion Chamber Outlet Duct and Combustor (Can) Positions.

The combustion chamber outlet duct, shown in figure 8.8, acts as a transition area to combine the gas flow from the eight combustion chambers and introduce it into the first stage nozzle vanes.

The fuel manifold consists of a secondary manifold within the primary manifold. Eight duplex fuel nozzles, one for each combustion chamber, are placed around the fuel manifold. Each nozzle has a primary orifice and a secondary orifice. Fuel sprays from the primary orifice during low pressure and from both orifices at high pressure. Fuel strainers in the primary and secondary passages of each nozzle prevent foreign matter from clogging the orifices. A fuel drain valve, at the bottom of the combustion case, automatically drains the combustion section after engine shutdown in the event of a false or aborted start.

8.5. GAS PRODUCER TURBINE SECTION

The gas producer turbine rotor assembly, illustrated in figure 8.9, consists of the turbine shaft and the first and second stage disks and blades. The turbine disks are attached to the turbine shaft hub with bolts and are separated from each other by the turbine rotor inner seal. The first and second stage turbine blades are shrouded. The shrouds form a continuous band which tends to reduce blade vibration, improve the airflow characteristics, and increase the efficiency of the turbine. The first and second stage turbine blades are placed in the fir-tree serrations of the disks and are held in place with rivets.

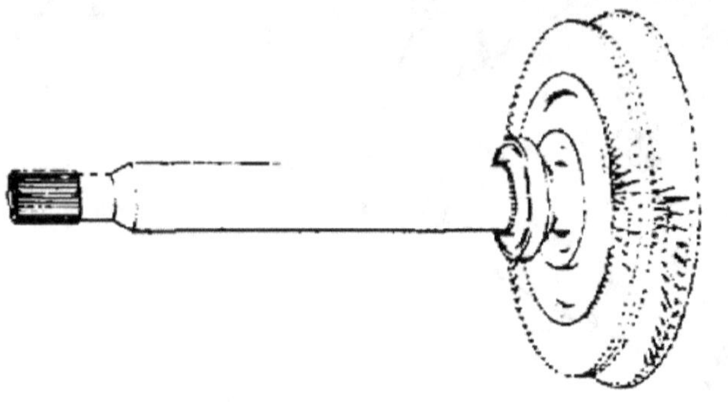

Figure 8.9. Gas Producer Turbine Rotor.

The turbine rotor is supported by the No. 3 bearing and by the splined end of the compressor rear hub. Installed on the turbine shaft is the No. 3 bearing seal assembly.

8.6. POWER TURBINE SECTION

The free-power turbine is a two-stage axial-flow turbine. The turbine inlet case, item 25, figure 8.2, is mounted on the rear flange of the gas producer turbine case. The power-turbine rotor turns counterclockwise while the gas producer rotor turns clockwise. The power turbine rotor assembly, items 16 and 18 in figure 8.2, consists of the turbine shaft, first and second stage disk and blades, turbine coupling, and accessory drive gear. Tiebolts attach the disk and blade assemblies to the front end of the turbine shaft. The shaft is supported at the rear by the No. 5 bearing and at the forward end by the No. 4 bearing.

The free-turbine inlet case houses the turbine inlet duct, the first and second stage vanes, and the free-stage turbine rotor.

The free-turbine case, item 24, figure 8.2, is mounted to the rear flange of the free turbine inlet case. The turbine case consists of the free-turbine outer case, inner front and rear case, four hollow struts and their Outer shrouds, twelve exhaust struts, and the No. 4 bearing housing. Housed in the case are the free-power turbine second-stage rotor, No. 4 bearing oil nozzle, and free turbine accessory drive gear and shaft assembly, item 20 in figure 8.2.

8.7. TURBINE EXHAUST DUCT

The free-turbine shaft and shaft cases are housed in the exhaust duct. The exhaust duct assembly consists of inner and outer assemblies, stiffeners, exhaust duct ring, and front and rear flanges. The exhaust duct bolts to the free-turbine case rear flange.

8.8. MAIN BEARINGS

All references to main bearings in this chapter are made by bearing number rather than by bearing nomenclature. Here are the bearing numbers with their corresponding nomenclatures and types.

Bearing Number	Nomenclature and Type
1	Compressor, front (roller)
2	Compressor, rear (ball)
3	Turbine (roller)
4	Free turbine, front (ball)
5	Free turbine, rear (roller)

8.9. ENGINE FUEL SYSTEM

The T-73 is equipped with a single-element, centrifugal-boost fuel pump. The pump is mounted on and driven by the component-drive gearbox. On the rear of the pump is a mounting pad for the engine fuel control. The fuel boost pump raises the fuel pressure by approximately 20 psi. Fuel then passes through an externally mounted fuel de-icing heater. From the de-icing heater, fuel returns to the pumps at the inlet fuel filter and is directed to the main pumping element. The main pumping element raises the pressure to approximately 800 psi, and fuel passes out of the pump to the engine fuel control.

The fuel control is a hydromechanical control designed to meter fuel to the engine. The control has a fuel metering system and a computing system. The metering system, subject to engine operating limitations, selects the rate of fuel flow supplied to the engine in accordance with the amount of power requested. The computing system allows the maximum engine performance available without exceeding operating limits. The fuel then flows through tubing to the right and left fuel manifolds and the eight fuel nozzles in the combustors.

8.10. INTERNAL COOLING AND PRESSURIZATION

Ninth stage air passes through holes in the compressor rotor rear hub, down through the inside of the rotor, and out through holes in the front hub. This air is used to pressurize the main bearing seals and to cool hot parts of the engine.

8.11. INLET GUIDE VANE ANTI-ICING SYSTEM

To prevent icing of the compressor inlet surfaces, an anti-icing air system is installed in the engine. Compressor bleed air is carried forward to the inlet case by an external tube on the left side of the engine. Air flow through this tube is controlled by a solenoid-operated valve. When icing conditions are encountered, the anti-ice switch in the cockpit is turned on to deenergize the solenoid which allows the valve to open and hot anti-icing air to pass forward.

Anti-icing air enters the compressor inlet outer case through the anti-icing air boss and into the cavity formed by the compressor-inlet outer case and the IGV outer, shroud. The air then passes through the hollow inlet-vanes and through openings in the inner shroud to the bullet nose cone.

The air passes through the bullet nose cone and is ejected out through louvers into the air-stream.

8.12. LUBRICATION SYSTEM

The engine lubrication system is illustrated in figure 8.10. Oil from the tank is fed to the inlet of the pressure section of the main oil pump. The pressure oil is then directed through the main oil filter, through the fuel oil cooler if cooling is required, and to the bearings. The oil passes through external tubing to the No. 1 and 2 bearing compartments. Oil to the No. 3 bearing compartment is supplied by an internal tube that connects with No. 2 bearing supply. Pressure oil flows through an external tube to the free turbine accessory drive gearbox. An internal line carries oil from this connector to the No. 4 and 5 bearings. Pressure oil flow is maintained by metering orifices, thus providing a relatively constant oil flow at

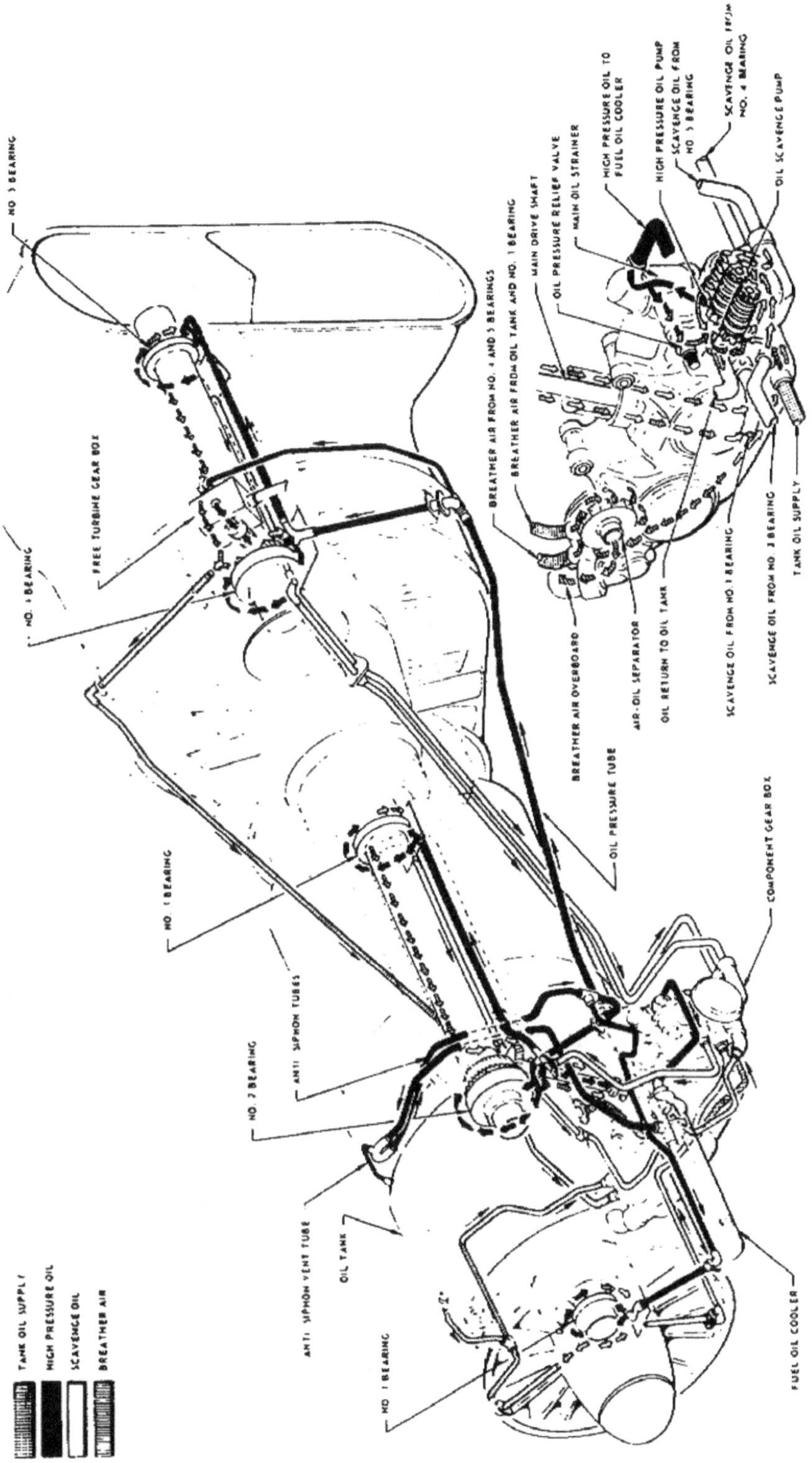

Figure 8.10. Engine Lubrication System.

all engine operating speeds. Oil pressure is controlled by the pressure relief valve. If the oil strainer becomes clogged, a bypass valve opens permitting oil flow and allowing engine operation to continue.

Scavenge oil from the five main bearings and gearbox is scavenged by five of the six gear stages in the main oil pump.

The first stage is a pressure pump. Stage two scavenges oil from No. 1 bearing; stage three from No. 3 bearing; stage four from the gearbox and No. 2 bearing; stage five from No. 4 bearing; and stage six from No. 5 bearing. The scavenge oil empties into a common tube that returns the oil to the tank. An air-oil separator in the gearbox removes oil from the breather air. Return oil passes through a de-aerator which removes most of the air. The oil tank contains baffles to prevent re-aeration of the oil in the tank.

Each of the bearing compartments and the oil tank are vented to the components drive gearbox. A rotary separator within the gearbox separates the majority of oil particles from the breather air. This air then exits through the overboard breather connector on the upper right side of the gearbox.

8.13. TORQUE SENSOR

The torque sensor, the ignition system, and the power turbine inlet temperature indicating system make up the engine electrical system.

The torque sensor consists of a torque shaft assembly, a three-pole magnetic torque sensor, and a torquemeter. It produces a visual indication of power transmitted from the engine to the main rotor gearbox by measuring the main drive shaft twist resulting from engine torque.

8.14. IGNITION SYSTEM

The ignition circuit is energized only during the starting cycle through operation of the starter circuit. The ignition system components consists of two identical, four-joule ignition exciters, one for each igniter plug. The ignition system operates satisfactorily with an input voltage of 14 to 30v dc. Operation of this system is the same as described in chapter 2, paragraph 2.22.

8.15. POWER TURBINE INLET TEMPERATURE INDICATING SYSTEM

This system consists of a dual-junction thermocouple cable and six thermocouples. Two connectors are provided in the thermocouple cable illustrated in figure 8.11. One is for connecting to an averaging indicator and the other is for individual checking of the thermocouples. This system functions the same as the exhaust gas temperature-measuring system described in chapter 2.

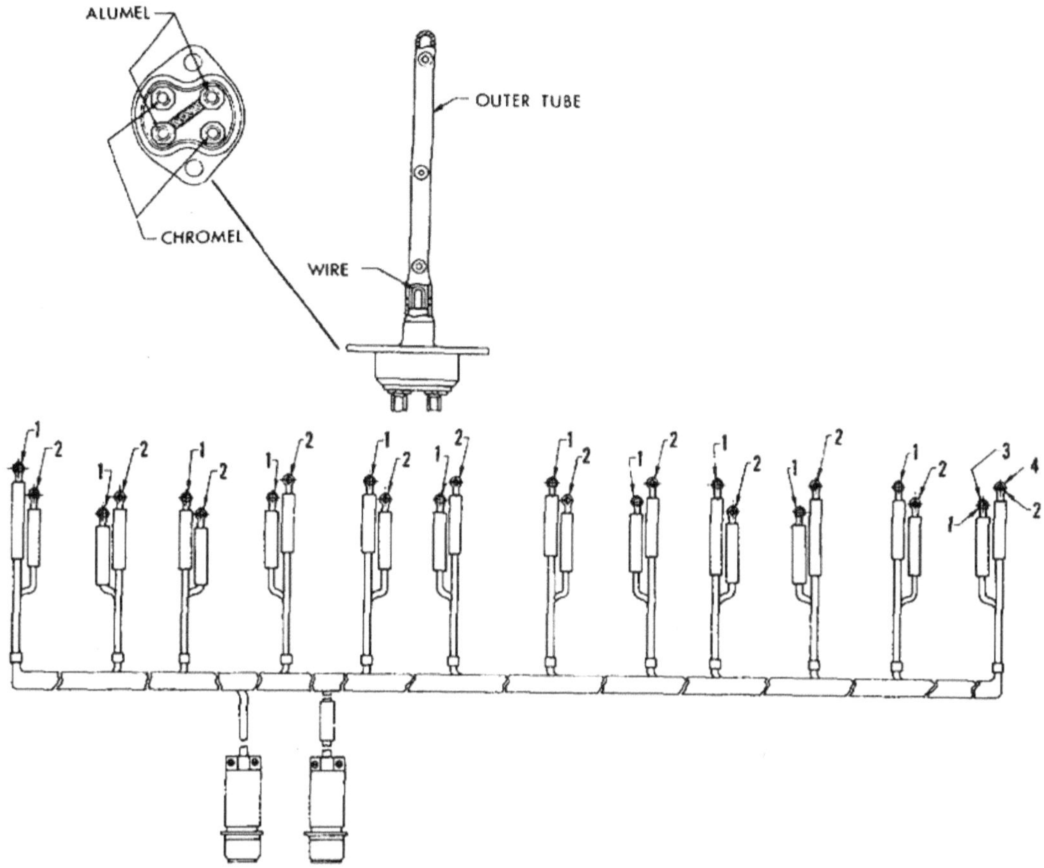

1. Individual lead
2. Averaging lead
3. Alumel terminal (green sleeve)
4. Chromel terminal (white sleeve)

Figure 8.11. Dual-Junction Thermocouple and Harness.

8.16. SUMMARY

The Pratt and Whitney T73-P-1 and the T73-P-700 are used to power the CH-54 flying crane helicopter. This engine is a straight-flow, axial-compressor, free-turbine powerplant. The axial flow compressor consists of nine stages. The combustion section has eight separate combustion chambers. The compressor is driven by a two-stage turbine mounted on the aft end of the compressor rotor shaft.

The power turbine section is also a two-stage turbine, which mounts on the front end of the turbine shaft. The turbine shaft extends out the exhaust end of the engine.

The engine fuel system has a hydromechanical fuel control with a metering and computing section that schedules fuel flow. Internal cooling and pressurization are maintained by compressor bleed air. Compressor bleed air is also used to prevent ice from forming on the engine inlet surfaces. All components of the lubrication system are mounted on the engine. The oil system is a dry-sump type.

The engine electrical system consists of the torque sensing system, ignition system, and turbine inlet temperature system.

Chapter 9

PRATT AND WHITNEY T74

9.1. INTRODUCTION

This chapter discusses the Pratt and Whitney T74 gas turbine engine. It is a reverse-flow, free-power turbine engine using a combination axial-centrifugal compressor assembly. The two models of the T74 are the T74-CP-700 and the T74-CP-702. They are used on the U-21 aircraft. Information in this chapter may be about either or both models.

Section I describes the engine operation, and section II discusses the major engine systems. Figure 9.1 shows what the engine looks like; in this case, it is the -700 model.

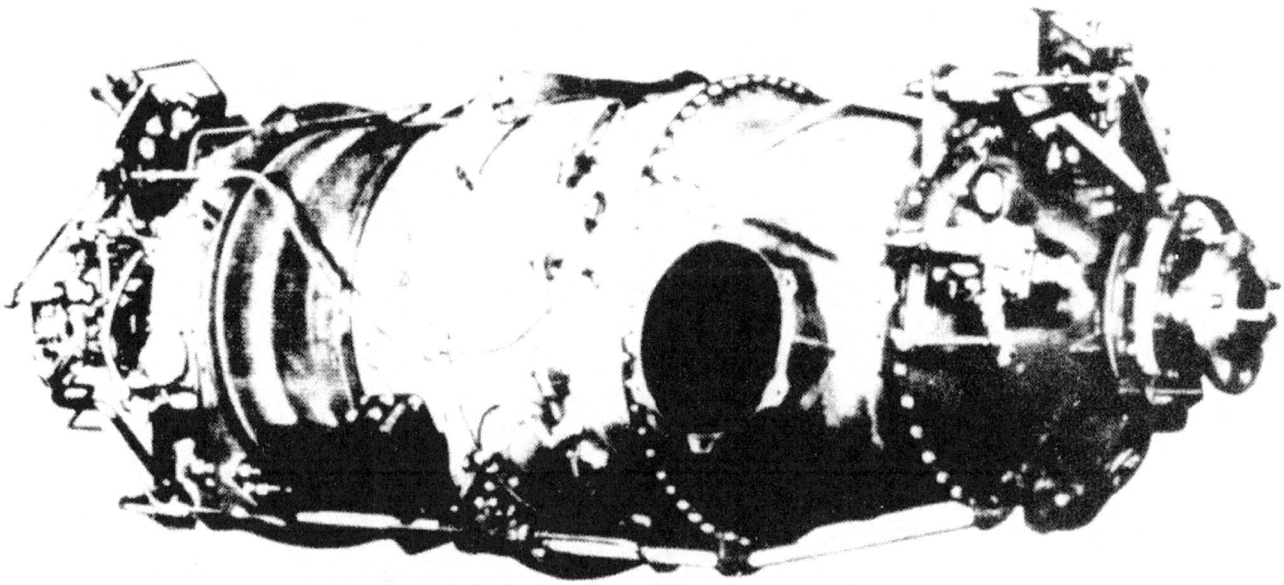

Figure 9.1. T74-CP-700.

Section I. Operational Characteristics and Description

9.2. GENERAL

This section discusses the operational characteristics and gives a description of the T74 engine. However, because these engines are undergoing continuous improvements in design and manufacture, the appearance of certain parts or details may change.

9.3. OPERATING CHARACTERISTICS

A cross-sectional view of the T74 showing the airflow through the engine is illustrated in figure 9.2. Inlet air enters the engine through a circular chamber formed by the compressor inlet case where it is directed to the compressor. The compressor consists of three axial stages combined with a single centrifugal stage.

A row of stator vanes, located behind each rotating disk, diffuses the air, raises its static pressure, and directs it to the next stage of compression. The compressed air passes through a diffuser, turns ninety degrees in direction, and is then led through straightening vanes into the combustion chamber liner.

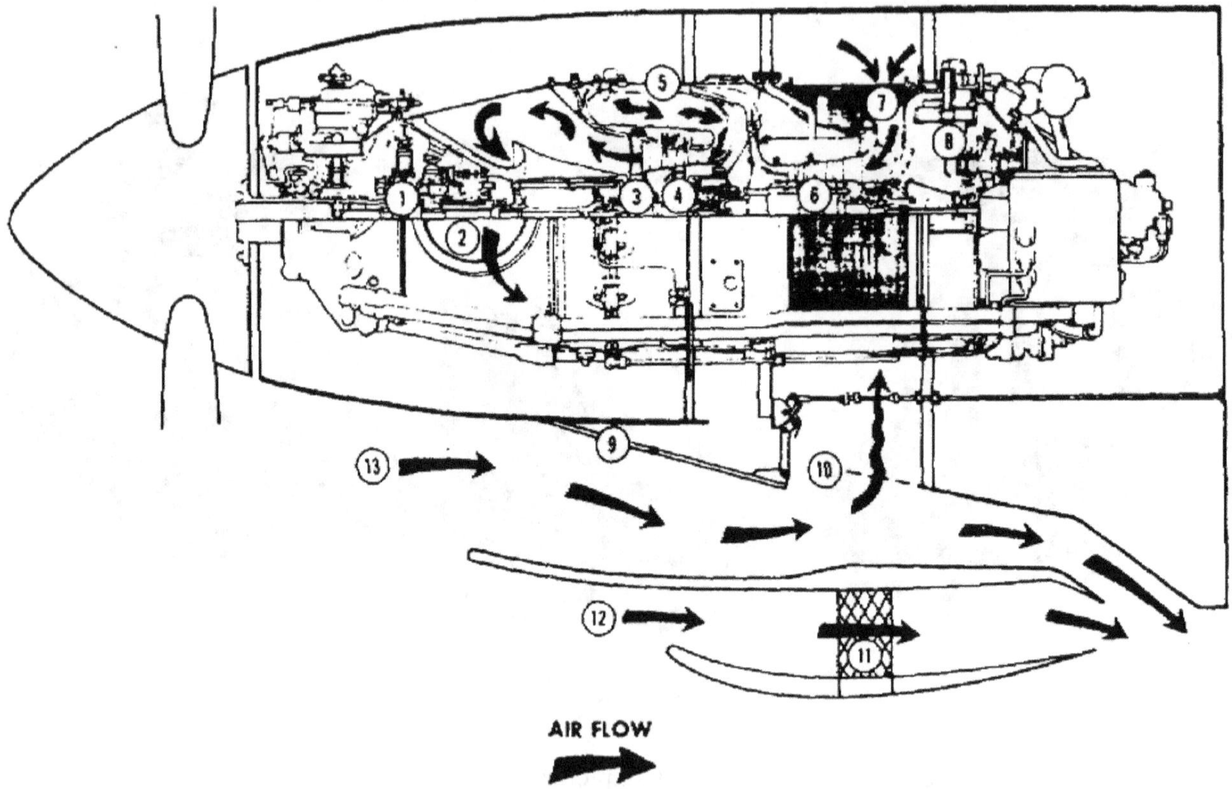

1. Propeller reduction gear
2. Exhaust outlet port
3. Power turbine (N2)
4. Compressor turbine (N1)
6. Compressor
7. Engine air inlet
8. Oil tank and accessory drive
9. Inertial separator anti-ice vane
10. Horizontal screen
11. Oil cooler
12. Oil cooler air intake
13. Engine air intake

Figure 9.2. Airflow.

The annular combustion chamber liner (ring shaped) has perforations which allow entry of compressed air. The flow of air changes direction to enter the combustion chamber liner, where it reverses direction and mixes with fuel. The location and shape of the combustion chamber liner eliminates the need for a long shaft between the compressor and turbine, reducing the overall length and weight of the engine.

Fuel is injected into the combustion chamber liner by fourteen simplex nozzles. The fuel-air mixture is ignited by two glow plugs which protrude into the combustion chamber liner. The expanding gas from the combustion chamber liner reverses direction and passes through the compressor turbine guide vanes to the compressor turbine. The gases then pass forward through the turbine guide vanes to drive the power turbine.

The compressor and power turbines, items 3 and 4 in figure 9.2, are located approximately in the center of the engine with their shafts extending in opposite directions. The exhaust gas from the power turbine is directed through an exhaust plenum to two exhaust ducts.

All engine-driven accessories, with the exception of the N_2 tachometer-generator and propeller governors, are mounted on the accessory gearbox and driven by the compressor. The oil tank is located forward of the accessory gearbox and forms part of the compressor inlet case. The tank has a total oil capacity of 2.3 gallons and has a dipstick and drain plug.

The power turbine drives a propeller through a two-stage planetary reduction gearbox located at the front of the engine. The torquemeter is located in the reduction gearbox.

9.4. STATIONS, FLANGES, AND SPECIFICATIONS

The engine stations and flanges are illustrated in figure 9.3. Stations, identified by numbers in the figure, are specific locations in the engine. Flanges, identified by letters in the figure, are rims or edges providing strength in the attachment of one engine section to another.

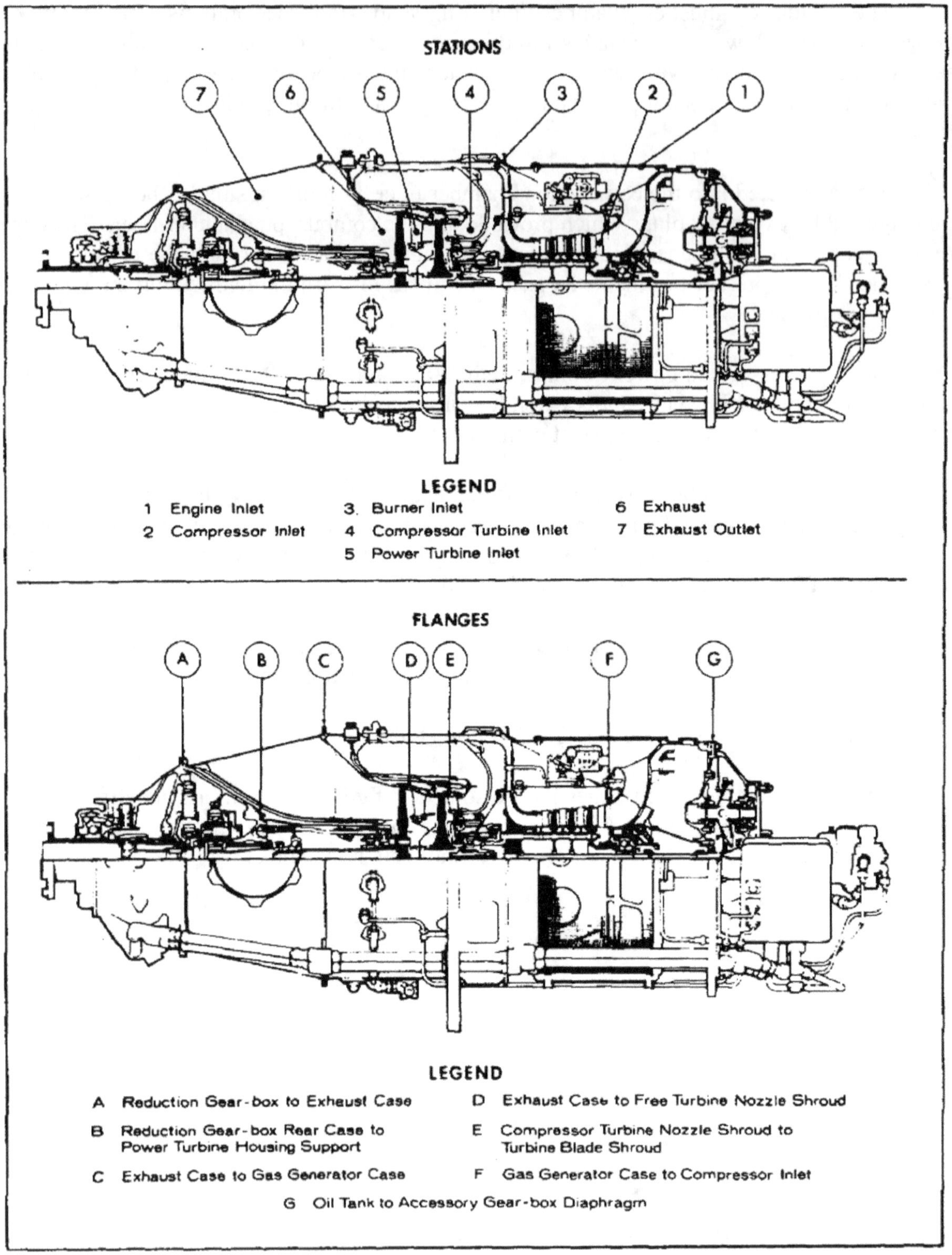

Figure 9.3. Stations and Flange Locations, -700.

Specifications for the T74-CP-702 engines used in Army aircraft are summarized in the following chart.

MODEL	T74-CP-702
COMPRESSOR	3 Axial, 1 Centrifugal
COMBUSTOR	Annular Reverse Flow
TURBINE (GP)	Single Stage
TURBINE (PT)	Single Stage
PROPELLER SHAFT GEAR RATIO	0.0668
ENGINE DIAMETER	19 in.
ENGINE LENGTH	62 in.
DRY WEIGHT	289 lb
TAKEOFF SHP	750
MILITARY SHP	700
NORMAL SHP	688
PROP rpm	2200
FUEL	MIL-T-5624 JP-4/5
OIL	MIL-L-23699
APPLICATION	Two in U-21 aircraft

9.5. INLET AND COMPRESSOR SECTIONS

The following subparagraphs discuss the inlet and compressor sections. An exploded view of the T74-CP-702 engine is shown in figure 9. 4.

a. The compressor inlet case, shown in figure 9.5, consists of a circular aluminum-alloy casting; the front forms a plenum chamber for the passage of compressor inlet air. The rear portion, which consists of a hollow compartment, is used to house the oil supply. A large circular steel screen (item 11 in figure 9.4) is bolted around the air intake and the rear of the gas generator case to prevent foreign object ingestion by the compressor.

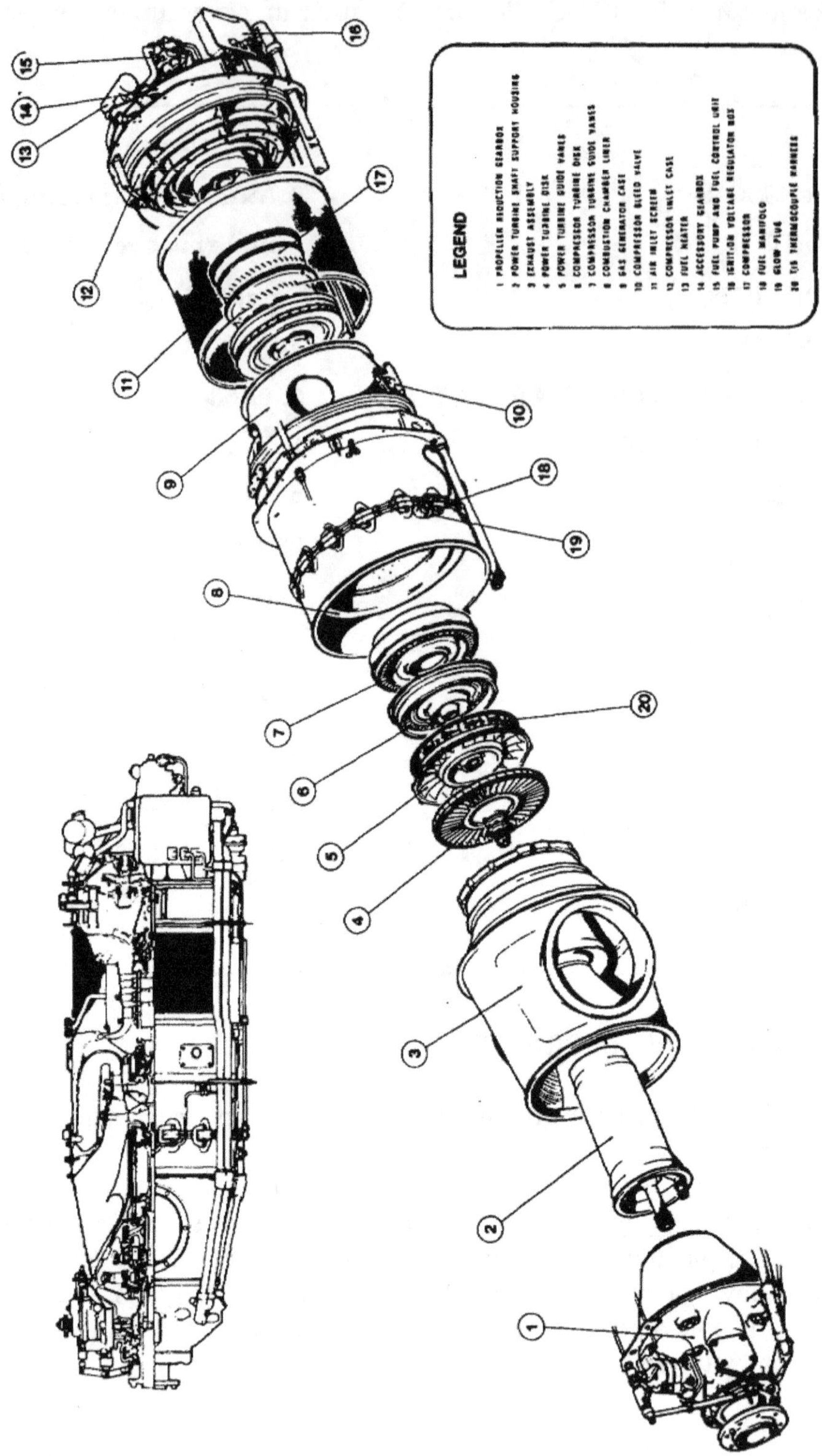

Figure 9.4. Exploded View of the T74-CP-702.

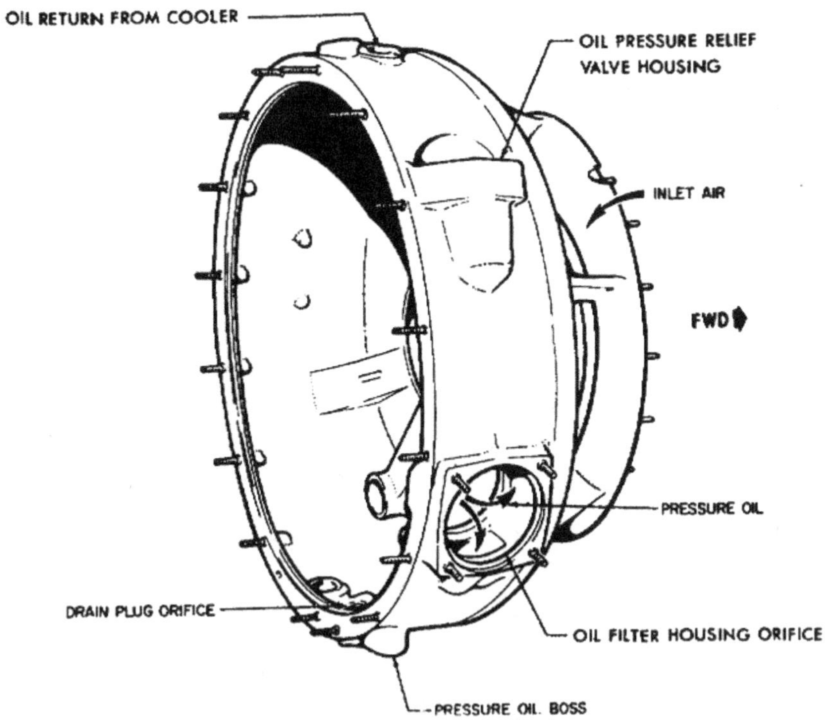

Figure 9.5. Compressor Inlet Case, -700 Engine.

The No. 1 bearing support is contained within the compressor inlet case centerbore. A conical tube is fitted in the centerbore of the oil tank compartment to provide a passage for the coupling shaft which extends the compressor drive to the accessories mounted on the rear accessory case. The pressure oil pump, driven by an accessory drive gear, is located in the bottom of the oil tank. The pressure oil relief valve and main oil filter, with check valve and bypass valve assemblies; are located on the right side of the inlet case at the 1 and 3 o'clock positions, respectively.

b. The compressor rotor and stator assembly, shown in figure 9.6, consists of a three-stage axial rotor, three interstage spacers, three stator assemblies, and a single-stage centrifugal impeller and housing. The compressor blades are made of stainless steel and attached, with limited clearance, to the compressor hub in dovetail grooves. This accounts for the metallic clicking heard during compressor run-down. Axial movement of compressor blades is limited by the interstage spacers located between the disks. The first stage compressor blade airfoil differs from those in the second and third stages which are identical. All three stages differ in

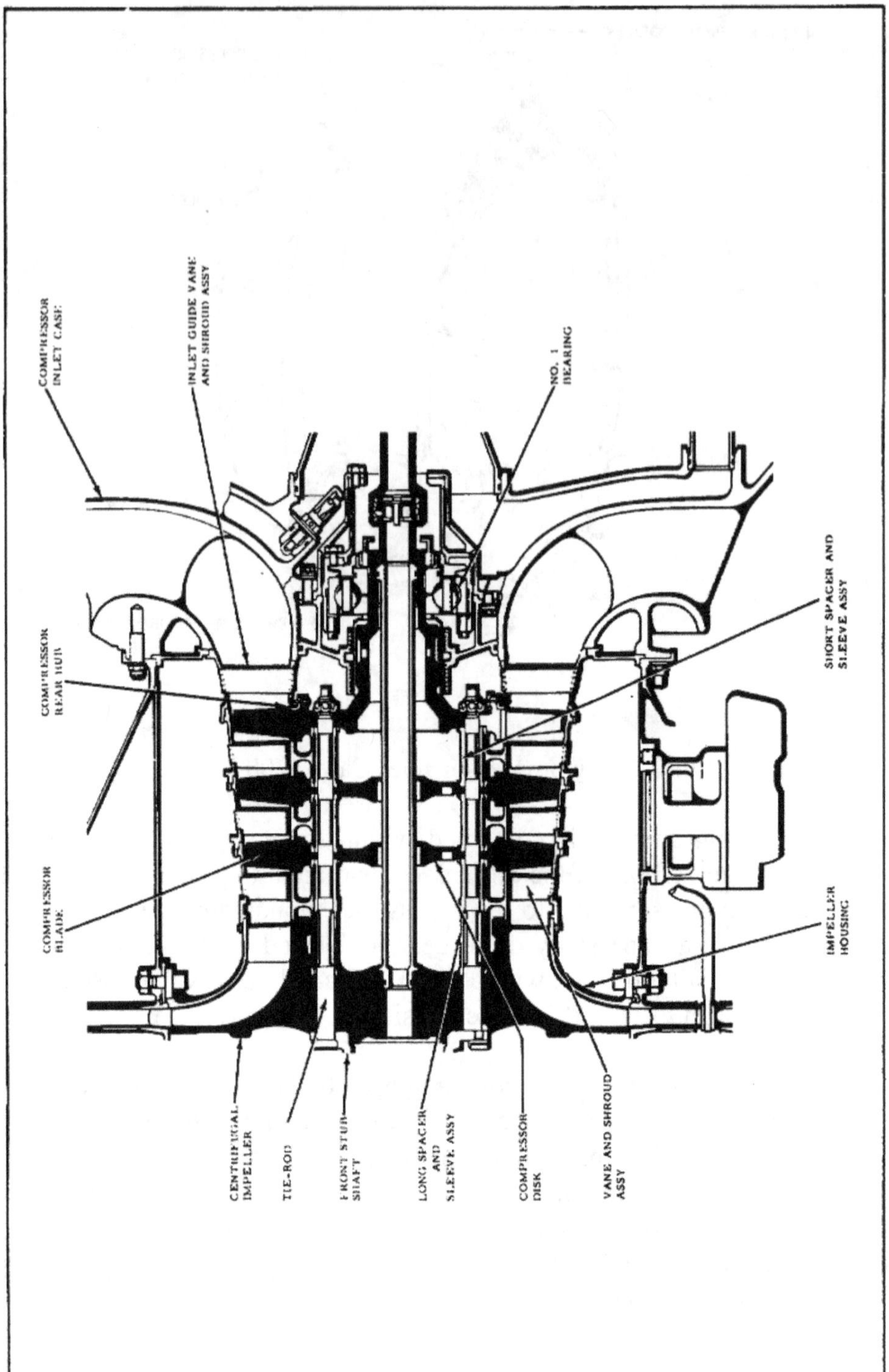

Figure 9.6. Compressor Assembly, -700.

length, decreasing from the first to the third stage. The No. 1 bearing is a ball bearing and supports the rear of the compressor assembly in the inlet case. The No. 2 bearing is a roller bearing and supports the front of the compressor and gas generator turbine.

c. The gas generator case is attached to the front flange of the compressor inlet case and consists of two stainless steel sections made into a single structure. The rear inner section supports the compressor assembly. The compressor bleed valve outlet port is located at the seven o'clock position forward of the inlet screen. The No. 2 bearing support is located in the centerbore of the case. The diffuser pipes or vanes brazed inside the center section of the gas generator case create a pressure increase in the compressor air as it leaves the centrifugal impeller. The compressed air is then directed through straightening vanes, located at the outlet of the diffuser, and out to the combustion chamber liner.

The front section of the gas generator case forms the outer housing for the combustion chamber liner. It consists of a circular stainless steel structure for mounting the 14 fuel nozzle assemblies and the manifold. Front and rear drain valves are mounted at the 6 o'clock position to allow residual fuel to drain overboard during engine shutdown after a false or aborted start. Two glow plugs protrude into the combustion chamber liner to ignite the fuel-air mixture. The engine is mounted on three flexible type mounts which are secured to mounting pads located on the outer circumference of the gas generator case.

9.6. COMBUSTION CHAMBER LINER

Located in the front section of the gas generator case is the combustion chamber liner. The liner is of the reverse flow type and consists primarily of an annular, heat-resistant steel liner open at one end. The combustion chamber is illustrated in figure 9.7.

The liner has perforations which allow air to enter the liner for combustion. The perforations insure an even temperature distribution at the compressor turbine inlet. The domed front end of the liner is supported inside the gas generator case by seven of the 14 fuel nozzles. The rear of the liner is supported by sliding joints which fit to the inner and outer exit duct assemblies.

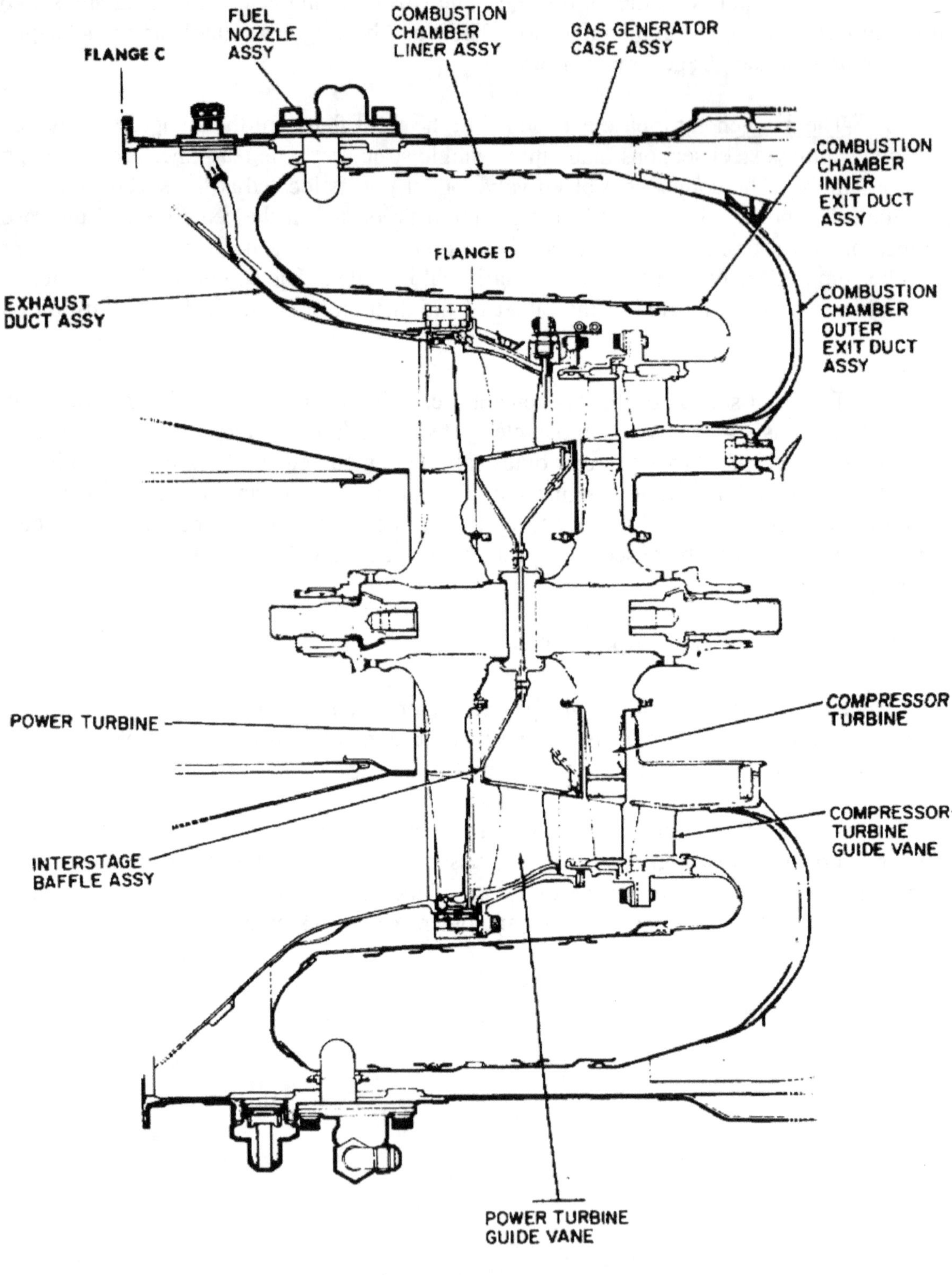

Figure 9.7. Combustion Chamber and Turbine Section, -702.

The exit duct forms an envelope which changes the direction of the gas flow by providing an outlet close to the compressor turbine guide vanes. The vanes ensure that the expanding gases are directed to the compressor turbine blades at the proper angle to drive the compressor.

9.7. TURBINE SECTION

The turbine rotor consists of two separate single-stage turbines located in the center of the gas generator case and completely surrounded by the annular combustion chamber liner. The following subparagraphs discuss the compressor and power turbines.

a. The compressor turbine consists of a turbine disk, blades, and weights. The turbine drives the compressor in a counterclockwise direction. The turbine assembly is splined to the compressor front hub and secured by a threaded centerlocking turbine bolt and washer. A master spline is provided to ensure that the disk assembly is always installed to a position to retain original balance. The disk has a circumferential reference groove to enable checking disk growth. The 58 cast steel alloy blades in the compressor turbine disk are secured in fir-tree serrations in the disk by individual tubular rivets.

The compressor turbine is separated from the power turbine by an interstage baffle. This baffle prevents dissipation of turbine gas and transmission of heat to turbine disk faces.

b. The power turbine disk assembly, consisting of a turbine disk, blades, and weights, drives the reduction gearing through the power turbine shaft in a clockwise direction. The power turbine guide vanes are located ahead of the power turbine rotor in the gas stream. The vanes direct the flow of gas at the most efficient angle to drive the power turbine. The power turbine disk is made to close tolerances and has a circumferential reference groove to permit taking disk growth measurements when required. A master spline insures that the turbine disk assembly is installed in a predetermined position to retain the original balance. The required number of weights is determined during balancing procedures and riveted to a special flange located on the rear of the turbine disk. The power turbine blades differ from those of the compressor turbine in that they are cast with notched and shrouded tips. The blades are secured by fir-tree serrations in the turbine disk. The blade tips rotate inside a double knife-edge shroud and form a continuous seal when the engine is running. This reduces tip leakage and increases turbine efficiency.

9.8. EXHAUST DUCT

The exhaust duct shown in figure 9.8 consists of a divergent, heat-resistant steel duct provided with two outlet ports, one on each side of the case. The duct is attached to the front flange of the gas generator case and consists of inner and outer sections. The outer conical section, which has two flanged exhaust outlet ports, forms the outer gas path and also functions as a structural member to support the reduction gearbox. The inner section forms the inner gas path and provides a compartment for the reduction gearbox rear case and the power turbine support housing. A removable sandwich-type heat shield insulates the rear case and support housing from the hot exhaust gases. A drain passage located at the 6 o'clock position at flange C leads to the gas generator case. This automatically drains the exhaust duct of any fluid accumulated during engine shutdown through the front drain valve on the gas generator case.

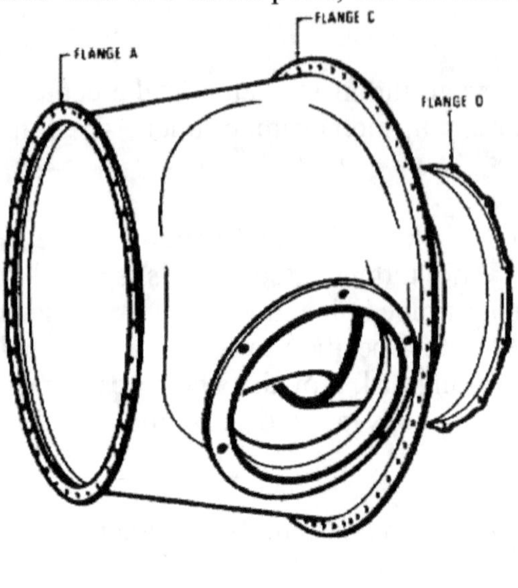

Figure 9.8. Exhaust Duct, -702.

9.9. REDUCTION GEARBOX

Located at the front of the engine is the reduction gearbox, which consists of two magnesium alloy castings bolted to the front flange of the exhaust duct. A cross-section view of the reduction gearbox is illustrated in figure 9.9.

The first stage of reduction is contained in the rear case. Torque from the power turbine is transmitted through the power turbine shaft to the first stage sun gear. The spur-gear end of the sun gear drives the three planetary gears in the first stage planet carrier. The first stage ring gear is located in helical splines in the rear case assembly. The torque developed by the power turbine is transmitted through the sun gear and planet gears to the ring gear. This results in rotation of the planetary carrier. The ring gear, though secured by the helical splines, is allowed to move axially between the case and three retaining plates. This movement is used in the torquemeter located inside the rear gearbox assembly, which is discussed in paragraph 9.16.

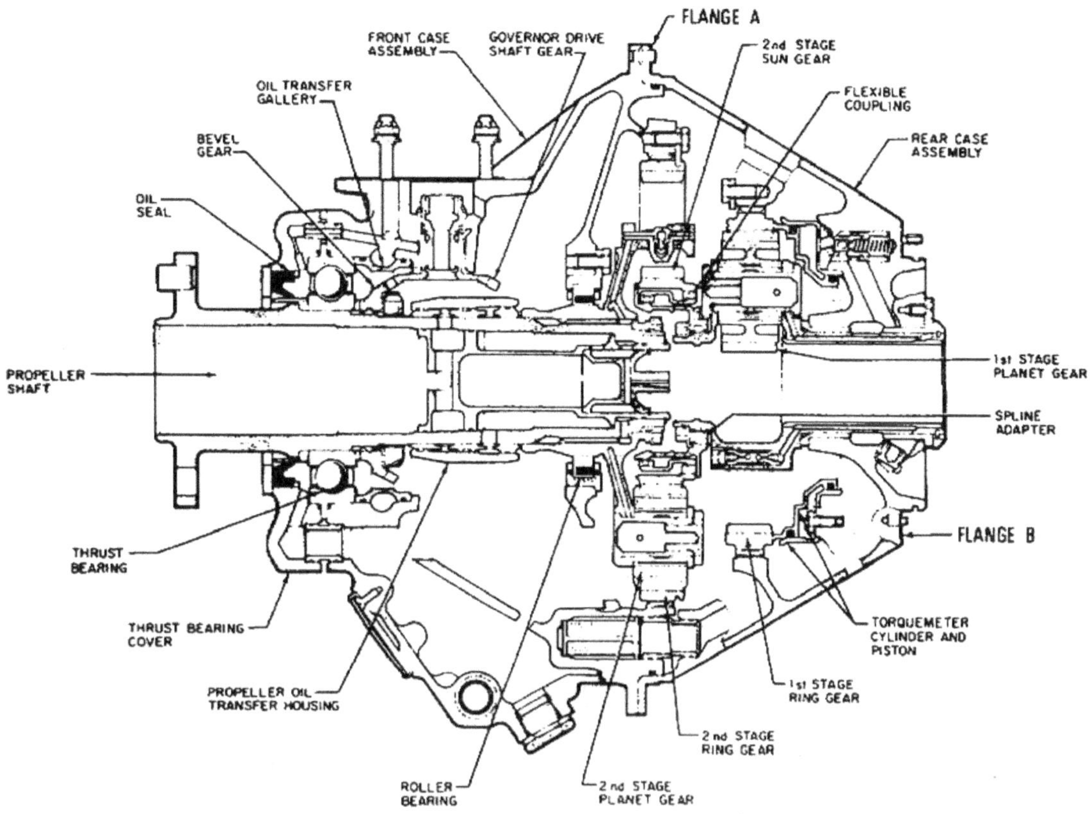

Figure 9.9. Reduction Gearbox.

The second stage of reduction is contained in the reduction gearbox front case. The first stage planet carrier is attached to the second stage sun gear by a flexible coupling which also dampens any vibrations between the two rotating masses. The second stage sun gear drives five planet gears in the second stage carrier. A second stage ring gear is fixed by splines to the reduction gearbox front case and secured by three bolted retaining plates. The second stage carrier is in turn splined to the propeller shaft and secured by a retaining nut and shroud washer.

The accessories located on the reduction gearbox front case are driven by a bevel drive gear, mounted on the propeller shaft behind the thrust bearing assembly. Propeller thrust loads are absorbed by a flanged ball bearing located in the front face of the reduction gearbox centerbore. The thrust bearing cover is secured to the front of the reduction gearbox, and it has a removable oil seal retaining ring for replacement of the oil seal.

9.10. ACCESSORY GEARBOX

Located at the rear of the engine is the accessory gearbox. It consists of two magnesium alloy castings attached to the rear flange of the compressor inlet case by 16 studs. The front casting, provided with front and rear O-rings, forms an oil-tight diaphragm between the oil tank compartment of the inlet case and the accessory drives. The accessory drive gearbox is illustrated in figure 9.10.

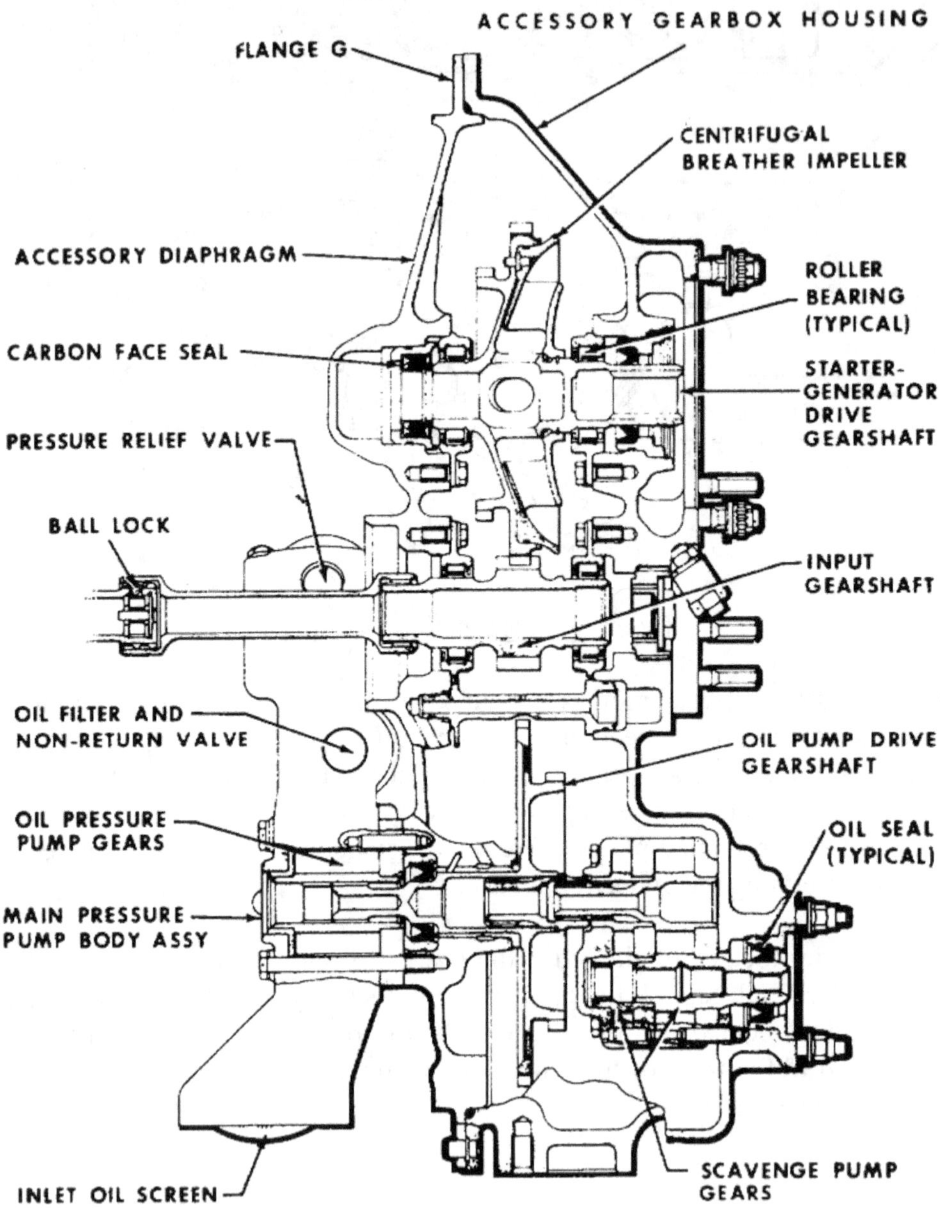

Figure 9.10. Accessory Gearbox Cross Section, -702.

The rear casting forms an accessory gearbox cover with support bosses for the accessory drive bearings and seals. The internal scavenge oil pump is secured inside the housing, and a second scavenge pump is externally mounted. Mounting pads and studs are located on the rear face for the combined starter-generator, the fuel control unit with the sandwich-mounted fuel pump, and the N_1 tachometer-generator. Three additional pads are available for optional requirements; see figure 9.11. An oil tank filler cap and dipstick are located at the eleven o'clock position on the rear housing. A centrifugal oil separator mounted on the starter-generator's drive gearshaft separates the oil from the engine breather air in the accessory gearbox housing.

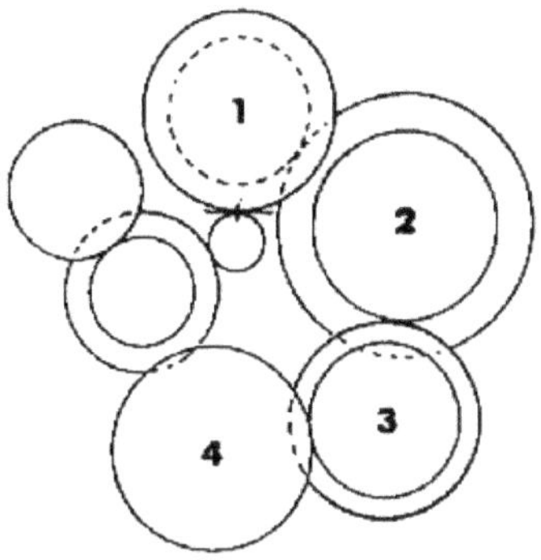

AS VIEWED FROM REAR

	DRIVE	RATIO	MAX. SPEED
	INPUT (100% Ng)	1:1	37,500
1	STARTER GENERATOR	293:1	10,991
2	FUEL PUMP & FCU	167:1	6,262
3	INTERNAL OIL SCAVENGE & PRESSURE PUMPS & TACHO-GENERATOR	112:1	4,203
4	EXTERNAL OIL SCAVENGE & OPTIONAL VACUUM PUMPS	103:1	3,821

Figure 9.11. Accessory Gearbox Train.

9.11. BEARING INSTALLATION

The compressor rotor assembly is supported and secured in the rear section of the gas generator case by two bearings. The outer race of No. 1 ball bearing is held in its flexible housings by a special nut and keywasher. The split inner race, spacer, and rotor seal are stacked against a shoulder on the compressor rear hub shaft and secured by a cup-washer and special nut. The outer flange of the No. 2 roller bearing is attached to the gas generator centerbore by four bolts and tablock washers.

The compressor turbine disk holds the plain inner race stacked in position between front and rear rotor seals on the compressor front stubshaft and a shoulder on the stubshaft. The bearing installation of the T74 engine is illustrated in figure 9.12.

The power turbine disk and shaft assembly is supported and secured in the power turbine shaft housing by the No. 3 and No. 4 bearings. Bearing No. 3 is a roller type, and No. 4 is a ball type.

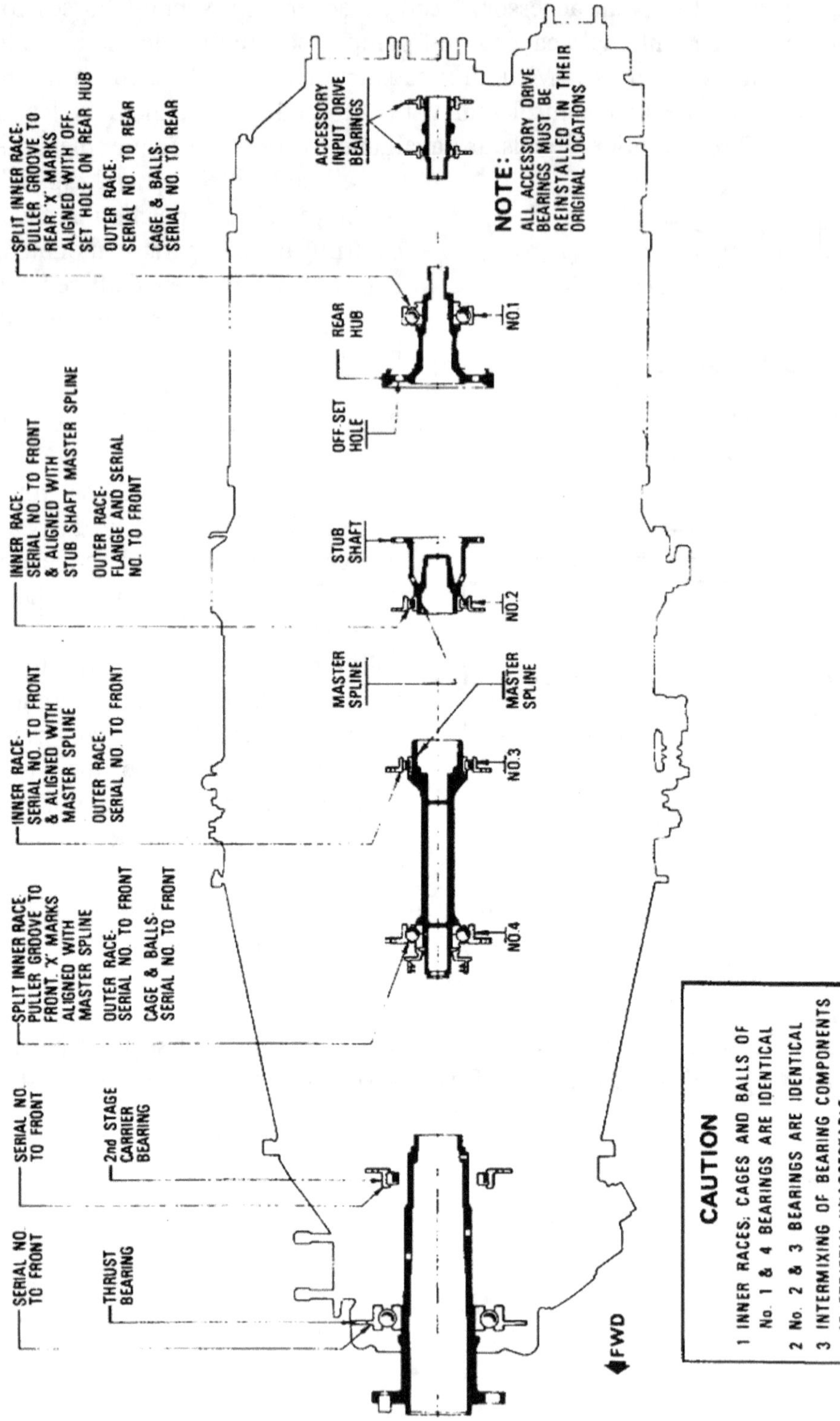

Figure 9.12. Bearing Installation.

9.12. SUMMARY

The T74 is a light-weight, free-power turbine engine designed for use in fixed-wing aircraft. It has two independent turbines, one driving a compressor in the gas generator section, and the second driving a reduction gearing for a propeller installation.

The different engine sections bolt together at flanges, which are identified by letters. Numbers are used to identify locations or stations on the engine.

The compressor is a three-stage axial rotor, which is housed in the gas generator case. Located forward of the compressor is the combustion chamber. The combustion chamber is of the annular reverse-flow type. Fourteen fuel nozzles are mounted at the front end of the combustion chamber liner. The two single-stage turbine rotors are located in the center of the annular combustion chamber. The compressor turbine assembly is splined to the compressor front hub. The power turbine rotor drives the reduction gearing through the power turbine shaft.

Ahead of the combustor and turbine assemblies is the exhaust duct. The duct is attached to the front flange of the gas generator.

Located at the front of the engine are the reduction gearbox and propeller shaft. The accessory drive gearbox is attached to the aft flange of the compressor inlet case at the rear of the engine.

The engine has four main bearings. Numbers 1 and 4 are ball bearings, and 2 and 3 are roller bearings.

Section II. Major Engine Systems

9.13. GENERAL

The T74-CP-702 has four major engine systems: the bleed air system, lubrication system, fuel system, and instrumentation and ignition system. Each of these systems is essential for safe engine operation. The section discusses each system in detail in the paragraphs that follow.

9.14. BLEED AIR SYSTEMS

The engine has three separate bleed air systems: a compressor bleed air control, a bearing compartment airseal, and a turbine disk cooling system. The engine is also equipped for cabin-pressurization air. The following subparagraphs describe the bleed air systems.

a. The compressor bleed air system. Automatically opening a valve in the gas generator case to spill interstage compressor air, the compressor bleed air system thereby prevents compressor stalls at low engine speeds. The valve closes gradually as higher engine speeds are attained.

b. Bearing compartment seals. Pressure air is used to seal the 1st, 2d, and 3d bearing compartments and also to cool both the compressor and the power turbines.

c. Turbine disk cooling system. The compressor and power turbine disks are both cooled by compressor discharge air bled from the straightening vane area of the diffuser. It is then metered through holes in the compressor turbine vane support into the turbine hub baffle, where it divides into three paths. Some of the air is metered to cool the rear face of the compressor turbine disk, and some to pressurize the bearing seals. The air is then led forward through passages in the compressor turbine hub to cool the front face of the compressor turbine. A portion of this cooling air is also led through a passage in the center of the interstage baffle to the rear face of the power turbine disk. The remaining air is used to cool the front face of the power turbine disk. The cooling air from both of the turbine disks is dissipated into the main gas stream flow to the atmosphere.

9.15. LUBRICATION SYSTEM

The T74 engine oil system is designed to provide a constant supply of clean lubricating oil to the engine bearings, reduction gears, torquemeter, propeller, and all accessory drive gears. The oil system consists of the components covered in the following subparagraphs. A schematic of the -702 engine lubrication system is illustrated in figure 9.13.

a. The oil tank is part of the compressor inlet case. The tank has a total capacity of 2. 3 gallons of which 1.5 gallons are usable. This capacity allows for oil expansion of approximately 0.8 gallons.

The oil tank has an oil filler neck and a quantity dipstick and cap which protrudes through the accessory gearbox housing at the 11 o'clock position. An anti-flooding and breather arrangement,

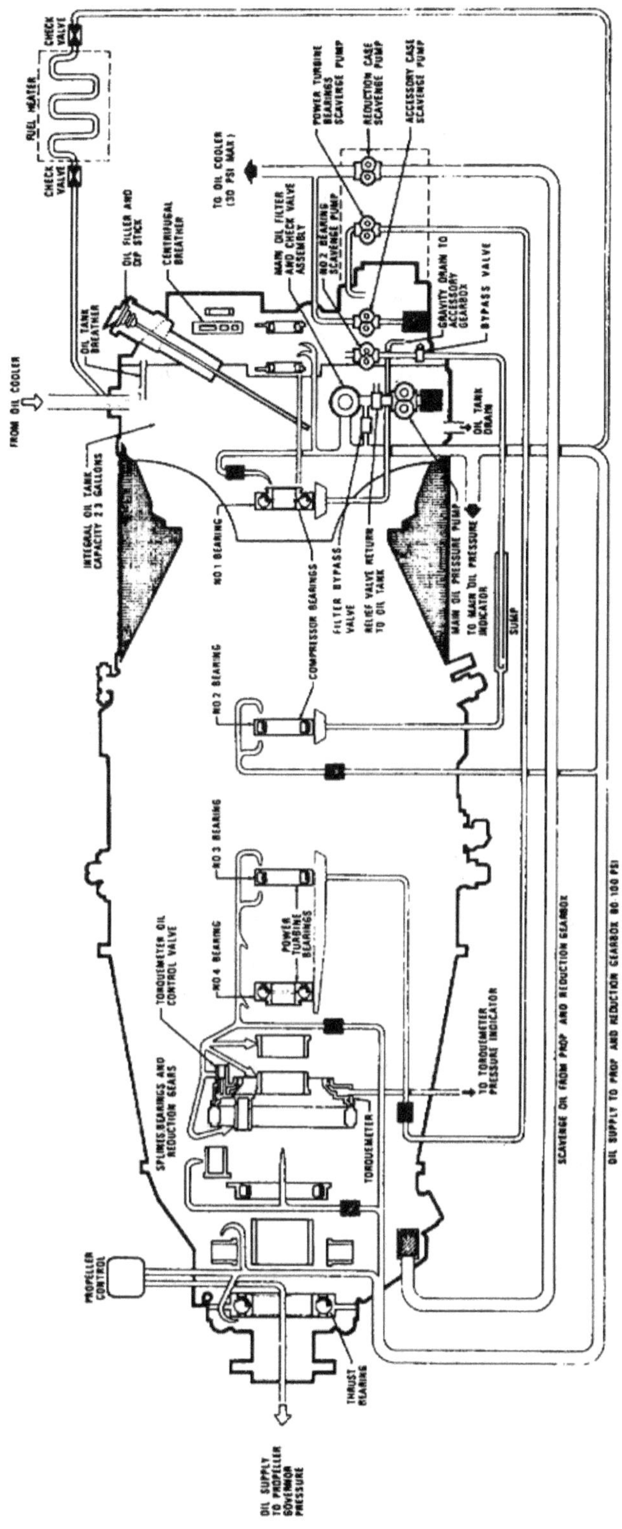

Figure 9.13. Engine Lubrication System Schematic, -702.

located in the highest point of the oil tank, prevents flooding of the accessory gearbox if the oil tank is overfilled. A drain plug is mounted in the bottom of the tank.

b. A gear-type oil pressure pump is located in the lowest part of the tank. The pump consists of two gears contained in a cast housing, bolted to the front face of the accessory diaphragm. It is driven by the accessory gearshaft which drives the internal double-element scavenge pump. The oil pump has an inlet filter screen, check valve, and relief valve.

c. The oil filter assembly consists of a disposable cartridge type filter element with a perforated flanged end, bypass valve, and check valve. The filter assembly housing is located in the compressor inlet case at the 3 o'clock position. If the filter becomes clogged, the increased pressure opens the bypass valve, and an alternate passage for unfiltered oil to flow through the engine is used. The check valve, positioned in the end of the housing, prevents gravity flow into the engine after shutdown and permits the filter element to be changed without having to drain the oil tank.

d. The centrifugal breather consists of a shrouded aluminum alloy impeller secured to the rear face of the starter-generator gearshaft by a retaining ring. Breather air flows radially inward through the rotating impeller housing where the oil particles are separated from the air mist by centrifugal force. The oil particles are thrown outward and drain freely to the bottom of the accessory gearbox. The air is then routed through a transfer tube to a breather boss on the rear face of the accessory housing where a connection for an overboard vent line is installed.

e. The oil-to-fuel heater assembly is essentially a heat exchanger which uses heat from the engine oil lubricating system to preheat the fuel in the engine fuel system. The heater has a Vermatherm element, which senses fuel temperature, and consists of a highly expansive material sealed in a metallic chamber. The expansion force is transmitted through a diaphragm and plug to a piston. The element senses outlet fuel temperatures and, at temperatures above 70° F, starts to close the core valve and simultaneously opens the bypass valve. At 90° F, the core valve is completely closed and the oil bypasses the heater core.

9.16. PRESSURE AND SCAVENGE OIL SYSTEMS

Oil for lubrication of engine parts is supplied under pressure by the pressure oil system. This oil is then returned to the oil tank by the scavenge oil system. The following subparagraphs discuss the pressure and scavenge oil systems.

a. The pressure oil system supplies oil at 85 to 95 psi to the reduction gearbox where it is divided into two branches. One branch is led to the first stage reduction gears, splines, torquemeter, and number 3 and 4 bearings. Pressure oil to the torquemeter is led through a metering valve which controls the flow into the torquemeter chamber. The bearings and gears are lubricated by oil spray jets. The second branch supplies oil to the propeller governor unit, the accessory drive gears, and propeller thrust bearing.

b. The scavenge oil system includes two double-element scavenge pumps connected by internal passages and lines to two main external transfer tubes. One pump is secured inside the accessory gearbox and the other is externally mounted. They are contained in separate housings and driven off accessory gearshafts.

The oil from the No. 1 bearing compartment is returned by gravity through an internal cored passage to the bottom of the compressor inlet case and then through internal passages in the oil tank and accessory diaphragm into the accessory gearbox. Number 2 bearing oil drains down from its compartment into an external sump leading rearward to the bottom of the tank. It is then scavenged to the scavenge pump which forces the oil into the accessory gearbox. The oil from the centrifugal breather and the input gearshaft and bearings drains to the bottom of the accessory gearbox. It is then scavenged from the gearbox together with the oil from No. 1 and 2 bearings by the rear element of the double-element internal scavenge pump. The internal scavenge pump is driven by a quillshaft from the main oil pump in the oil tank. The external scavenge pump scavenges any reduction gearbox oil which drains rearward when the engine is in extreme climbing attitudes. Oil from the propeller governor, front thrust bearing, reduction gear, and torquemeter bleed orifice drains into the reduction gearbox sump. The oil is then scavenged by the external scavenge pump through the external transfer tube to the accessory gearbox housing. The oil from both internal and external scavenge pumps is forced through a T fitting into the airframe oil cooler, where the oil is cooled and passed on to the oil tank. The normal oil operating temperature is 74° to 80° C.

9.17. TORQUEMETER

The torquemeter is a hydromechanical torque-measuring device located inside the first stage reduction gear housing. The torquemeter gives an accurate indication of the torque being produced by the power turbine. The torque pressure value is obtained by tapping the two outlets on the top of the reduction gearbox case. The pressure differential between the two outlets is then read on an instrument which indicates the correct torque pressure.

9.18. FUEL SYSTEM

The T74 basic fuel system consists of a single, engine-driven, fuel pump; p fuel control unit, temperature compensator, and starting control; and a dual fuel manifold with 14 simplex fuel nozzles. Two drain valves are mounted on the gas generator case to insure drainage of residual fuel after engine shutdown after a false or aborted start. The -702 fuel system is illustrated in figure 9.14.

a. The fuel pump is a positive displacement gear-type pump and is driven off the accessory gearbox. Fuel from a booster pump enters the fuel pump through a screen filter and then on to the pump gear chamber, from where the fuel is pumped at high pressure to the fuel control unit. The filter screen is spring loaded and should it become blocked, an increase in fuel pressure differential overcomes the spring, lifts the screen from its seat, and allows unfiltered fuel to flow into the system. The fuel control unit returns bypassed fuel from the fuel control unit to the pump inlet; see the internal line in figure 9.14.

b. The fuel control unit is mounted on the engine-driven fuel pump and is driven at a speed proportional to compressor turbine speed (N_1). The control determines the proper fuel schedule for the engine to produce the power required as established by controlling the speed of the compressor turbine (N_1). Engine power output is directly dependent upon compressor turbine speed. The fuel control governs the N thereby governing the power output of the engine. Control of N_1 is accomplished by regulating the amount of fuel supplied to the combustion chamber.

c. The temperature compensator is mounted on the compressor case with the bimetallic disks extending into the inlet air stream. The temperature compensator is illustrated in figure 9.15.

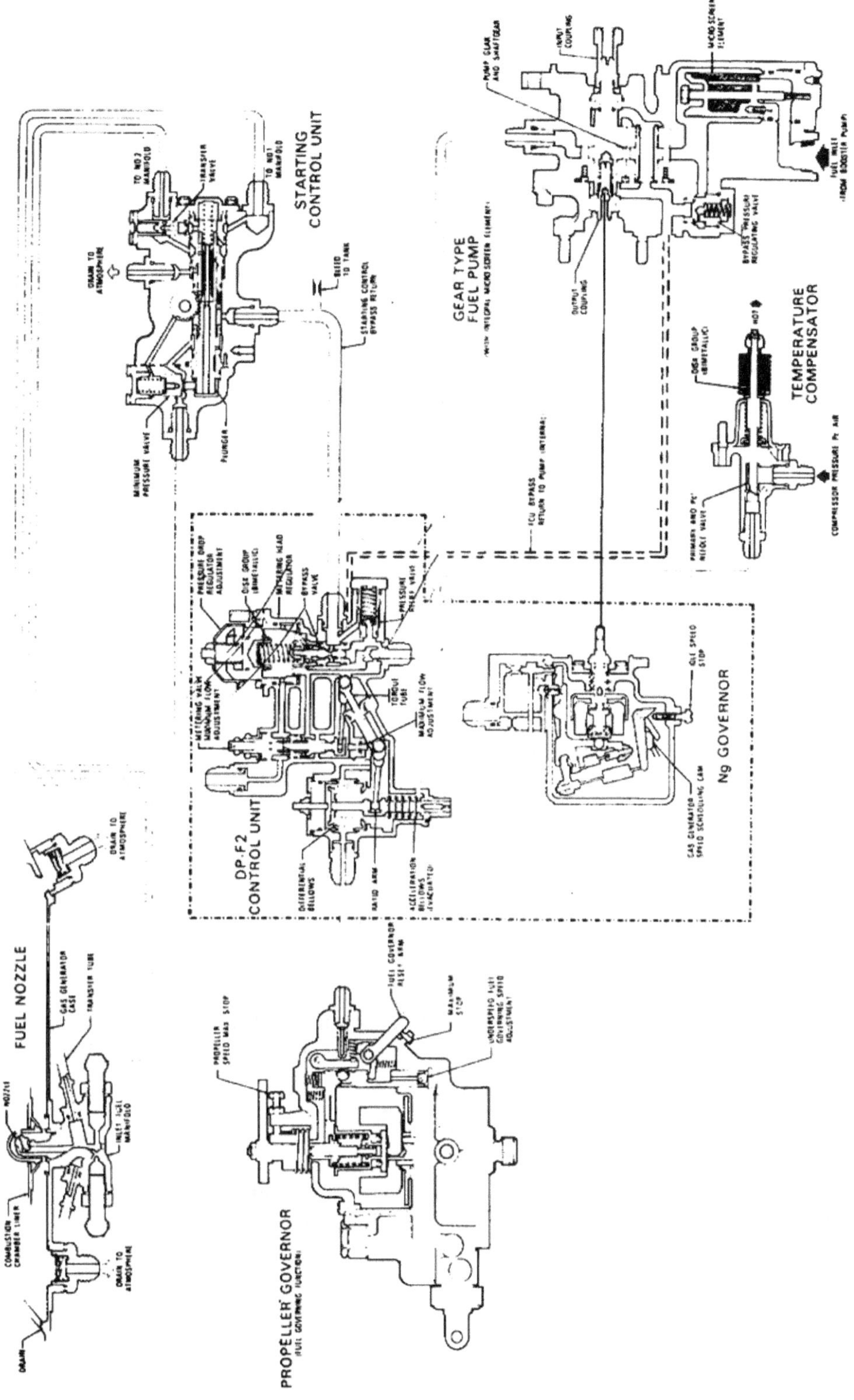

Figure 9.14. Engine Fuel System, -702.

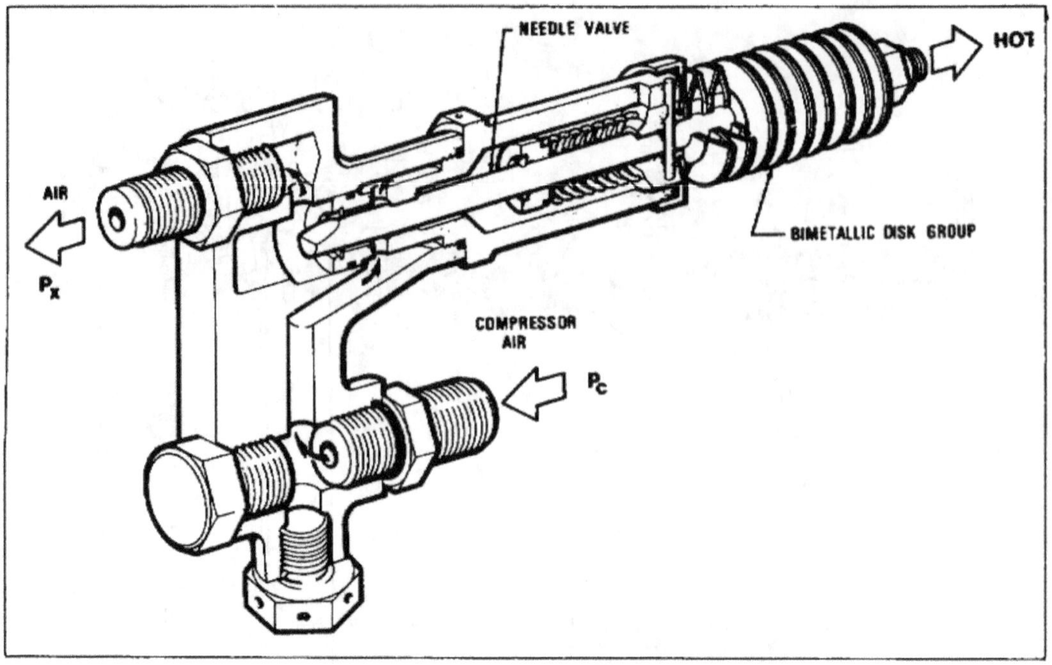

Figure 9.15. Temperature Compensator, -700.

Compressor discharge pressure (P_C) is applied to the compensator. This pressure source is used to produce a pressure signal to the fuel control unit. The compensator changes the pressure signal to the fuel control to produce an acceleration schedule, based on inlet air temperature, to prevent compressor stall or excessive turbine temperature.

d. The starting control consists of a ported plunger sliding in a ported housing. A schematic of the starting control fuel flow is shown in figure 9.16.

Rotational movement of the input lever is converted to a linear movement of the plunger through a rack and pinion engagement. A pressurizing valve, located at the inlet to the control, maintains a minimum pressure in the fuel control to insure correct metering. This valve permits the primary manifold to fill initially for lightup, and as pressure increases in the control, the transfer valve opens. This allows fuel to flow into secondary manifold.

e. The fuel manifold assembly delivers a constant supply of high pressure fuel from the starting control to two sets of seven fuel manifold adapters with simplex nozzles. The dual manifold consists of 28 short fuel transfer tubes fitted with O-rings at each end and interconnected by 14 fuel manifold adapters. Locking plates

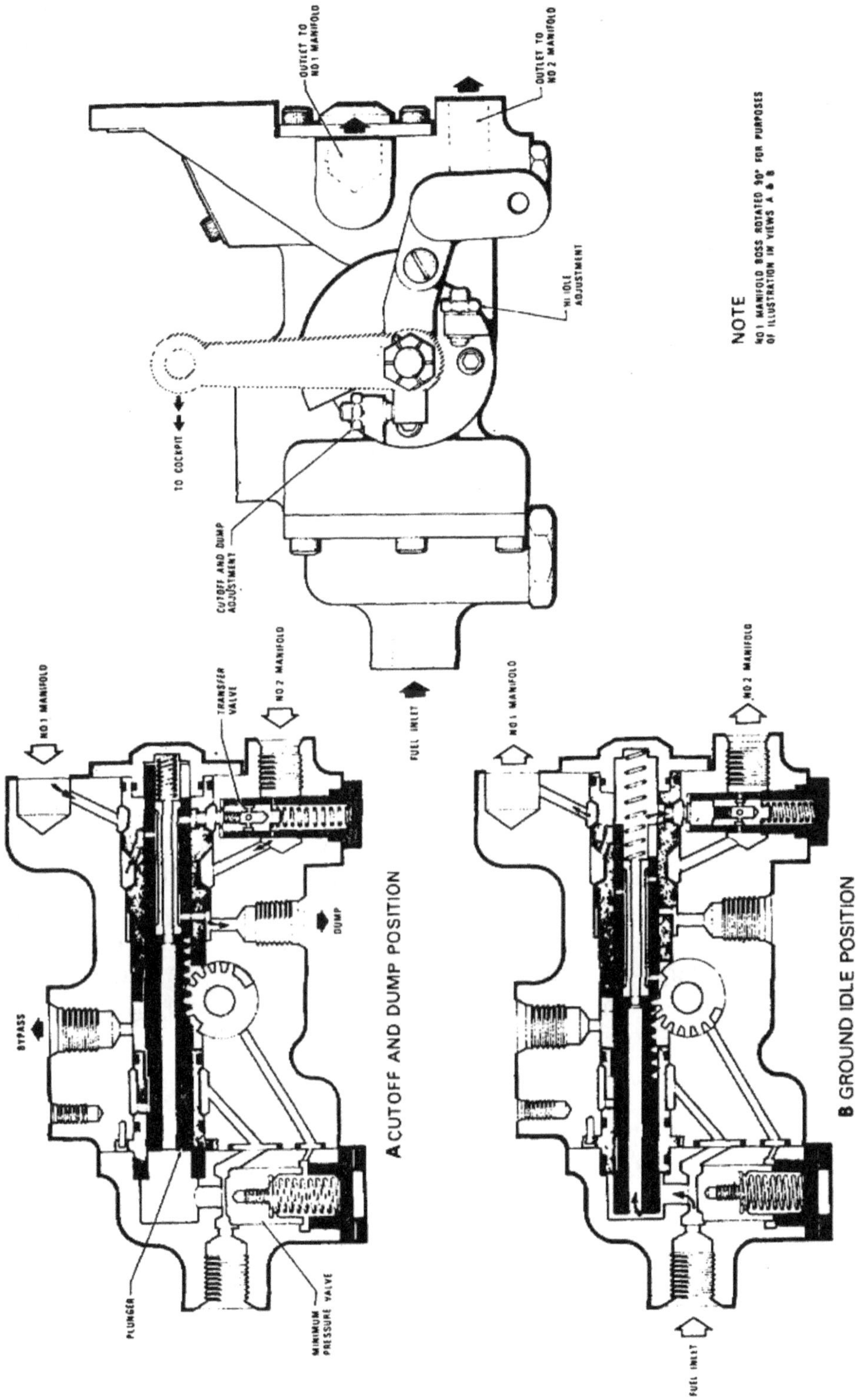

Figure 9.16. Starting Control Fuel Flow, -702.

keep the transfer tubes in proper alignment, secured by two bolts to a mounting boss on the circumference of the gas generator case. Individual transfer tubes or fuel nozzles can be removed and replaced without necessarily disconnecting the remainder of the set. The fuel manifold assembly is illustrated in figure 9.17.

9.19. IGNITION SYSTEM

The glow-plug type ignition system is used on the T74 engine for quick light-offs (starts), even at extremely low ambient temperatures. The basic system consists of a current regulator and two sets of tubes, two shielded plug leads, and two glow plugs. The following subparagraphs discuss the components in the ignition system.

a. The current-regulator unit is normally secured to the accessory gearbox housing but can be remotely mounted if required. The regulator contains four electron tubes, shown in figure 9.18. Each tube has a pure iron filament surrounded by helium and hydrogen gas and enclosed in a glass envelope sealed to an octal base. The iron filament, having a negative co-efficient of resistance (resistance decreases with temperature increase caused by current flow), stabilizes the current flow across the tubes to a nearly constant value over a wide range of voltages. Each glow plug is wired in series with two parallel connected ballast tubes. Either glow plug may be selected for starting the engine. The tubes provide an initial current surge when switched on, which stabilizes to a constant value in approximately 30 seconds. The system heats the glow plugs for fast light-offs.

b. The glow plug consists of a heating element fitted into a short conventional type plug body. A cross-section illustration of a glow plug is shown in figure 9.19.

The plugs are secured to the gas generator case in threaded bosses. The heating element consists of a helically wound coil which lies slightly below the end of the plug body. During starting procedures, the fuel sprayed by the fuel nozzles runs down along the lower wall of the combustion chamber liner and into the helical coil in the glow plug body. The fuel is vaporized and ignited by the hot coil element which heats up to approximately 2,400° F. Three air holes in the plug body allow compressor discharge air from the gas generator case into the plug body, then past the hot coil in the combustion chamber liner to produce a hot streak or torching effect which ignites the remainder of the fuel. The air also serves to cool the coil elements when the engine is running with the glow plugs switched off.

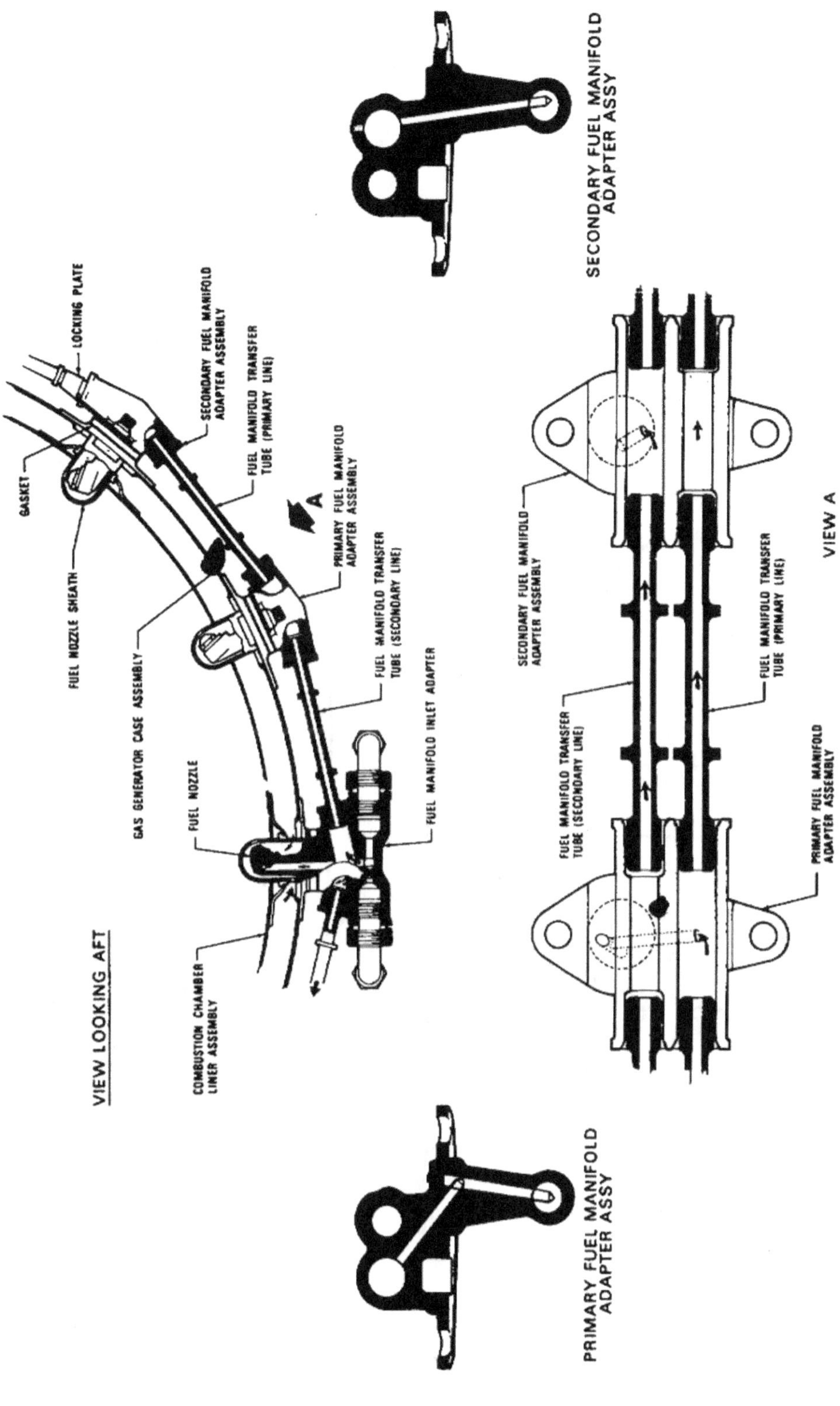

Figure 9.17. Fuel Manifold Assembly, -702.

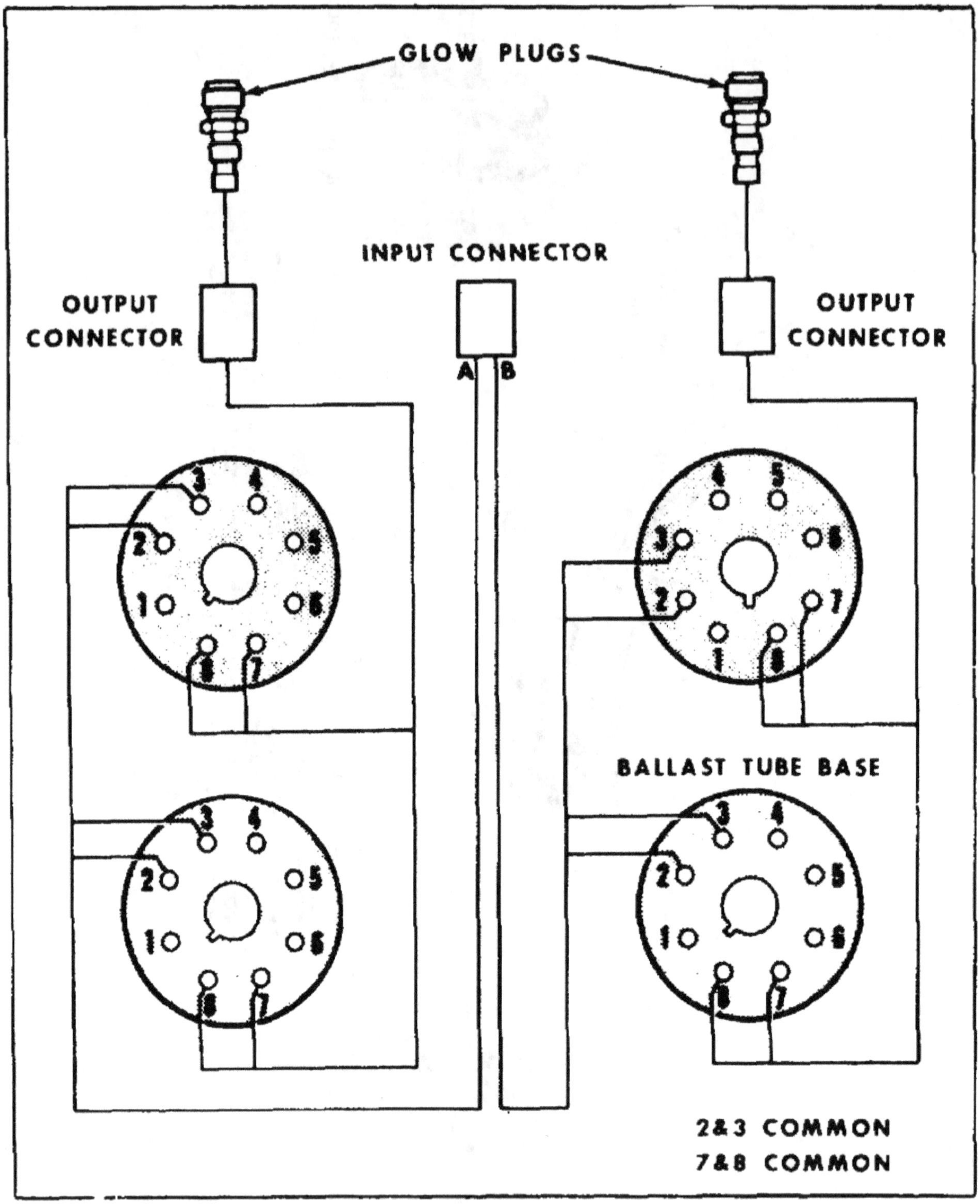

Figure 9.18. Current Regulator Circuit.

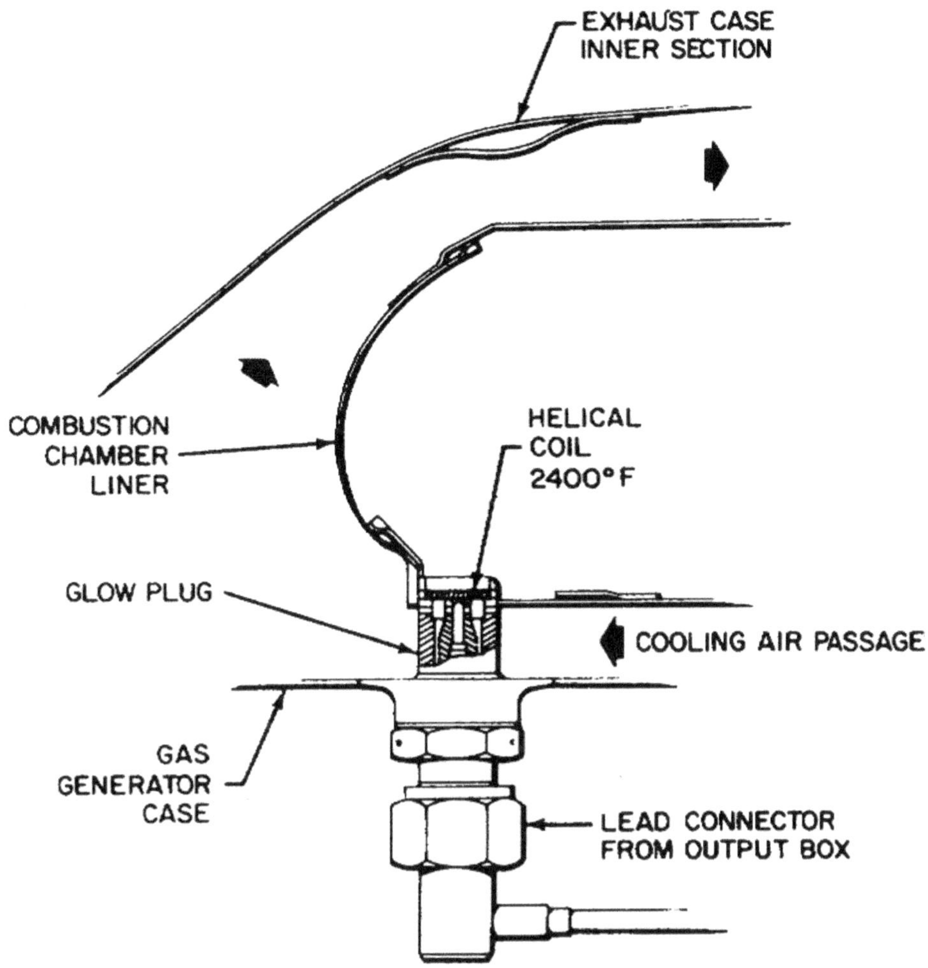

Figure 9.19. Glow Plug.

9.20. INSTRUMENTATION

The following instrumentation is considered necessary for normal operation.

a. The interturbine temperature sensing system is designed to provide the pilot with an accurate indication of engine operating temperature taken between the gas generator and power turbines. The system shown in figure 9.20 consists of twin leads, two busbars, and 10 individual chromel-alumel thermocouple probes connected in parallel. Each probe protrudes through a threaded boss on the power turbine stator housing into an area adjacent to the leading edge of the power turbine vanes. This system generates its own electric current.

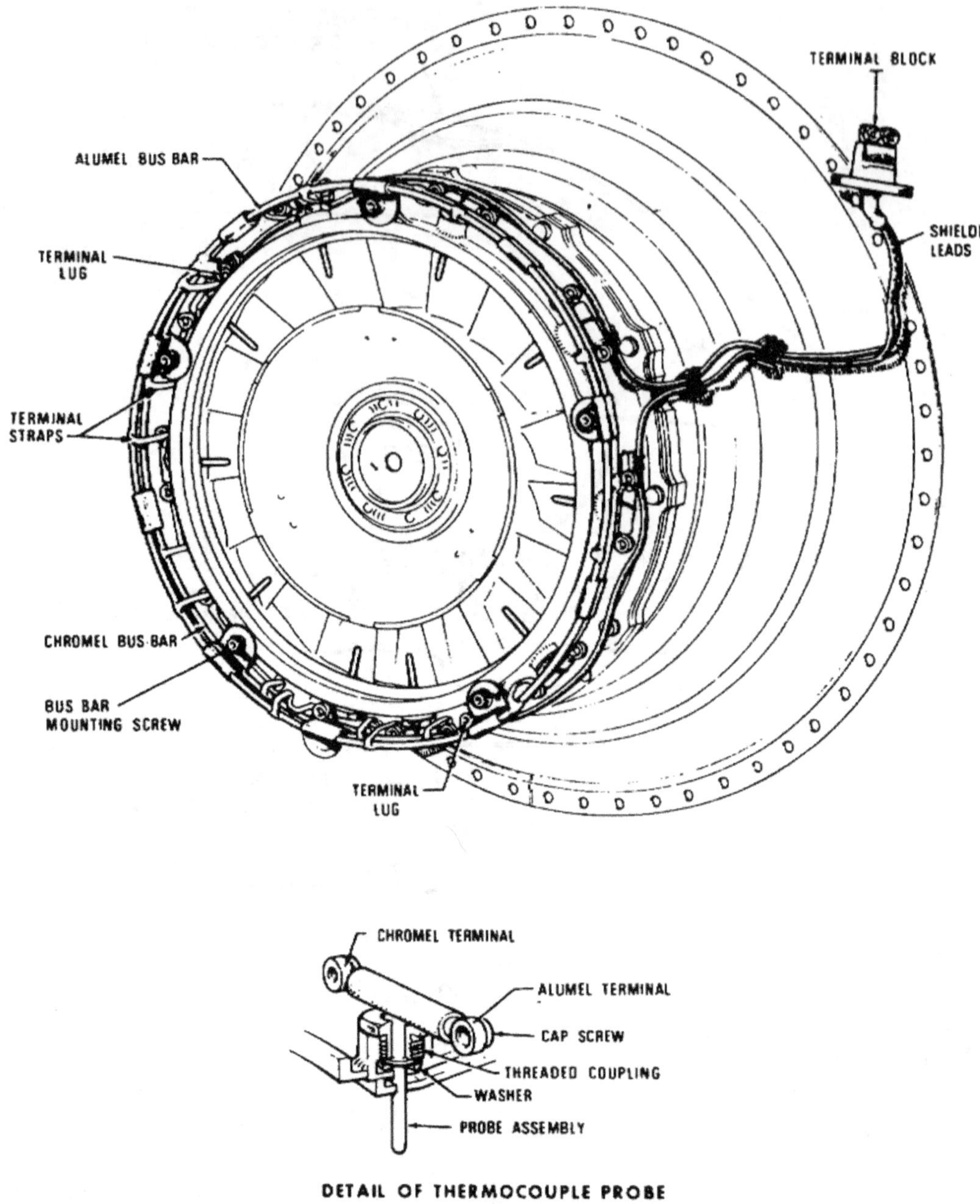

Figure 9.20. Interturbine Temperature Thermocouple Assembly, -700.

b. The oil temperature indicator measures the temperature of the oil as it leaves the delivery side of the oil pressure pump. The oil temperature bulb is mounted on the accessory gearbox housing. This system uses 28v dc from the aircraft electrical system.

c. The oil pressure indicator measures oil pressure in psi at the delivery side of the oil pump. The oil pressure transmitter is mounted just above the oil temperature bulb on the accessory gearbox housing.

d. The turbine tachometer registers compressor turbine speed (N_1) as a percentage of maximum rpm. Propeller shaft speed is registered in hundreds of rpm. The tachometer-generator for the N_1 is mounted at the 5 o'clock position on the accessory gearbox and is driven from the internal scavenge pump. The propeller tachometer-generator (N_2) is mounted on the right side of the reduction gearbox front case and is driven by a bevel gear on the propeller shaft. The N_1 and N_2 tachometer-generators produce their own electrical current.

e. The torquemeter indicating system registers engine output power in psi of torque. The transmitter converts oil pressure to an electrical signal that registers engine torque on a gage in the pilot's cockpit.

9.21. SUMMARY

The Pratt and Whitney T74-CP-702 has four major engine systems. The compressor bleed-air system prevents the compressor from stalling during low engine speeds. Compressor bleed-air is also used for bearing compartment seals and turbine cooling.

The lubrication pressure system produces a constant supply of clean oil to the engine bearings, reduction gears, torquemeter, propeller, and accessory gears. The oil is then transferred back to the oil tank by the internal and external scavenge pumps.

The basic fuel system consists of a single engine-driven pump, fuel control unit, temperature compensator starting control, fuel manifold, and 14 simplex fuel nozzles. The fuel control determines the fuel flow to the engine to produce the power required. The temperature compensator sends an air pressure signal to the fuel control to prevent compressor stall or excessive turbine temperature. The starting control permits the primary manifold to fill for engine starts. The fuel manifold delivers fuel to two sets of seven simplex nozzles.

The glow plug ignition system is capable of quick light-offs at extremely low ambient temperature. The fuel is vaporized and ignited by the hot coil element in the glow plug. The engine is equipped with instrumentation to monitor engine operation during flight measures temperature in the turbine section.

Turbine tachometers register N_1 speed as a percentage of maximum gas generator rpm. The propeller tachometers (N_2) register shaft speed in hundreds of rpm. Each indicator is directly responsive to a tachometer-generator unit attached to a corresponding engine section.

Oil pressure is taken from the delivery side of the main oil pressure pump and registered in psi.

Oil temperature is taken from the delivery side of the oil pressure pump and registered in degrees centigrade.

The torquemeters indicate torque in psi applied to the propeller shaft.

Chapter 10

GENERAL ELECTRIC T700-GE-700

10.1. INTRODUCTION

General Electric is developing the T700-GE-700 turboshaft engine for the U. S. Army Utility Tactical Transport Aircraft System (UTTAS). The T700 is designed to operate in combat with improved reliability, easy maintenance, low fuel consumption, and extended operating life.

The T700 will operate with no visible smoke and at low noise levels. External lines and leads have been reduced in number and grouped for armoring protection. The engine also contains its own lubrication and electrical systems to reduce dependence on the airframe systems.

10.2. GENERAL DESCRIPTION

The T700 is an axial-flow, free-power, turboshaft engine rated at 1,500 shp. It has a five-stage axial, single-stage centrifugal compressor; an annular combustor; a two-stage gas generator (N_1) turbine; and a two-stage power turbine. An engine-driven inlet particle separator is located ahead of the compressor. For field maintenances, the engine breaks down into four modules: controls and accessories, cold section, hot section, and power turbine.

The engine is constructed of corrosion-resistant steel except for a titanium axial-compressor casing, aluminum inlet-separator frame, and magnesium gearbox case. The engine dimensions in inches are length 47, height 23, and width 25. A cutaway illustration of the T700-GE-700 is shown in figure 10.1.

10.3. CENTRIFUGAL INLET SEPARATOR

Figure 10.2 shows the centrifugal inlet separator and a schematic of its operation. The inlet air passes through the fixed separator swirl vanes which swirl the air, and centrifugal force throws the particles to the separator collection scroll. The scroll is scavenged by an engine-driven blower mounted on the accessory gearbox.

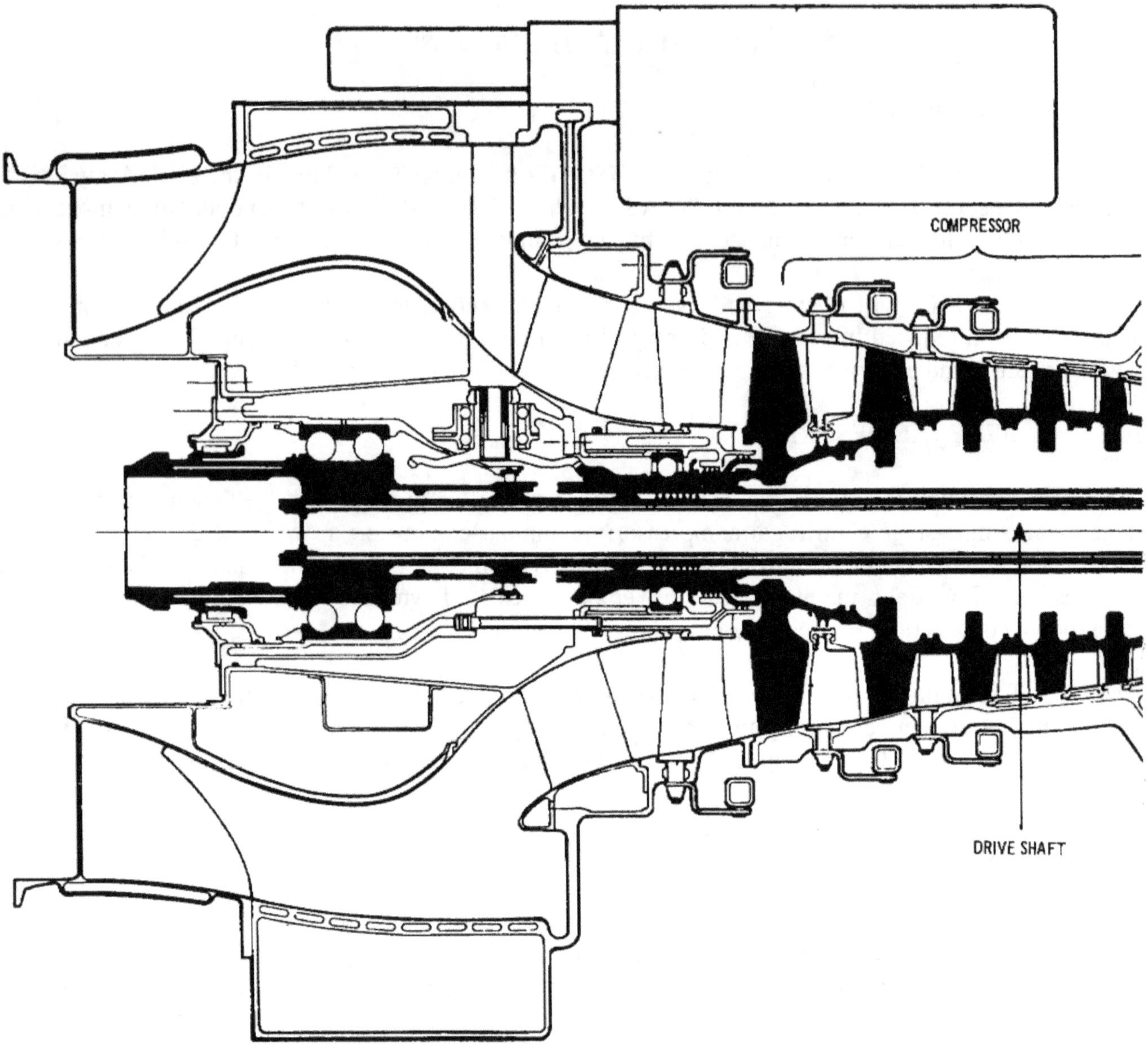

(continued on next page)

Figure 10.1. Cutaway View of the T700-GE-700.

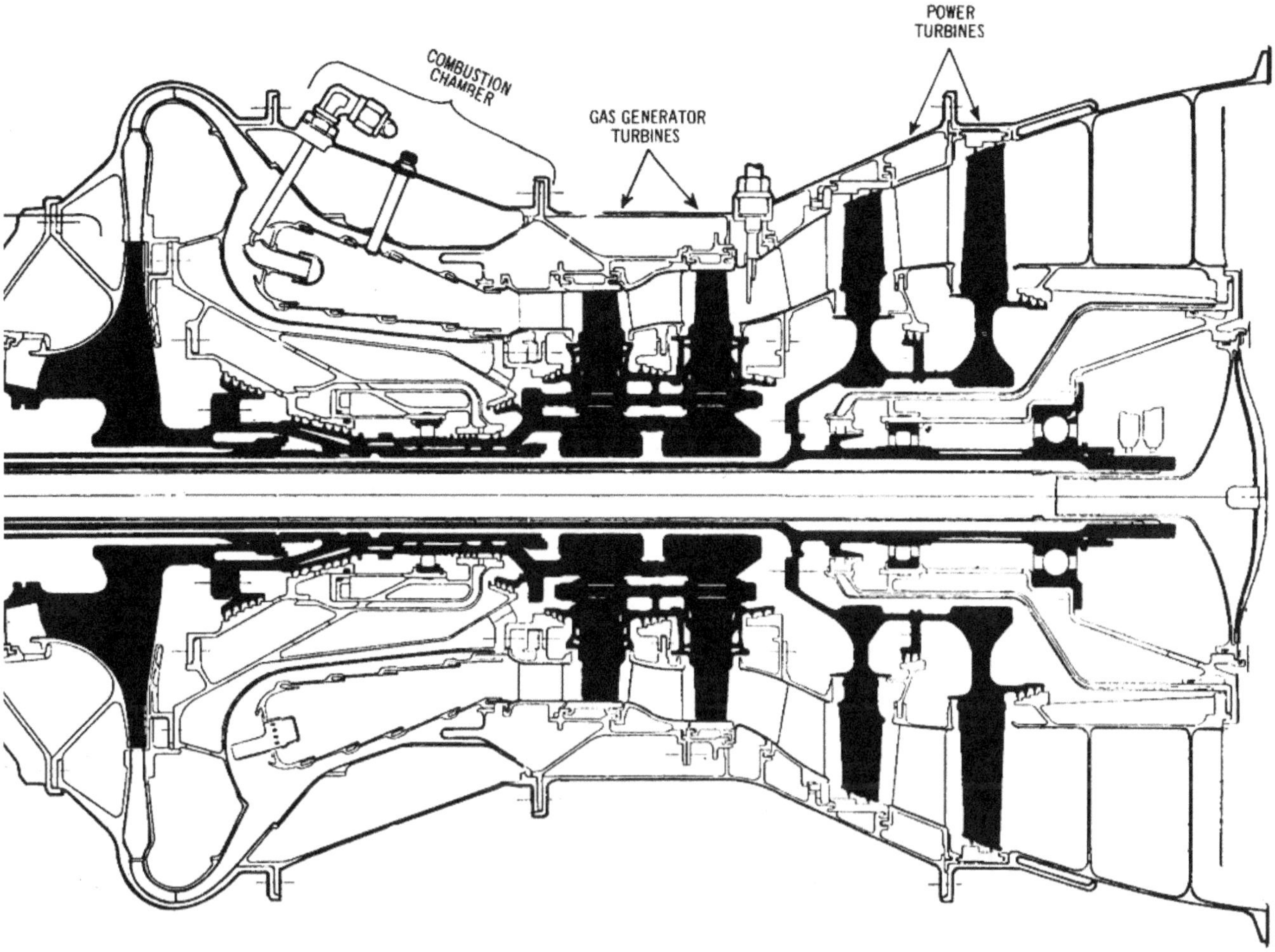

(continued from previous page)

Wexford
2008

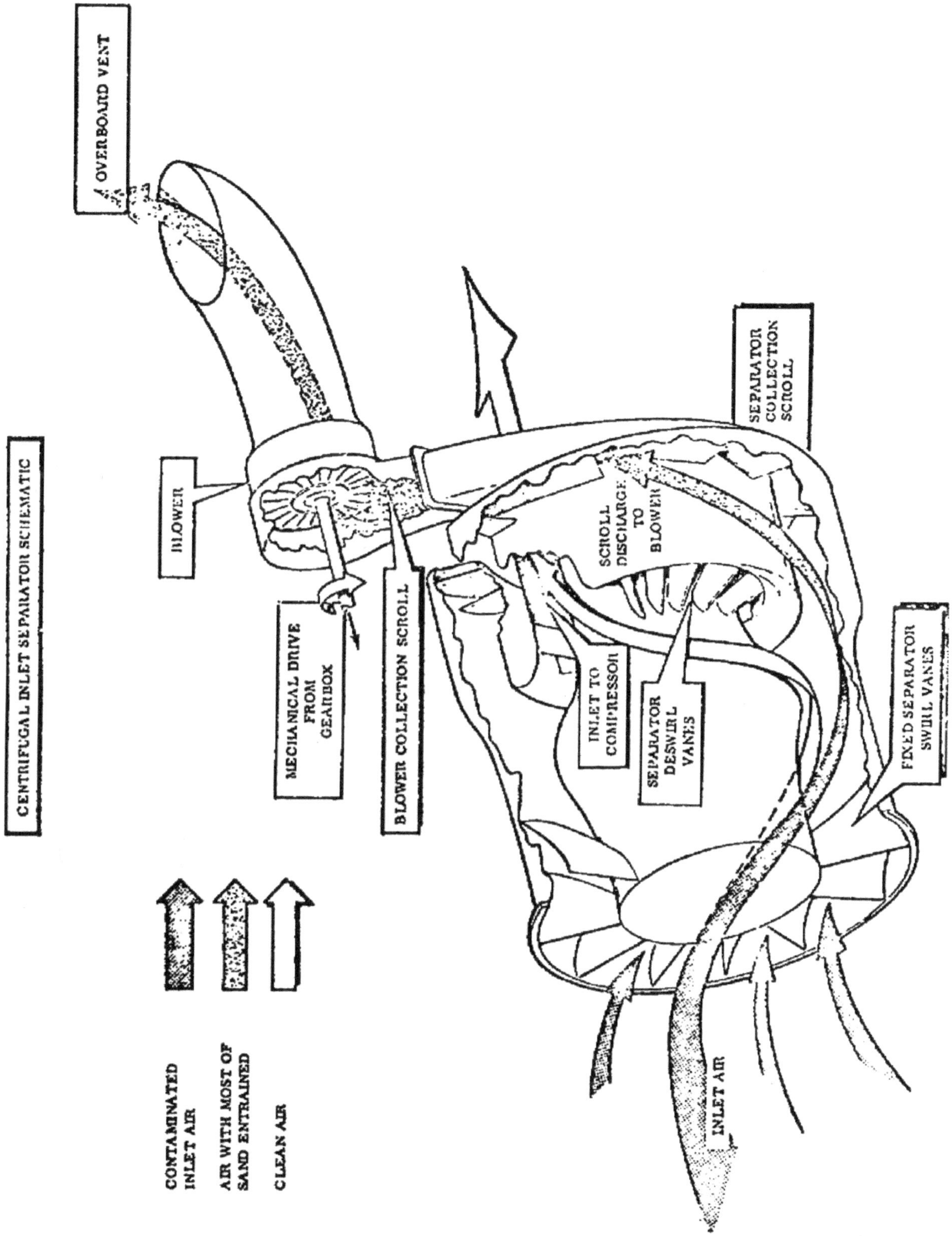

Figure 10.2. Centrifugal Inlet Separator.

10.4. COMPRESSOR

The compressor has five axial stages and one centrifugal stage. The inlet guide vanes and the first and second stage stator vanes in the compressor vary their angle of attack according to compressor speed. The centrifugal impeller is mounted aft and on the same shaft as the axial compressor. One of the major design differences of the T700 is that the axial-compressor rotor stage blades and disks are one-piece castings. This means better rotor integrity and also easier buildup and overhaul. The compressor has very few blades, 332 total. This is about one-third the number in comparable current engines. Stages one and five have borescope ports for inspections. Just aft of the centrifugal impeller is the diffuser passage.

10.5. COMBUSTOR

The annular combustion chamber is located aft of the diffuser. Engine durability and time between overhauls (TBO) are directly related to the heat of the combustor. If the combustor can develop the power required at a lower temperature, component service life can be increased.

The combustor has twelve vaporizing fuel injectors and two atomizing start-fuel nozzles. The axial type vaporizing combustor operates with a lower peak temperature factor. Axial flow ensures minimum liner area and hence minimizes the amount of secondary cooling air to be mixed into the hot gas stream. The annular combustion chamber can be seen in figure 10.1.

10.6. GAS GENERATOR TURBINE

The T700 gas generator turbine is a two-stage, air-cooled, high-performance, axial-flow turbine. The modular concept has been used in the gas generator turbine design for ease of assembly and maintenance. The turbine is divided into easily accessible submodules consisting of the first stage turbine nozzle and the turbine rotor.

The first stage nozzle submodule consists of 24 air-cooled cast vane segments brazed in pairs. These twelve pairs are assembled to the inner support and held in place by a retaining ring with bolts.

The rotor submodule consists of the turbine rotor combined with the second nozzles, supports, and shrouds. Both first and second stage disks, cooling plates, and turbine blades are clamped by five short tiebolts. Five longer tiebolts clamp this rotor assembly to the forward shaft through the forward coupling. Loosening these five longer tiebolts does not disturb the rotor assembly itself, permitting simple maintenance without special tools. Thirteen second-stage nozzle segments of two vanes each are held by the outer support, which is also assembled with the first stage shroud segments and second stage shroud support.

All airfoils (vanes) are internally cooled. The first stage nozzle leading edge is air-cooled, with the air exiting through holes in the vane airfoils. The midchord region is convection-cooled, with cooling air exiting both through pressure holes and trailing edge slots. Cooling air for the second stage nozzle is bled from the centrifugal compressor exit and piped back through the turbine casting. The air enters the second stage nozzle through bushings and cools the nozzles by internal airflow exiting through trailing edge and inner band holes. The turbine blades are air-cooled through radial holes. Air enters through the dovetail and exits at the tip. The first stage blades also employ trailing edge holes for cooling.

10.7. POWER TURBINE

This component is a two-stage uncooled tip-shrouded design with replaceable turbine blades and nozzle segments. The output shaft governing speed range is from 17,000 to 23,000 rpm with a maximum rated speed of 24,000 rpm.

The power turbine is a self-contained module which can be disassembled and reassembled to the gas generator without special tools.

The rotor assembly consists of the third and fourth stage disks mounted on a drive shaft supported by four bearings. The third stage disk is secured to the drive shaft by a flange, allowing quick removal from the drive shaft without removal of the aft sump or rear frame. The third stage disk has 46 tip-shrouded blades attached to the disk through conventional dovetails and retained axially by bent locking strips. A similar arrangement is used on stage four where 50 blades are employed.

The drive shaft is a hollow one-piece unit splined to the short output shaft at the forward end. An integral feature of the drive shaft is the torque sensor that mechanically displays the total twist of the drive shaft, which is electrically sensed and processed. This mechanical display of total twist is accomplished by a reference shaft that is pinned to the front end and extends back to the aft end, where it is free to rotate relative to the drive shaft. The relative rotation is due to transmitted torque, and the phase angle between the reference surfaces is electronically sensed by two pickups, one sensing five teeth on the reference shaft, the other sensing five teeth on the drive shaft.

Both stages of each turbine nozzle are cast in segments. The third stage has six segments of six vanes each, and the fourth stage has 10 segments of five vanes each. The static tip shrouds are replaceable in four segments each and are of open cell honeycomb arrangement.

10.8. BEARING ARRANGEMENT

Figure 10.1 shows the T700-GE-700 bearing and frame arrangement. The four bearings and shaft arrangement at the front of the axial compressor isolate the driveshaft from the engine.

Circumferential carbon seals are used on the power turbine shaft, and labyrinth seals are used on the gas generator shaft. The front frame has eighteen deswirl vanes which are part of the structure.

10.9. LUBRICATION SYSTEM

A schematic of the lubricating system is shown in figure 10.3. The oil tank is built into the inlet particle separator. All scavenge lines have magnetic chip detection for fault isolation. An emergency lubrication system has been built into the design for assuring continued bearing operation after loss of oil from any cause. Small oil reservoirs are included in each bearing sump, and are kept full during normal operation by the oil pressure pumps. Oil is always bleeding out of these reservoirs at a slow rate. Air jets, also continuous, act as "foggers" for this oil bleed and provide oil mist lubrication at all times. This continues for at least six minutes even if the oil supply fails. A fuel-oil heat exchange is a self-contained oil cooler and fuel de-icer.

10.10. MAINTAINABILITY

The engine has been designed for ease of servicing and maintenance. Borescope inspections can be made in critical areas where sand and dust erosion and clogging, corrosion, and foreign object damage can occur and in other critical areas where inspections are necessary.

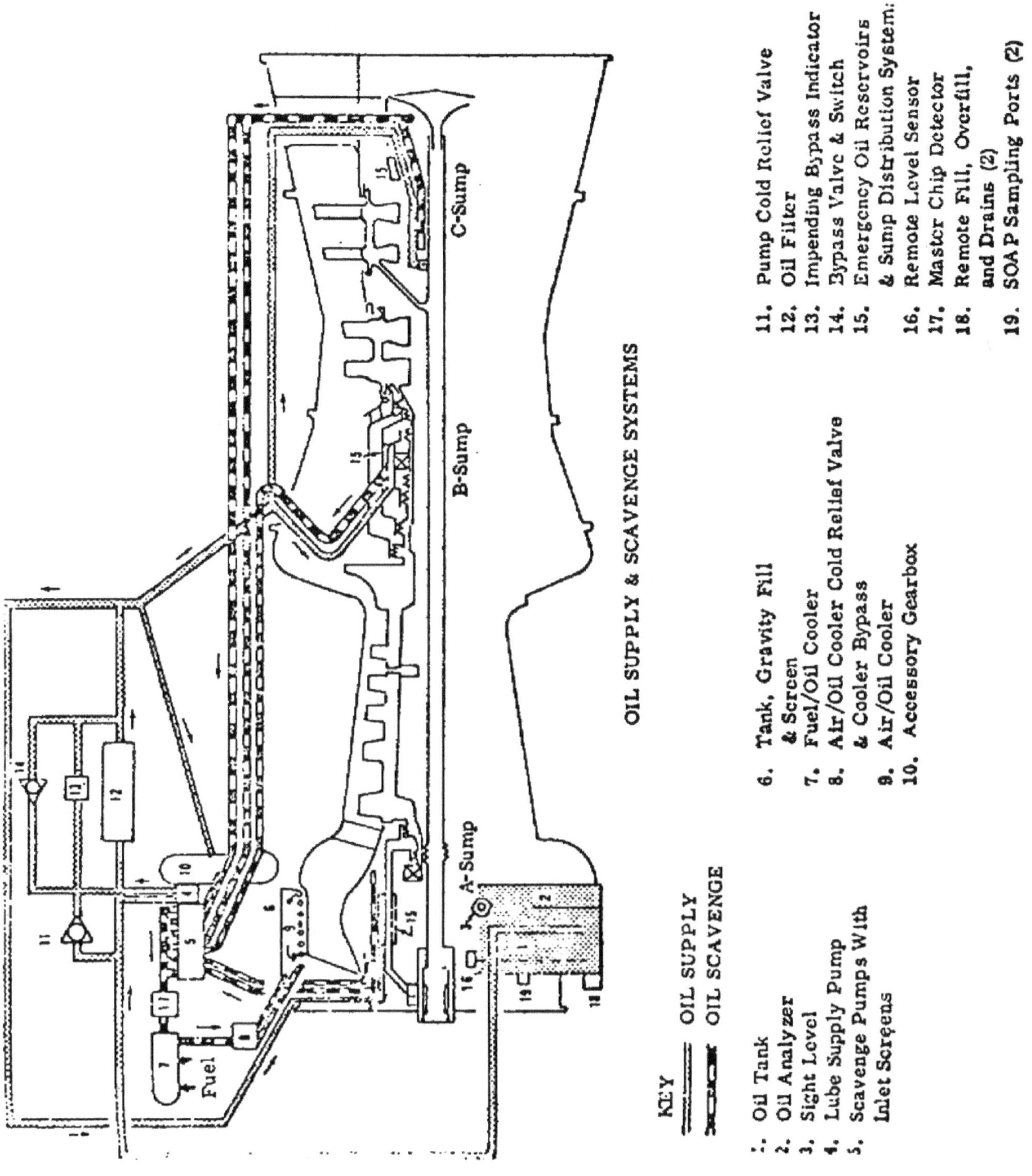

Figure 10.3. Lubrication System Schematic.

The T700 design offers the following maintenance benefits: no special tools required for module replacement, no critical measurements required for module replacement, no component replacement at direct support. Any maintenance action requiring special tools is accomplished at depot level.

10.11. SCHEDULED MAINTENANCE

The T700-GE-700 engine has been designed to meet the required design life of 5,000 hours. The T700 is a durably designed engine and does not require frequent scheduled checks or adjustments. An example of this feature is that the fuel control requires no field adjustments.

10.12. SUMMARY

The General Electric T700-GE-700 turboshaft engine is presently being developed to power the Utility Tactical Transport Aircraft System (UTTAS). The engine is an axial-flow free-power turbine design, rated at 1, 536 shp. The engine is made up of four modules: accessory gearbox, cold section, hot section, and power turbine.

Mounted on the front of the engine is a centrifugal inlet particle separator to keep foreign objects from entering the compressor. The compressor has five axial stages and one centrifugal stage. The annular combustion chamber has 12 vaporizing fuel injectors and two atomizing start-fuel nozzles. The gas generator turbine and power turbine are two-stage axial flow turbines. An emergency oil system has been designed to ensure limited engine operation after loss of oil. The oil tank is built into the inlet separator.

The engine has been designed for improved reliability, easy maintenance, low fuel consumption, and extended operating life. All field maintenance can be performed without special tools. The engine can be internally inspected by the borescope method, rather than by engine teardown. The engine will have a design life of 5,000 hours.

Appendix I

REFERENCES

Army Regulations

AR 310-25	Dictionary of United States Army Terms
AR 310-50	Authorized Abbreviations and Brevity Codes

Technical Manuals

TM 55-406	Fundamentals of Aircraft Powerplant Maintenance
TM 55-2840-229-24/T.O. 2J-T53-16	Maintenance: Organizational, Direct Support, and General Support, Engine, Shaft Turbine

U. S. Army Transportation School Publications

333-335	T53-L-13 Gas Turbine Engine Familiarization Manual
333-337	T74-CP-700 Gas Turbine Engine Familiarization Manual
333-339	T63-A-5A/700 Gas Turbine Engine Advance Sheet
333-374	JFTD-12A-1/T-73-P-1 Gas Turbine Engine Familiarization Manual
333-375	T55-L-11 Turboshaft, Gas Turbine Engine Familiarization Manual
333-385	Jet Cal Analyzer
333-394	T55-L-7, T55-L-7C Gas Turbine Engine Familiarization Manual
333-397	T62 Gas Turbine Engine Familiarization Manual

Appendix II

GLOSSARY

Acceleration lag -- in the turbine engine, delay between the time instant power is requested and when power is available. The time it takes the engine to accelerate and give the required power increase.

Aerodynamic drag -- force which thrust must overcome to move an aircraft forward. Design can lessen aerodynamic drag through streamlining. Drag increases with increased speed.

Airbleed actuator -- device that operates the interstage bleed system, to improve compressor acceleration characteristics by unloading small amounts of compressed air.

Air density -- total mass of air per given volume, the weight of a given volume of air. Air is denser at lower altitude, at lower temperature, and lower humidity.

Air-fuel ratio -- 15 parts of air to 1 part of fuel by weight, the mixture to be burned in the combustion chamber.

Air inlet -- large, smooth aluminum duct to conduct the air into the compressor.

Ambient air -- surrounding air.

Angle of attack -- the acute angle formed by the direction of the relative wind and some longitudinal reference axis of the aircraft.

Annular combustion chamber -- two-part combustion chamber made up of an annular liner and a housing assembly. The compressed air goes into a ring-shaped space formed by the annular liner around the turbine shaft rather than into individual combustion chambers. The space between the outer liner wall and the housing assembly allows the flow of cooling air. Used with axial-flow and dual compressors.

Annular reverse-flow -- type of gas turbine engine most commonly used in Army aircraft. Air flow direction is reversed in the combustion area.

Anodizing -- putting a protective oxide film on a light metal by an electrolytic process.

Anti-icing system -- device that supplies hot air under pressure to prevent icing of the inlet housing areas and inlet guide vanes. Hot scavenged oil is also circulated through internal passages in the walls and struts.

Army Spectrometric Oil Analysis Program (ASOAP) -- periodic oil analysis for microscopic metal particles. This takes place at an oil analysis laboratory.

Atmospheric pressure -- barometric pressure exerted by the atmosphere as a result of gravitational attraction above the point in question.

Atomizer -- nozzle that creates a highly atomized and accurately shaped spray of fuel suitable for rapid mixing and combustion.

Axial-flow compressor -- one in which the air is compressed parallel to the axis of the engine. It is made up of a series of alternating rotor and stator vane stages.

Bleed system -- device that unloads small amounts of air to relieve pressure.

Boss -- raised rim around a hole; e.g., axle hole in a wheel. Circular projection on a casting, usually serving as the seat for a bolt head or nut.

Brayton cycle -- constant pressure cycle, with four basic operations which it accomplishes simultaneously and continuously for an uninterrupted flow of power. The turbine engine operates on this cycle.

Can-annular combustion chamber -- one with characteristics of both the can and annular types. It has an outer shell and a number of individual cylindrical liners.

Can combustion chamber -- one made up of individual combustion chambers in which the air from the compressor enters each individual chamber through the adapter.

Caustics -- substances that can burn, corrode, or destroy animal or other organic tissue by chemical action.

Centrifugal-axial flow compressor -- combination of the centrifugal-flow and the axial-flow compressors. It usually consists of a five- or seven-stage axial-flow compressor and one centrifugal-flow compressor. Also called the dual compressor.

Centrifugal-flow compressor -- one with an impeller (rotor), stator, and compressor manifold. The rotor revolves at high speed, drawing air into the blades. Centrifugal force accelerates the air, and it moves through the stator and through the manifold.

Combustion -- process of burning the fuel-air mixture in a gas turbine engine.

Combustion chamber -- part of a turbine engine in which the propulsive power is developed by combustion of the injected fuel and the expansive force of the resulting gases.

Combustion chamber liner -- engine part usually constructed of welded high-nickel steel, subjected to flame of extremely high temperature. It is behind the compressor and receives the compressed air which is mixed with fuel and ignited. The combustor is where the combustion takes place.

Combustor -- the combustion chamber of a gas turbine engine with its associated burners, igniters, and injection devices.

Compressor -- that section of an engine that produces an increase in air pressure. It is made up of rotating and stationary vane assemblies. It is the gas producer, or it may be thought of as an air pump.

Compressor rotor -- impeller, may be thought of as an air pump. It accelerates the air rearward into the first stage vane assemblies.

Compressor stall -- separation of the airflow from the suction surface of the fixed or rotating blades of a compressor. Any degree of stall reduces airflow.

Concave -- pressure side of an airfoil.

Conduction -- transfer of heat through material by communication of kinetic energy from particle to particle rather than by a flow of heated material.

Convergent area -- place where the cross-sectional area of a duct becomes smaller.

Convergent exhaust duct -- used on fixed-wing aircraft, formed by tapering toward the rear of the duct.

Convex -- suction side of an airfoil.

Crossover tube -- duct carrying flame to the individual cylindrical liners of the can-annular combustion chamber.

Diffuser -- aft structural member of an engine. It receives high velocity air from the centrifugal impeller and decreases velocity and increases air pressure. In the combustor, a diffuser forms a divergent flow path for the exhaust gases.

Diffusion -- process by which gases intermingle as the result of their spontaneous movement caused by thermal agitation.

Directional references -- specific definitions of terms referring to gas turbine engines to identify front and rear, right and left, bottom and top.

Divergent area -- place where air flows from a smaller into a larger area.

Divergent exhaust duct -- used on helicopters, device to diffuse the exhaust gases rearward and to eliminate thrust.

Dry-cleaning solvent -- cleaning compound that may be used for all metal parts.

Dry-sump engine -- one in which the oil is stored separate from the engine.

Dual compressor -- see centrifugal-flow, axial-flow compressor.

Duplex nozzle -- dual-orifice channel through which highly atomized and accurately shaped sprays of fuel go into the combustion chamber.

Engine airflow path -- route of the airflow through the engine. See paragraph 4. 4a for a detailed description of airflow through a T53-L-13.

Engine oil pressure indicating system -- device that gives continuous readings of engine oil pump pressure in psi.

Engine oil temperature indicating system -- device electrically connected to the 28v dc system which transmits temperature readings to the indicator in degrees centrigrade.

Engine speed notation -- capital letter N represents the rotational speed of the engine. When a number is placed after the N, as in N_1, it indicates a specific system on the engine.

Engine stations -- specific locations on the engine designating temperature- or pressure-measuring locations. For example, T3 means the third temperature pickup on the engine.

Engine surge -- result of compressor stall. The complete engine in stall.

Exhaust -- hot gases discharged from the engine through the exhaust diffuser section.

Exhaust diffuser -- section composed of an inner and outer housing, separated by hollow struts across the exhaust passage. It forms a divergent flow path for the exhaust gases.

Exhaust gas temperature indicator -- sensitive millivoltmeter calibrated in degree centigrade, activated by an electrical force generated by its thermocouple.

Fir tree mount -- manner of attaching the blades to the disk in the turbine rotor assembly. The root of the blade where it is attached to the disk is shaped like a fir tree. See figures 1.27 and 1.28 on page 42.

Foreign object damage -- commonly called FOD, harm or destruction to the turbine engine caused by foreign objects sucked into the inlet area of the engine with the required air.

Forged -- shaped by hammering. Only the malleable metals are worked successfully. The application of heat increases plasticity.

Free-power turbine engine -- the turbine engine used by the Army. Sixty percent of the energy produced by combustion is extracted by the gas producer turbine to drive the gas producer rotor. The rest of the energy is converted to shaft horsepower to drive the out-put shaft of the engine.

Frictional loss -- resistance to the relative motion of air flowing along a duct.

Frontal area -- front part of a gas turbine engine, smaller than that of a reciprocating engine, therefore producing less drag.

Front of engine -- end from which power is extracted. An exception is the T73 engine on the CH-54, in which the power is extracted at the end where the exhaust gas is expelled.

Fuel-air ratio -- see air-fuel ratio.

Fuel atomizer -- See atomizer.

Fuel controls -- devices to control fuel flow. They are usually hydromechanical and include speed governors, servo systems, valves, metering systems, and sensing pickups.

Fuel divider -- device that meters fuel to the engine nozzles according to a predetermined schedule of secondary flow versus primary flow.

Fuel nozzle -- device to inject fuel into the combustion chamber in a highly atomized and accurately shaped spray.

Fuel pressure indicating system -- device that gives continuous readings in psi of fuel pressure in the main fuel supply line.

Gas producer (N_1) -- compressor in a free-power turbine engine.

Gas turbine engine -- aircraft powerplant that is adaptable for both airplanes and helicopters.

Gerotor pump -- modified gear-type pump with two moving parts, an inner toothed element and an outer toothed element. The inner one has one less tooth than the outer.

Glow plug -- device that consists of a heating element in a short conventional-looking spark plug.

Heat exchanger -- fuel-oil cooler, to help cool the oil. The exchanger is a cylindrical oil chamber surrounded by a jacket through which the fuel passes. Heat from the oil is transferred to the fuel by conduction.

Humidity -- amount or degree of moisture in the air. If humidity increases, air density is decreased. Humidity has little effect on density, however, in comparison with temperature and pressure changes.

Igniter plugs -- spark plugs which function only during starting and cut out of the circuit as soon as combustion is self-supporting.

Imbalance -- uneven distribution of weight resulting in rotating parts being out of balance. Measured in inch-grams or inch-ounces.

Impeller rotor -- rotor in a compressor that revolves at high speed, drawing air into the blades.

Inlet guide vanes -- devices positioned by the inlet guide vane actuator pilot valve. They are located in front of the first compressor rotor, and they control the angle of incidence of the inlet air, thus ensuring a compressor surge margin.

Inlet housing assembly -- forward structural support of the engine.

Jetcal analyzer -- device used to check the exhaust gas temperature during periodic maintenance inspections or when abnormally high or low temperatures are noted.

Jet propulsion -- propulsion of a body by means of a jet or stream of gas, obtaining all or most if its thrust by reaction to the ejection of the combustion products (gas).

Joule -- unit of energy or work used in rating gas turbine ignition systems. A joule is equal to the amount of energy expended in one second by an electric current of one ampere through a resistance of one ohm.

Kinetic energy -- work energy associated with motion.

Labyrinth seal -- device for preventing leakage of gas on the gas generator shaft in a turbine. A labyrinth consists of a series of projections on the rotating element running in close contact with grooves on the stationary element.

Maintenance allocation chart -- chart in a -20 TM that assigns maintenance tasks to the lowest level capable of doing them, based on experience, skills, tools, and time available.

Manifold -- component in which air or gases are collected for intake or expulsion.

Micron --one millionth of a meter.

Mil -- unit of length equal to 1/1000 inch. Unit of angular measurement equal to 1/6400 of the circumference of a circle.

N_1 system -- gas producer.

N_2 system -- power turbine and shaft.

Nacelle -- enclosed housing for an aircraft engine, outside the fuselage.

Nozzle -- channel through which gas is conveyed to the rotor vanes of a turbine. Its purpose is to convert pressure into velocity.

Orifice -- opening having a closed perimeter through which a fluid may discharge. It may be open to the atmosphere, or it may be partially or completely submerged in the discharged fluid.

Oscillograph -- instrument that produces a record of variations in an electrical quantity.

Oscilloscope -- instrument that shows the presence and form of an electric current.

Outside air temperature -- commonly abbreviated as O. A. T, the temperature of the air outside the engine.

Otto cycle -- a constant volume cycle, with four distinct operations performed intermittently. Reciprocating engines operate on this cycle.

Overspeed governor, N_2 -- gearbox mounted on engine inlet housing and driven from the power shaft.

Overspeed governor, fuel control -- part of the torquemeter system, an individual pumping unit which, with the tachometer drive assembly, sets the torquemeter oil pressure.

PD 680 -- cleaning solvent for exterior of engine and its attached components.

Pneumatic -- something moved or worked by air pressure.

Power-to-weight ratio -- relationship between power and weight. Turbine engines produce more power for weight than reciprocating engines.

Power turbine (N_2) -- turbine that is free and independent of the gas producer system. It develops rotational shaft power.

Pressure oil system -- method of supplying oil under pressure to engine parts.

Pressure pumps -- devices to put oil into the system.

Pressurizing and drain dump valve -- device to prevent flow of fuel to the nozzle until enough pressure is built up in the fuel control. One also drains the fuel manifold at engine shutdown and traps fuel in the upper portion of the system to keep the fuel control primed for faster starts.

Primary air -- air that mixes with fuel in the combustion chamber, to form a combustible mixture. The ratio is 15 parts of air to 1 part of fuel.

Radial inflow turbine -- type of turbine made by some manufacturers, not used in any Army aircraft today, even though it is rugged and simple, relatively inexpensive, and easy to manufacture. Similar in design and construction to the centrifugal-flow compressor.

Ram air pressure -- free stream air pressure, provided by the forward motion of the engine.

Rear of engine -- end of engine from which exhaust gas is expelled.

Reciprocating engine -- device which produces motion in which the power originates in pistons and cylinders.

Reverse flow -- change in direction of airflow in the combustion chamber of a gas turbine engine.

Rotational direction -- direction of movement of the rotating part, determined by viewing the engine from the rear.

Rotor -- in a gas turbine engine, the turbine wheel.

Scavenge oil system -- method of returning oil from the engine to the oil tank, for cooling and reuse.

Scavenger pumps -- those that drain oil from the sumps at various parts of the engine, return it through the oil cooler, and back to the oil tank.

Secondary air -- large surplus of air that cools the hot sections of a gas turbine engine to lower temperatures.

Shaft horsepower (SHP) -- energy used to drive the compressor and accessories in a turbine engine.

Shroud -- device used with turbine rotor to prevent blade tip losses and excessive vibrations. The shrouded blades can be thinner than unprotected ones.

Simplex nozzles -- single-orifice channels through which highly atomized and accurately shaped sprays of fuel go into the combustion chamber.

Solvent immersion -- cleaning method in which parts are immersed in solvent to remove carbon, gum, grease, and other surface contaminants.

Splines -- teeth in a gear.

Speed governor -- device to relieve the pilot from resetting the power lever when outside air temperature and pressure change. Consists of flyweights balanced by a spring.

Standard day conditions -- 59°F, sea level barometric pressure (29.92 inches of mercury).

Stator -- part of assembly that remains stationary with respect to a rotating part. Stator vanes are a stationary set of airfoils in a compressor.

Tachometer -- device that gives the pilot a continuous indication of engine rpm.

Tachometer generator -- device that supplies power at a frequency proportional to the driven speed which drives the synchronous motors in the indicator.

TBO -- time between overhauls. This is established by the Army and the engine manufacturer.

Test cell -- building, usually concrete, that contains both a control room and an engine room, used for testing engines. The test cell is at the manufacturer's; a mobile engine-test unit is used in the field.

Thermocouple -- device composed of two pieces of metal or wire joined where heat is to be applied and the free end connected to an electrical measuring instrument.

Thermodynamic cycle -- succession of processes which involve changes in temperature, pressure, and density in which the substance acts as a means of transformation of energy. See Otto and Brayton cycles.

Thrust -- pushing or pulling force developed by an aircraft engine.

Torquemeter -- hydromechanical torque-measuring device located in the reduction-gear section of the inlet housing. The measurement is read as torque oil pressure in psi.

Torquemeter indicating system -- pressure indicator for continuous readings of engine output-shaft torque.

Transducer -- device actuated by power from one system and supplying power to a second system.

Turbine -- rotary engine actuated by the reaction of a current of fluid (gas in this case) subject to pressure. The turbine is usually made with a series of curved vanes on a central spindle arranged to rotate.

Turbine nozzle -- stationary nozzle which discharges a jet of gas against the blades on the periphery of a turbine wheel.

Turbine rotor -- rotating portion of a turbine engine. It is made of specially alloyed steel because of severe centrifugal loads, the result of high rotational speeds.

Turbine section -- part of the turbine engine that extracts the kinetic energy of the expanding gases and transforms it into shaft horsepower.

Turbofan -- compressor section of a turbine engine.

Turbojet -- engine most commonly used in high-speed, high-altitude aircraft.

Turboprop engine -- one in which a turbine rotor converts the energy of the expanding gases to rotational shaft power, to provide power for a propeller.

Turboshaft -- engine in which a turbine rotor converts the energy of the expanding gases to rotational shaft power, to provide power for a helicopter transmission.

Vapor blasting -- abrasive method used to clean combustor parts. Not to be used on ceramic, magnesium, painted, or aluminum surfaces.

Vapor degreasing -- cleaning method used on unpainted metal parts or aluminum-painted steel parts.

Vaporizing tubes -- devices used instead of fuel nozzles in a T53-L-11 engine.

Variable inlet guide vanes -- devices located in front of the first compressor rotor to guide the angle of incidence of the inlet air to the first compressor rotor.

Velocity -- speed or rate of motion in a given direction, and in a given frame of reference.

Venturi -- short tube with flaring ends connected by a constricted middle section forming a throat. It depends for operation upon the fact that as the velocity of a fluid increases in the throat, the pressure decreases. It is used for measuring the quantity of a fluid flowing in connection with other devices for measuring airspeed and for producing suction, especially for driving aircraft instruments by means of a branch tube joined at the throat.

Vermatherm element -- device which senses outlet fuel temperature and closes the core valve and opens the bypass valve.

Vibration -- periodic motion of a body or medium in alternately opposite directions from the position of equilibrium. Vibration meter--device for measuring vibrations.

Weldment -- unit formed by welding together an assembly of pieces, as in gear housing.

INDEX

	Paragraph	Page
ASOAP	3.5	72
Abbreviations used for gas turbine engines	Table I	23
Accessory drive assembly		
T53	4.9	99
T62	6.5	178
Accessory gearbox, T74	9.10	237
Accessory gearbox section, T63	7.5	193
Advantages of turbine engines	1.12	18
Air-bleed system, T63	7.10	206
Air density effect on compressor efficiency	1.19	27
Air flow through T74 engine	9.3	225
Air inlet section	1.18	25
Annular-type combustion chamber	1.21a	36
Anti-icing system		
T53	4.17	124
T55	5.15	165
T63	7.10	206
T73	8.11	219
Army spectrometric oil analysis program (ASOAP)	3.6	72
Assembly of engine	3.11b	81
Automatic fuel control system	2.3a	50
Axial-flow compressor	1.19b	29
Axial-flow turbine	1.22a	39
Bearing arrangement, T700	10.8	262
Bearing compartment airseal, T74	9.14b	241
Bearing installation	9.11	238
Bernouilli's Principle	1.3	3
Bleed air systems, T74	9.14	240
Boyle's Law	1.3	3
Brayton cycle	1.8	12
Can-annular-type combustion chamber	1.21c	38
Can-type combustion chamber	1.21b	37
Centrifugal-axial-flow compressor	1.19c	30
Centrifugal breather, T74	9.15d	243
Centrifugal-flow compressor	1.19a	28
Centrifugal impeller	1.19c	30
Centrifugal inlet separator, T700	10.3	256
Charles' Law	1.3	3

	Paragraph	Page
Check valve assembly and accumulator,		
T63	7.9d	206
Chip detectors, T55	5.19c	169
Classifications of turbine engines	1.7	12
Cleaning compressor rotor blades	3.10	78
Cleaning engine assembly	3.9	77
Combustion section	1.21	35
T55	5.12	156
T73	8.4	215
Combustor, T700	10.5	260
Combustor assembly, T62	6.6	181
Combustor chamber housing	4.14a	113
Combustor chamber liner, T74	9.6	232
Combustor turbine assembly (N_2)	4.12	107
Compressor		
construction	1.20	33
efficiency.	1.19	27
housing, T53	4.10	99
principle	1.5	6
rotor.	1.19	27
section.	1.19	27
section, T63	7.4	191
stall	1.19d	32
turbine, T74	9.7a	234
Compressor, T700	10.4	260
Compressor assembly		
T53	4.10	99
T55	5.10	154
Compressor bleed-air system, T74	9.14a	241
Compressor inlet case, T74	9.5a	228
Compressor rotor and stator as-		
sembly, T74	9.5b	230
Compressor section, T73	8.3	213
Constructinn of compressors	1.20	33
Current-regulator unit, T74	9.19a	249
Description		
fuel system, T53	4.13	111
fuel system, T55	5.14	158
lubrication system, T53	4.18	124
T55 engine	5.3	146
T62 APU	6.2	174
T73-P-700 engine	8.2	210
T700-GE-700 engine.	10.2	256
Diffuser housing, T73	4.11	101

	Paragraph	Page
Diffuser section, T55	5.11	154
Directional references	1.16a	22
T53-L-13	4.17	124
T55 engine	5.6	151
Disadvantages of turbine engines.	1.13	19
Disassembly of engine	3.11a	81
Dry-cleaning solvent.	3.9d	78
Dual compressor	1.19c	30
Duplex nozzle	2.7b	53
Einstein's Law of Conservation of Energy	1.3	3
Electrical system, T53.	4.20	130
Electrical system, T55.	5.21	171
Electrical system, T62.	6.11	186
Engine		
fuel system, T73	8.9	218
instrumentation	2.24	65
maintenance precautions	3.7	73
model designations	1.17	23
oil filters, T55	5.19d	171
oil pressure indicating system	2.25c	66
oil temperature indicating system	2.25d	66
speed notation	1.16c	23
station notation	1.16b	22
stations, T53-L-13	4.7	97
stations, T55-L-11	5.6	151
surge	1.19d	32
terminology	1.16	21
vibration test	3.3	69
Exhaust diffuser	1.24,	43
	4.14c	115
Exhaust duct, T74	9.8	235
Exhaust ducts	1.24	43
Exhaust gas temperature harness		
T53	4.20b	134
T55	5.21d	173
Exhaust gas temperature indicating system	2.25e	66
Exhaust section.	1.24	43
Extended storage	3.13c	86
Flyable storage	3.13a	83
FOD, disadvantage of turbine engines.	1.13a	20
Free-turbine engine cycle of operation	1.9	12
Fuel-air mixture	1.21	35
Fuel atomizer, T53	4.15a	119

	Paragraph	Page
Fuel control	2.3	48
T55	5.14a	161
T63	7.9	201
T74	9.18b	245
Fuel filters	2.8	55
Fuel flow, T53 engine	4.15a	117
Fuel flow control in turbine engines	2.3	48
Fuel flow divider, T53	4.15a	117
Fuel manifold, T53	4.15a	117
Fuel manifold assembly, T74	9.18e	247
Fuel nozzle, T63	7.9e	206
Fuel nozzles	2.7	52
Fuel-oil cooler	2.10	57
Fuel pressure indicating system	2.25f	66
Fuel pump	2.4	51
Fuel pump, T74	9.18a	245
Fuel pump assembly, T63	7.9a	201
Fuel system		
T62	6.8	182
T74	9.18	245
Fuel system operation, T62	6.9	182
Gas generator case, T74	9.5c	232
Gas generator turbine, T700	10.6	260
Gas producer, N_1	1.5	6
Gas producer fuel control (N_1), T63	7.9b	201
Glow plug, T74	9.19b	249
High cost, disadvantage of turbine engines	1.13e	21
Hot air solenoid valve	4.17	124
Hour meter, T62	6.11f	187
Hydromechanical fuel control	2.3	48
Igniters	2.22	62
Ignition exciter		
T55	5.21b	172
T62	6.11a	186
Ignition lead and coil assembly	5.21c	172
Ignition system		
T53	4.20a	130
T63	7.11	207
T73	8.14	221
T74	9.19	249
Ignition unit	2.21	62
Impeller	1.19a	28
Impeller housing	4.10	99

	Paragraph	Page
Inlet housing assembly		
T53	4.10	99
T55	5.9	153
Instrumentation, T74	9.20	252
Internal cooling and pressurization system		
T53	4.14	112
T55	5.15	165
T73	8.10	219
Internal cooling system.	2.23	64
Interstage air-bleed system		
T53	4.16	121
T55	5.18	166
Interturbine temperature sensing system, T74	9.20a	252
Jetcal analyzer	3.4	72
Jet propulsion principle	1.4	5
High fuel consumption, disadvantage of turbine engines	1.13d	21
High temperatures, disadvantage of turbine engines	1.13b	20
Liquid-to-liquid cooler, T55	5.14c	161
Lubricating oil precautions	3.7a	74
Lubrication flow.	4.20 a	130
Lubrication system		
T55	5.19	168
T62	6.10	184
T63	7.7	194
T73	8.12	219
T74	9.14	240
T700	10.9	262
Lubrication systems.	2.13	58
Main bearings, T73	8.8	218
Main electrical cable assembly, T55 . . .	5.21a	171
Main fuel system	2.6	52
T55	5.14d	161
Main oil pressure supply system	4.20a	130
Maintainability, T700	10.10	262
Maintenance allocation chart	3.12	81
Maintenance procedures	3.8	77
Manual fuel control system	2.3b	51
Manual fuel flow, T53	4.15b	121
Model comparison of T55 engines	5.4	149
Multiple-rotor turbine	1.22c	40

	Paragraph	Page
N_1, T53 engine	4.4	91
N_1 accessory drive gearbox assembly . . .	4.11a	106
N_2 T53 engine	4.4	91
N_2 overspeed governor assembly	4.11b	106
Newton's First Law of Motion	1.3	3
Newton's Second Law of Motion	1.3	3
Newton's Third Law of Motion	1.3	3
Oil cooler	2.18	61
Oil filters	2.17	61
Oil pressure indicator, T74	9.20c	254
Oil scavenge system, T53	4.20b	137
Oil temperature indicator, T74	9.20b	253
Oil-to-fuel heater assembly, T74	9.15e	243
Oil filter assembly, T74	9.15c	243
Oil flow, T55	5.19b	169
Oil level indication system, T55	5.19a	168
Oil pressure pump, T74	9.15b	243
Oil tank, T74	9.15a	241
Oil tanks	2.14	60
Operating characteristics, T74 engine . . .	9.3	225
Operation		
T53 engine	4.4	91
T62 engine	6.3	174
T63 engine	7.2	188
Operation of lubrication system	4.20	130
Otto cycle	1.8	12
Output reduction carrier and gear assembly	4.10a	101
Overspeed governor in torquemeter system	4.21	137
Power-to-weight ratio	1.12a	18
Power turbine, N_2.	1.5	6
Power turbine, T700.	10.7	261
Power turbine disk assembly, T74	9.7b	234
Power turbine governor (N_2), T63	7.9c	203
Power turbine inlet temperature . . . . . indicating system, T73.	8.15	222
Power turbine section		
T55	5.13	157
T73	8.6	217
Precautions for engine maintenance	3.7	73
Pressure and scavenge oil system, T74 . .	9.16	243
Pressure and velocity control	1.3	3
Pressure oil system, T74	9.16a	244

	Paragraph	Page
Pressure pumps	2.15	60
Pressurizing and drain dump valves	2.9	57
Propeller reduction carrier and gear assembly	4.10b	101
Protective devices for T62	6.4	178
Radial-flow turbine	1.22a	36
Reduction drive, T62	6.5	178
Reduction gearbox, T74	9.9	235
Repair of engine parts	3.11c	81
Scavenge oil, T53	4.20b	137
Scavenge oil system, T74	9.16b	244
Scavenger pumps	2.16	61
Scheduled inspections	3.5	72
Scheduled maintenance, T700	10.11	264
Simplex nozzle	2.7a	53
Single-rotor turbine	1.22b	37
Slow acceleration, disadvange of turbine engines	1.13c	20
Solvent immersion	3.9b	78
Spark plugs, T62	6.11b	186
Special inspections	3.5	72
Specifications for engines		
T53	4.6	96
T55	5.5	150
T63	7.3	190
T74	9.4	226
Speed switch, T62	6.11c	186
Spray oil jets	2.19	61
Start fuel system, T55	5.14d	161
Starting control fuel flow, T74	9.18d	247
Starting fuel systems	2.5	51
Stator vanes	1.19b	29
Storage and preservation	3.13	82
Symbols used in gas turbine engines	Table I	23
T53		
engine description	4.3	89
engine operation	4.4	91
model comparison	4.5	95
T53-L-13 and -701 specifications	4.6	96
T55 model comparison	5.4	149
T73-P-700 engine description	8.2	210
T700-GE-700 gas turbine engine	10.2	256
TBO	3.11	81
TOT thermocouple assembly, T63	7.11	207

	Paragraph	Page
Tachometer drive assembly in torquemeter system	4.21	137
Tachometer generator, T62	6.11e	187
Tachometer system	2.25a	65
Temperature compensator, T74	9.18c	245
Temporary storage	3.13b	86
Test cell	3.2	68
Theory		
compressor	1.5	6
gas producer (N_1)	1.5	6
jet propulsion	1.4	5
N_1 and N_2 systems	1.5	6
reciprocating engines	1.5	6
Thermocouples, T53	4.20b	137
Thermodynamic cycles	1.8	12
Time between overhauls	3.11	81
Torquemeter indicating system.	2.25b	66
T74	9.20e	254
Torquemeter system		
T53.	4.19	127
T55	5.20	171
T63	7.8	201
T74.	9.17	244
Torque sensor, T73.	8.13	221
Turbine assembly.	4.15b	113
T62.	6.7	181
Turbine-cooling blood-air system, T74 . .	9.14c	241
Turbine construction	1.23	41
Turbine engine		
advantages	1.12	18
classifications	1.7	12
disadvantages	1.13	19
troubleshooting.	3.14	86
Turbine exhaust duct, T73.	8.7	218
Turbine exhaust thermal switch, T62. . .	6.11d	187
Turbine outlet temperature measurement system, T63	7.11	207
Turbine section	1.22	39
T63.	7.6	194
T73.	8.5	217
T74.	9.7	234
Turbine tachometer, T74	9.20d	254
Turbojet engine	1.10	14
Turbojet engine cycle of operation	1.9	12

	Paragraph	Page
Turbojet engine principle.	1.5	6
Turboprop engine	1.11	15
Turboshaft engine	1.11	15
VIGV.	4.15	115
Vapor blasting	3.9c	78
Vapor degreasing	3.9a	78
Vaporizing tube	2.7c	55
Variable inlet guide vane system, T53 . . .	4.15	115
Variable inlet guide vane system, T55. . .	5.16	165
Venturi Principle	1.3	3
Vibration equipment	3.3	69
Vibration sources	3.3	69

www.ingramcontent.com/pod-product-compliance
Lightning Source LLC
LaVergne TN
LVHW080033210926
848481LV00011B/3174
9781934939130